Research on River Engineering

Research on River Engineering

Topic Editors

Vlassios Hrissanthou
Mike Spiliotis
Konstantinos Kaffas

MDPI

Basel • Beijing • Wuhan • Barcelona • Belgrade • Novi Sad • Cluj • Manchester

Topic Editors
Vlassios Hrissanthou
Department of Civil Engineering
Democritus University of Thrace
Xanthi
Greece

Mike Spiliotis
Department of Civil Engineering
Democritus University of Thrace
Xanthi
Greece

Konstantinos Kaffas
Department of Science
Roma Tre University
Rome
Italy

Editorial Office
MDPI AG
Grosspeteranlage 5
4052 Basel, Switzerland

This is a reprint of the Topic, published open access by the journals *Hydrology* (ISSN 2306-5338) and *Water* (ISSN 2073-4441), freely accessible at: https://www.mdpi.com/topics/9LC3P5IML6.

For citation purposes, cite each article independently as indicated on the article page online and as indicated below:

Lastname, A.A.; Lastname, B.B. Article Title. *Journal Name* **Year**, *Volume Number*, Page Range.

ISBN 978-3-7258-4755-6 (Hbk)
ISBN 978-3-7258-4756-3 (PDF)
https://doi.org/10.3390/books978-3-7258-4756-3

Contents

About the Editors

Vlassios Hrissanthou

Vlassios Hrissanthou is an emeritus professor at the Department of Civil Engineering, Democritus University of Thrace, Greece, with extensive academic experience and achievements in hydrology, hydraulics, and hydraulic structures. He holds a Civil Engineering diploma from Aristotle University of Thessaloniki, along with postgraduate certificate and a Dr.-Ing. in soil erosion and sediment transport from Germany's Karlsruhe University (KIT). He has also conducted postdoctoral research in hydraulics at the University of the Armed Forces in Munich. He taught graduate and postgraduate courses, including hydraulics, engineering hydrology, advanced engineering hydrology, and sediment transport, imparting knowledge to students in related fields. His academic output includes 51 journal papers, 91 conference proceedings, 14 book chapters (in English, Greek, and German), 3 edited English books, 2 Greek books, and 5 guest-edited English Special Issues. His work has accrued 1144 citations on Google Scholar and 1134 on ResearchGate. He reviews for 60 international journals, contributing to the advancement of academic quality, engaging in exchanges, and having notable influence in the fields of international hydrology, hydraulics, and hydraulic structures.

Mike Spiliotis

Mike Spiliotis is an associate professor in the Department of Civil Engineering at Democritus University of Thrace, Greece. His in-depth research focuses on areas such as water resources management. He studied civil engineering at Democritus University of Thrace, Greece, and then did postgraduate studies in water resources technology at the School of Civil Engineering of the National Technical University of Athens. He also completed doctoral research in water resources management at the School of Rural and Surveying Engineering of the same university. His doctoral research direction was the application of fuzzy system theory in strategic water resources management. In addition, he also conducted postdoctoral research in adaptive water resources management at the National Technical University of Athens. He is mainly responsible for teaching graduate courses, including basic courses such as water resources management, open channel flow, irrigation, and drainage systems, and advanced courses such as hydraulics and water resources management and restoration. Through these courses, he imparts professional knowledge and skills related to water resources to civil engineering graduate students. He has a fruitful research output, with 56 papers published in English in scientific journals, 57 papers published in English and Greek in conference proceedings, and 5 book chapters in English and Greek. His research results have received widespread attention, with 1148 citations on Google Scholar. He also co-authored the book *Conventional and Fuzzy Regression: Theory and Engineering Applications*, which was published by Nova Science Publishers in 2018. Mike Spiliotis has served as a co-guest editor of three English Special Issues, participated in the editing of academic journals, and plays an active role in facilitating academic exchange and the dissemination of results in related fields.

Konstantinos Kaffas

Konstantinos Kaffas is an environmental engineering researcher with a lot of expertise. He currently works at the Faculty of Science, at the University of Rome Tre (Italy), and is affiliated with Democritus University of Thrace (Greece), where he earned his 2007 Environmental Engineering degree, postgraduate qualification in Hydraulic Engineering and Water Resources Management, and PhD in Basin-Scale Hydromorphological Modeling in 2017. As a researcher at Democritus

University's Civil Engineering Department and a Greek Association of Civil Engineers member since 2008, he has carried out postdoctoral work at Italy's Free University of Bolzano and the University of Florence. His research focuses on soil erosion, sediment transport, hydrological/hydraulic modeling, watershed management, and flood risk assessment. Since 2014, he has taught postgraduate courses like Advanced Engineering Hydrology at Democritus University. Kaffas has 23 English journal papers, 25 English/Greek conference publications, and 6 English book chapters (including a chapter in *Handbook of Climate Change Impacts on River Basin Management*), with 709 Scopus citations. He reviews for 15 international journals and has guest-edited three Special Issues, promoting academic progress in his field.

water

MDPI

Editorial

Research on River Engineering

Vlassios Hrissanthou [1,*], Mike Spiliotis [1] and Konstantinos Kaffas [2]

1 Department of Civil Engineering, Democritus University of Thrace, 67100 Xanthi, Greece; mspiliot@civil.duth.gr
2 Department of Science, Roma Tre University, 00146 Rome, Italy; konstantinos.kaffas@uniroma3.it
* Correspondence: vhrissan@civil.duth.gr

1. Introduction

River engineering is one of the most important subjects of hydraulic engineering. The main scientific fields necessary for understanding the basic principles of river engineering, include hydrology, hydraulics, and geomorphology.

Using hydrologic rainfall–runoff models, the river inflows originating from rainfall-induced overland flow can be calculated. River floods have to be routed during intense storms. Flood routing can be calculated using hydrologic or hydraulic models. Hydraulic models are based on water mass and momentum conservation equations, which are hyperbolic-type partial differential equations solved using numeric methods, e.g., finite difference schemes.

Soil erosion products, due to rainfall on the surrounding basins, are transported by the overland flow into the rivers and constitute the so-called wash load. The river bed can be eroded by the river flow, or suspended sediment can be deposited onto the river bed. Numerous computational models for the bed load and the total load have been developed in the past. To take into account sediment transport in rivers, the sediment continuity equation should be added to the water mass and momentum conservation equations. Sediment transport is mainly influenced by unsteady turbulent flows, which are normal physical condition in rivers. The vegetation on river banks also influences river flow and sediment transport.

The hydrologic and geomorphologic conditions in reservoirs and lakes are different from those in rivers. Generally, hydraulic structures, e.g., dams, modify the hydraulic and geomorphologic conditions in rivers.

2. Overview of the Topic "Research on River Engineering"

The above topic includes thirteen articles submitted to *Water* and one article submitted to *Hydrology*. The article authors are active at universities, research institutes, and water management authorities from eleven different countries: Bangladesh, Canada, China, Egypt, France, Iran, Italy, Japan, Mexico, Peru, and the USA. One article resulted from the cooperation of three universities in Iran, Italy, and the USA, respectively, while another article resulted from the cooperation of a university and a technical bureau in the USA and a research institute and a water management authority in Bangladesh. Generally, most authors are active at universities.

The fourteen articles can be divided into three categories: Category A: "Hydraulic geometry, bed roughness and sediment particle size of rivers"; Category B: "Hydraulic structures"; Category C: "Water quality recovery". Articles [1–6] belong to Category A, articles [7–12] belong to Category B, and articles [13,14] belong to Category C.

The contents of the articles of Category A are summarized below:

In the article by Qin et al. [2], the well-known empirical relationships of regime theory, in the form of a power law in alluvial channels, that express river width, average flow depth and flow velocity as functions of discharge, were determined for six major exorheic rivers located in the Qinghai–Tibet Plateau (China). The included equations were also modified so that the discharge frequency was incorporated into these relationships.

The study by Takata et al. [5] attempted to back-calculate Manning's roughness coefficients by repeating a two-dimensional flow simulation to fit the spatially and temporally dense river water level data observed in Japan's Yamatsuki River, a typical mountainous river with an average riverbed gradient of 1/50 and an average river width of 17.9 m. The software Nays2DH was used for the river flow calculations. The flow field calculation model used a general curvilinear coordinate system, which allowed for complex boundaries and riverbed topography to be directly considered. The roughness coefficient during flooding was found to be correlated with the channel slope and step height (H)–step length (L)–channel slope (S) ratios (H/L/S), and a corresponding regression equation was proposed.

The study by Gabr et al. [3] assessed the maintenance condition of the main surface drains (Baloza and Elfarama) located in the Tina Plain (50,000 acres) and a portion of the Southeast Elkantara region (25,000 acres) in North Sinai, Egypt, based on the values of the Discharge Capacity Ratio (DCR) and Manning's roughness. DCR is defined as the ratio of actual discharge to the projected or design discharge.

The study by Dehkordi et al. [6] developed an empirical relation capable of estimating the median sediment particle size in gravel river bends. Field data were collected from different cross-sections placed at the bend apex and crossovers in various rivers (Iran). The Buckingham π-theorem was applied to identify the effective dimensionless numbers, such as the Shields number, Reynolds particle number, Froude number, submerged specific gravity of sediment, and aspect and curvature ratios. Three regression techniques, containing the power function approach, the general additive model (GAM), and the multivariate adaptive regression spline (MARS), were chosen to achieve the best relation between the above-given dimensionless variables. It was found that two parameters, the curvature ratio and the Shields number, were the most important in affecting the median size of bed sediment at the bends of meandering rivers.

The study by Sarker et al. [1] provided detailed and quantitative insights into the properties of planform complexity and dynamics of channel patterns. This was achieved by investigating the applicability of anastomosing classification on the Brahmaputra River's planform (Bangladesh) and computing the disorder/unpredictability and complexity of fluctuations using the notion of entropy and uniformity of energy conversion rate by the channels using a power spectral density approach.

In the article by Yi et al. [4], the effects of flooding characteristics, namely flooding depth and flooding duration, on riparian plants were studied in the case of Three Gorges Reservoir (China). The results of this study show that the riparian plant diversity and functional diversity varied by season. A significant negative relationship between plant diversity and flooding depth was observed, while the flooding duration was not a significant predictor in different seasons.

The contents of the Category B articles are reported in the following paragraphs:

According to the article by Cabarello et al. [7], the maximum scour that occurred at the Carrizal River hydraulic control structure (Tabasco, Mexico) was assessed experimentally and numerically. The physical model was built at a scale of 1:60 (without distortion) at the Engineering Institute of the National Autonomous University of Mexico. The maximum experimental scour was compared to the results of a 2D free surface numerical model and to the estimations of four empirical equations: Breusers, Farhoudi and Smith, Negm, and Dietz. The numerical model consisted of a mass conservation equation for water, Saint-Venant equations, a mass conservation equation for sediment (Exner equation) regarding bed load transport, and an advection–diffusion equation regarding suspended load transport. Only the Breusers method provided values close to the measured values.

In the study by Zhang et al. [8], a quasi-stump group structure was proposed and placed upstream of the bridge piers to mitigate the scour of water flow on the riverbed. Both experimental and numerical simulations using FLOW-3D were employed to study the protective effect of this structure. The experiments were carried out in the hydraulic laboratory at Zhengzhou University's School of Water Conservancy and Civil Engineering (China). The numerical model consists of the continuity equation for turbulent flow and the Reynolds-averaged Navier–Stokes equations. FLOW-3D uses the RNG (Renormalization Group) k-ε turbulence model to solve the latter equations. A separate equation, including the critical shear stress, was employed to calculate the volumetric bed load transport rate. The numerical results were in good agreement with the experimental findings. It was found that the quasi-stumps group could effectively reduce the flow velocities around the bridge piers, thereby promoting the deposition of suspended sediment.

The study by Tu et al. [10] adopted a flume model experiment to investigate the scouring and deposition around geotube groins and mixed clay–geotube groins. The motivation of the study was the fact that during the cofferdam construction of the toe reinforcement project at the Qiantang River estuary (China), the scouring of the riverbed at the groin head often led to the collapse of geotube groins due to strong tidal currents. Results indicated that the influence of tidal surges on geomorphic changes surrounding the groins was more pronounced during spring tides than neap tides. Under the same flow conditions, the scour depth at the head of the geotube groin was notably deeper than that of the mixed clay–geotube groin. Additionally, sediment silting behind the mixed clay–geotube groin was significantly greater than that behind the geotube groin.

The study by Girolami et al. [9] focused on the mechanisms that trigger erosion of the pervious foundation of flood protection dikes. The origin of these permeable areas is generally attributed to the presence of a paleo-valley and paleo-channels filled with gravelly sandy sediments beneath the riverbed and dikes. These layers may extend into the protected area. The possibility of internal erosion must also be considered as the cause of leaks, sand boils, and sinkholes. Two classical geophysical methods (Electromagnetic Induction and Electric Tomography) were applied to Agly dikes (France) for the geological observation of the foundation soils. A three-dimensional finite element numerical model for internal flows was used for quantifying the soil permeability.

The study of Chenari et al. [11] investigated the problem of low efficiency and lack of water supply at the Hemmat Water Intake, Iran, where severe sediment accumulation was observed at the intake mouth. The FLOW-3D software was used to simulate the flow patterns under various scenarios of hydraulic regimentation works. The numerical model uses the Volume-of-Fluid (VOF) method on a gridded domain to solve the Navier–Stokes–Reynolds equations for three-dimensional analysis of incompressible flows. The considered parameters include the following: (i) three alternative locations of the spur dike; (ii) four spur dike lengths; and (iii) five spur dike deviation angles. Overall, the main flow of the river with the highest velocity and depth and best directed towards the water intake occurs for the placement of the longest spur dike in front of the inlet and for a spur dike deviation angle of 135 degrees.

In the study by Zhang et al. [12], flume experiments were conducted to study the upstream water level rise of submerged rock weirs. The flume was installed in the Key Laboratory of Hydraulic and Waterway Engineering of the Ministry of Education at Chongqing Jiaotong University (China). The main findings were as follows: The recorded water level along the flume showed that for submerged conditions, the rock weirs primarily rose the upstream water level while exerting minimal influence on the changing tail-water level. For a given tail-water depth and void ratio of rock weirs, the upstream water level rise increased with increased discharge. However, this response became insignificant as tail-water depth increased. Furthermore, as weir void ratio increased, the water level rise upstream of the weir was expected to decrease for a given discharge and tail-water depth. Based on the experimental data and observations, a predictor including the effects of Froude number before building the weir, weir submergence, and weir void ratio was proposed for estimating the water level rise upstream of I-shaped rock weirs for submerged conditions.

The contents of the Category C articles are described below:

In the article by Mori-Sánchez et al. [13], a two-dimensional numerical model (Iber model) was applied to the Lurin River to improve the water quality. Lurin is one of the main sources of water for the city of Lima (Peru). The main water quality parameters were Dissolved Oxygen (DO), Biochemical Oxygen Demand at 5 days (BOD_5), and Escherichia Coli (E. Coli). The Iber model consisted of two submodels: a hydrodynamic submodel (Saint-Venant equations) and a convection–diffusion submodel for each polluting substance. Using this model, the authors found where the greatest contamination occurred. It was proposed to improve the river by optimizing the San Bartolo WWTP (Waste Water Treatment Plant) and building a new WWTP in Pachacámac to avoid diffuse contamination.

The study by MacKenzie et al. [14] presented a novel methodology to evaluate the relative contributions of the stream corridor and upland watershed contributions to total sediment and phosphorus loads in receiving watercourses. The new method could be used to develop a cost-optimized mitigation plan for the impacted watersheds, including the implementation of Low-Impact Development (LID) for urban areas and Best Management Practices (BMPs) for rural areas of the watershed, as well as stream restoration projects for degraded stream sections. The estimation of stream banks and bed erosion was based on the concept of stream power. The new method was applied to the Tannery Creek watershed (Canada), which has experienced significant urbanization in recent decades.

Acknowledgments: We acknowledge the contributions of all authors of the 14 papers in this Topic.

Conflicts of Interest: The authors declare no conflicts of interest.

References

1. Sarker, S.; Sarker, T.; Leta, O.T.; Raihan, S.U.; Khan, I.; Ahmed, N. Understanding the Planform Complexity and Morphodynamic Properties of Brahmaputra River in Bangladesh: Protection and Exploitation of Riparian Areas. *Water* **2023**, *15*, 1384. [CrossRef]
2. Qin, C.; Wu, B.; Xue, Y.; Fu, X.; Wang, G.; Wang, G. Multifrequency Downstream Hydraulic Geometry of Alluvial Mountain Rivers Located on the Qinghai-Tibet Plateau. *Water* **2023**, *15*, 2139. [CrossRef]
3. Gabr, M.E.; Fattouh, E.M.; Mostafa, M.K. Determination of the Canal Discharge Capacity Ratio and Roughness to Assess its Maintenance Status: Application in Egypt. *Water* **2023**, *15*, 2387. [CrossRef]
4. Yi, X.; Huang, Y.; Jiang, Y.; Ma, M.; Chen, Q.; Wu, S. Flooding Depth and Flooding Duration with the Zonation of Riparian Plant Communities in the Three Gorges Reservoir of China. *Water* **2023**, *15*, 3228. [CrossRef]
5. Takata, H.; Obata, S.; Sato, T.; Shimatani, Y. Back-Calculation of Manning Roughness Coefficient by 2D Flow Simulation and Influence of In-Channel Physical Parameters in a Mountain River, Japan. *Water* **2024**, *16*, 320. [CrossRef]
6. Dehkordi, A.N.; Sharafati, A.; Mehraein, M.; Hosseini, S.A. An Empirical Relation for Estimating Sediment Particle Size in Meandering Gravel-Bed Rivers. *Water* **2024**, *16*, 444. [CrossRef]

7. Caballero, C.; Mendoza, A.; Berezowsky, M.; Jiménez, A. Numerical-experimental Study of Scour in the Discharge of a Channel: Case of the Carrizal River Hydraulic Control Structure, Tabasco, Mexico. *Water* **2023**, *15*, 2788. [CrossRef]
8. Zhang, Y.; Wang, J.; Zhou, Q.; Cai, Y.; Tang, W. The Investigation of Local Scour around Bridge Piers with the Protection of a Quasi-stumps Group. *Water* **2023**, *15*, 2858. [CrossRef]
9. Girolami, L.; Bonelli, S.; Valois, R.; Chaouch, N.; Burgat, J. On Internal Erosion of the Pervious Foundation of Flood Protection Dikes. *Water* **2023**, *15*, 3747. [CrossRef]
10. Tu, X.; Gao, H.; Fan, K.; Xu, J.; Li, J.; Xia, C.; Pang, X. Comparative Experimental Study of Geotube Groin and Mixed Clay-Geotube Groin under Various Flow Conditions. *Water* **2023**, *15*, 3844. [CrossRef]
11. Chenari, S.A.; Nadian, H.A.; Ahadiyan, J.; Valipour, M.; Oliveto, G.; Sajjadi, S.M. Enhancing Hydraulic Efficiency of Side Intakes Using Spur Dikes: A Case Study of Hemmat Water Intake, Iran. *Water* **2024**, *16*, 2254. [CrossRef]
12. Zhang, W.; Liu, X.; Gan, B. Experimental Study on Upstream Water Level Rise of Submerged Rock Weirs. *Water* **2024**, *16*, 2136. [CrossRef]
13. Mori-Sánchez, O.L.; Ramos-Fernandez, L.; Llué-Chero, W.E.; Pino-Vargas, E.; del Pino, L.F. Application of the Iber Two-Dimensional Model to Recover the Water Quality in the Lurin River. *Hydrology* **2023**, *10*, 84. [CrossRef]
14. MacKenzie, K.; Auger, S.; Beitollahpour, S.; Gharabaghi, B. The Role of Stream Restoration in Mitigating Sediment and Phosphorous Loads in Urbanizing Watersheds. *Water* **2024**, *16*, 363. [CrossRef]

water

MDPI

Article

Enhancing Hydraulic Efficiency of Side Intakes Using Spur Dikes: A Case Study of Hemmat Water Intake, Iran

Saman Abbasi Chenari [1], Hossein Azizi Nadian [2], Javad Ahadiyan [3,4], Mohammad Valipour [5,*], Giuseppe Oliveto [6,*] and Seyed Mohsen Sajjadi [3,4]

1 Department of Water Sciences and Environmental Research Center, Islamic Azad University, Shoushtar Branch, Shoushtar 64517-41117, Iran; samanabasi@yahoo.com
2 Department of Civil Environmental Architectural Engineering and Mathematics, University of Brescia, Via Branze 43, 25123 Brescia, Italy; hossein.azizinadian@unibs.it
3 Faculty of Water and Environmental Engineering, Shahid Chamran University of Ahvaz, Ahvaz 61357-83151, Iran; j.ahadiyan@scu.ac.ir (J.A.); m.sadjadi@scu.ac.ir (S.M.S.)
4 Center of Excellence of the Irrigation and Drainage Networks Improvement and Maintenance, Golestan Blvd, Ahvaz 61357-43311, Iran
5 Department of Engineering and Engineering Technology, Metropolitan State University of Denver, Denver, CO 80217, USA
6 School of Engineering, University of Basilicata, Viale dell'Ateneo Lucano 10, 85100 Potenza, Italy
* Correspondence: mvalipou@msudenver.edu (M.V.); giuseppe.oliveto@unibas.it (G.O.)

Abstract: This study investigates the problem of low efficiency and the lack of a water supply at the Hemmat Water Intake, in Iran, where severe sediment accumulation was observed at the intake mouth. The Flow-3D software was used to simulate the flow patterns under various scenarios of hydraulic regimentation works. The considered parameters include: (i) three alternative locations of the spur dike (i.e., a spur dike placed on the opposite side of the intake inlet and aligned with the upstream edge of the intake, to be regarded as a witness spur dike; a spur dike at a distance D_S of 7 m downstream of the witness spur dike, which implies a dimensionless distance D_S/b_{i1} of 1/3, with b_{i1} being the intake opening width; and a spur dike at a distance of 7 m upstream of the witness spur dike with a dimensionless distance, still, of 1/3); (ii) four spur dike lengths, L_S/B_r, with L_S being the effective spur dike length and B_r the approach river width; and (iii) five spur dike deviation angles of 75, 90, 105, 120, and 135 degrees (the deviation angle is the angle between the spur dike axis and the original river-bank line from which the spur dike extends). The results showed that, with the increase in the relative spur dike length (L_S/B_r), the velocity of the flow entering the water intake increases by 11%. A spur deviation angle of 135 degrees increases the flow depth at the intake inlet by 9% compared to a smaller deviation angle of 75 degrees. In addition, the spur dike increases the flow shear stresses at the intake inlet by up to 50%. Overall, the main flow of the river with the highest velocity and depth, and best directed towards the water intake, occurs for the placement of the longest spur dike (i.e., L_S/B_r = 0.46) in front of the inlet (i.e., witness spur dike) and for a spur dike deviation angle of 135 degrees. The spur dike increases the shear stress at the intake entrance by more than five times with respect to the case of its absence. In general, the presence of a spur dike on the opposite bank and with a deviation angle in the direction of the intake inlet well directs the main flow towards the canal intake. Moreover, it reduces the possibility of sedimentation in the canal inlet by increasing the flow velocity. Therefore, the results of this study could also be useful in increasing the hydraulic efficiency of lateral intakes by reducing the sedimentation phenomena.

Keywords: diversion channels; Flow-3D software; hydraulic efficiency; side intakes; spur dike

1. Introduction

Lateral intakes are hydraulic structures that divert water from rivers for domestic, agricultural, and industrial purposes [1]. The structure and performance of the water intake structures lead, in their vicinity, to the deviation of the streamlines near them, which, together with the flow-pressure gradient and centrifugal forces, leads to the formation of some vortices at the structure entrance, creating complex three-dimensional flow patterns at lateral intakes [2]. At any time, the condition of a water supply with a maximum (diverted) discharge and minimal settling should be considered as the optimal operating condition for an intake structure [3,4]. Determining the intake flow depth is one of the most important issues in lateral intake design, and affects the diversion flow ratio [5]. However, artificial structures such as spur dikes can increase the water intake efficiency and reduce sedimentation loads [6,7]. Intake structures also have some disadvantages, such as flow separation in the upstream wall of the intake channel, the generation of secondary currents and eddies that prevent the flow from entering the intake channel, and the formation of a stagnant flow area prone to the settling of sediments near the outer wall, causing a reduction in the width of the passing flow and the efficiency of the intake [8–10]. Moreover, the flow diverted toward the intake leads to changes in the hydraulic conditions and distribution of the flow velocity in the river and at the intake inlet [6,7]. Therefore, a complete understanding of the processes that occur after intake construction can help to optimize the water diversion mechanism and improve the performance of the intake structures. Below, a short review on various research studies focused on increasing the hydraulic efficiency of lateral intakes is presented.

Some researchers have suggested some key points to increase the efficiency of lateral intakes. As examples: the use of a 90-degree water intake, the application of intakes with different angles, and placing the intake location on the outer bend of the river. However, many of these configurations need to be analyzed especially with numerical models. Abolghasemi et al. [11] evaluated water intakes with 90- and 52-degree angles. They reported that water intakes with a 52-degree deviation angle exhibit better efficiency with regard to the transfer of sediment and return flows at the water intakes. Kashyap et al. [12] showed the entrainment sediment capacity of the flow depending on the positions and sizes of regions of high bed shear stress. Azimi et al. [13] discovered that a secondary circulation cell develops downstream of the intake channel based on the simulation results for lateral intakes with a vertical diversion. Montaseri et al. [2] investigated a 90-degree-angle water intake in a 180-degree river bend. They concluded that sedimentation would occur along the inner bound curve if they installed the intake structure in the second half of the bend. On the other hand, if the intake were placed in the first half of the bend, sediments would accumulate along both the inner and outer bounders of the bend, and more sediments would enter the intake.

Based on a laboratory study on lateral intakes in a 180-degree bend, Masjedi and Foroushani [5] showed that the intake discharge ratio is directly related to the increase in flow depth. In branching channel flow, the discharge ratio is the ratio between the branch channel discharge and the main channel upstream discharge. Jalili et al. [9], using a laboratory study and the Sediment Simulation in Intakes with Multiblock option (SSIIM) model, showed that the amount of sediments entering an intake increases with an increase in the intake–discharge ratio. Babagoli Sefidkoohi et al. [14] concluded that the Re-Normalisation Group (RNG) turbulence model in Flow-3D leads to results with a minimal error in flow simulations in a river with lateral intakes. Safarzadeh and Khaiatrostami [15] concluded that, along the diversion channel, the presence of highly turbulent dividing surfaces induces an instability of shear layers matched by strong instantaneous vertical motions and, consequently, bed shear stresses. Tavakoli et al. [16] considered laboratory and numerical models of lateral intakes in 180-degree bends. They showed that the minimum sediment load entering the intake occurs for an intake deviation angle of 50 degrees, and the best location of the intake in the bend is at an angle of 120 degrees for all discharge ratios. In addition, Sayed [17] showed, in his experimental work, that the largest discharge ratio occurs at the branching angle of 45 degrees, while the smallest at the branching angle of 90 degrees. Moreover, the laboratory results by Alomari et al. [4] showed that reducing the intake diversion angle would lead to a reduction in sediment transport in the diversion channel; a 30-degree diversion angle can reduce the sediment transport by up to 64%. Heidari Rad et al. [18] concluded that, when diverging the flume to sizes of 0.75 and 0.50, the diversion sediment to the intake decreased by 35.4 and 49.9%, respectively. They also compared the experimental results to those from CCHE2D (i.e., an integrated software package developed at the National Center for Computational Hydroscience and Engineering, the University of Mississippi) and Flow-3D numerical models, noting good accuracy. The models founds that the diversion flow decreases by 21.9 and 31.8%, respectively. Furthermore, Montaseri et al. [2], using the Ansys Fluent (version 6.2) software, showed that sedimentation occurs along the inner bank of a bend and enters the intake, from the downstream edge, in the case of intakes in the second half of the bend. Still considering research based on three-dimensional flow simulations, Meshkati and Salehi [19] found that the modelling errors in the main channel and intake were about 7.3% and 19.7%, respectively, when considering lateral intakes in a 180-degree bend. Niknezhad et al. [20] concluded that the RNG turbulence model is more accurate than the standard k-ε turbulence model.

Various structures, including submerged plates and vanes, are used to increase the efficiency of lateral intakes. With regard to previous laboratory studies, it was shown that submerged plates could change the flow patterns in front of an intake so that the incoming sediments could be reduced by 31% [21]. By applying Flow-3D (version 10.1), Sarhadi and Jabbari [22] confirmed that the parallel arrangement of plates with an angle of 60 degrees does not provide suitable conditions for increasing the relative flow rates in the lateral channel; whereas the plaid and linear arrangement offers the best conditions for transferring the flow to the intake channel. The Flow-3D numerical model simulations by Firozjaei et al. [23] revealed that the circulation area between the submerged plates and the lateral walls of the main channel increases the intake efficiency and reduces the amount of sediments entering the lateral intake. Baltazar et al. [24] found that, in comparison to the case without elements, the plates changed their near-velocity field by creating vortices and increasing the dimensions of the separation zone inside the intake channel. Al-Zubaidy and Ismaeil [10] found that secondary eddies resulting from submerged plates depend on the arrangement of the latter, which causes a change in the velocity distribution patterns and, as a result, reduces sedimentation and sediment entry into the intake channel. Further, the physical modelling of Moghadam et al. [25] revealed that using parallel and staggered submerged vanes at 10- and 30-degree angles would reduce shear stresses at the intake in comparison to the case without vanes.

With regard to the use of spur dikes in order to increase the hydraulic efficiency of lateral intakes, the numerical results of Shamloo et al. [26] showed that the performance of the vanes may be increased using a proper spur dike, thus mitigating the bed-load sediment rate into the intake channel. Analyzing the flow velocity and scour around a T-shaped spur dike in a 90-degree bend, Daneshfaraz et al. [27] noticed that, in the bend without a spur dike, the scour rate was five times higher than in the case with a spur dike. Based on 3D numerical models and laboratory experiments, Karami et al. [28] found that shear stresses increase in the main channel toward the entrance of the diversion channel. Their research considered the presence of spur dikes and vanes. They also found that sedimentation occurs in the vortex areas within the diversion channel and behind the dike in the main channel. The placement of a spur dike perpendicular to the approach flow, upstream of the intake inlet and on the opposite bank with $L_S/W_c = 0.2$ (where L_S is the spur dike length and W_c is the main channel width), not only reduced the sediment load to the intake channel but also doubled the diverted discharge. Moreover, shear stresses decreased in the main channel due to moving away from the entrance of the diversion channel as a result of the gradual weakening of secondary flows and the reduction in discharge and velocity in the main channel. Geravandi et al. [29] investigated the placement of L-shaped spur dikes upstream and downstream of lateral intakes in an internal river bend. They concluded that the flow velocity at the intake inlet increased. Hence, the best spur dike deflection angle, among the angles of 30, 45, 60, and 90 degrees, was investigated, looking at the maximum inlet flow rate. The best angle was found to be equal to 60 degrees. Zamani et al. [30] experimentally investigated the effect of a spur dike in increasing the intake discharge and minimizing the turbulence, erosion, and sedimentation. Their results showed that the spur dikes in front of the intake inlet and just downstream of it involved the highest discharge ratio. However, the smallest rate of erosion and sedimentation was achieved when the spur dike was located upstream of the intake channel. Finally, Moradinejad et al. [31] performed an experimental study on the flow patterns and sedimentation at intake structures using spur dikes and skimming walls. They showed that a trench is created towards the intake structure when using a skimming wall, which in turn increases the intake efficiency up to 66% in the case of a skimming wall only and up to 81% when also a spur dike is also placed.

However, the effects of spur dikes on the hydraulic efficiency of side intakes are not clear. This study aims to contribute to this field. The location, effective length, and deviation angle of a spur dike are considered to investigate the hydraulic efficiency of side intakes in terms of water depths, flow velocities, shear stresses, and sedimentation. This is carried out with contextualization to the real case of the Hemmat Water Intake, in Iran. Several numerical simulations were performed by using the Flow-3D (version 11.0.4) Computational Fluid Dynamics (CFD) software in this regard. All the scenarios and simulations were based on the lowest flow rate of 12 m^3/s to focus on simple (but substantially ordinary) conditions.

2. Materials and Methods

2.1. Survey of the Area under Study

The lateral intake of the Shahid Hemmat Dam is located on the Jarahi River in Khuzestan province, upstream of Shadegan City in Iran (Figure 1). This small dam is intended to raise the water level and provide the necessary water depth for the suction pond of the pumping station, which is 28.8 m long on the river. An intake facility with a capacity of 10 m^3/s was built on the right bank of the river to supply the required water to the pumping station. The inlet angle of the intake is 60 degrees with the central axis of the river. During periods of water scarcity (i.e., dry season)—and therefore of low flows of the Jarahi River—the inlet flow depths at the lateral intake are quite low and the water heads at the pumping station are not enough. Moreover, the approaching flow velocities are very low, leading to sedimentation at the entrance and inside the lateral intake. These are the main factors causing low efficiency and sediment accumulation at the Hemmat Water Intake. Hence, finding a solution to these issues and increasing the intake hydraulic efficiency performing numerical simulations of different scenarios is the main aim of this research. The Flow-3D model was employed for these purposes.

The area under study was carefully surveyed. The river characteristics were gathered from 160 m upstream of the dam to 25 m downstream of the dam. Topographical surveys to get all the physical features of the structures were also carried out. The topography map of the station and its upstream and downstream river bed, with a scale of 1:500, was prepared using the direct ground method (Ahadiyan et al. [32]). The studied area was mapped, and a number of points were obtained. Then, the basic information of bed elevation measurements according to the original scale was entered into AutoCAD software, the 3D shape was output, and the same file was defined for Flow-3D. The effective width of the river was B_r = 30 m and the width of the intake inlet was b_{i1} = 21 m. Based on the information from the hydrometric station, the flow depths of 1.35, 2.20, 2.80, 3.70, and 4.38 m were identified for the flow rates of 12, 32, 62, 100, and 143 m^3/s, respectively (Ahadiyan et al. [32]). Topographic and bathymetric data can be made available upon direct request to the authors of the present paper. Table 1 provides some key physical features as well as some target parameters, such as the spur dike length and position. As previously reported, the parameters which have been considered include: (i) three alternative locations of the spur dike (i.e., a spur dike placed on the opposite side of the intake inlet and aligned with the upstream edge of the intake inlet, to be regarded as a witness spur dike; a spur dike placed at a distance D_S of 7 m downstream of the witness spur dike placement, which implies a dimensionless distance D_S/b_{i1} of 1/3, with b_{i1} being the intake opening width; and a spur dike at a distance of 7 m upstream of the witness spur dike placement with a dimensionless distance, still, of 1/3); (ii) four relative spur dike lengths, L_S/B_r, of 0.24, 0.32, 0.40, and 0.46, with L_S being the effective spur dike length and B_r the approaching river width; and (iii) five spur dike deviation angles of 75, 90, 105, 120, and 135 degrees. All these configurations, almost equally spaced between them, would cover all possible arrangements of a spur dike at the Hemmat Water Intake. Further configurations outside the ranges here considered would be ineffective or impracticable. Figure 2 shows a 3D view of the area under study as introduced in the Flow-3D model.

Figure 1. Photographs of the lateral intake structure of the Shahid Hemmat Dam. (**a**) Aerial view of the Hemmat Dam, lateral intake, and pumping station; (**b**) intake inlet from the top of the pumping station; (**c**) sedimentation at the intake inlet; (**d**) pumping station.

Table 1. Some physical features of the area under study and target parameters for spur dikes.

Distance of the Spur Dike Base from the Witness Spur Dike D_S [m]	Effective Spur Dike Length L_S [m]	Intake Inlet Width b_{i1} [m]	Effective River Width B_r [m]
7	7.2, 9.6, 12, 14	21	30

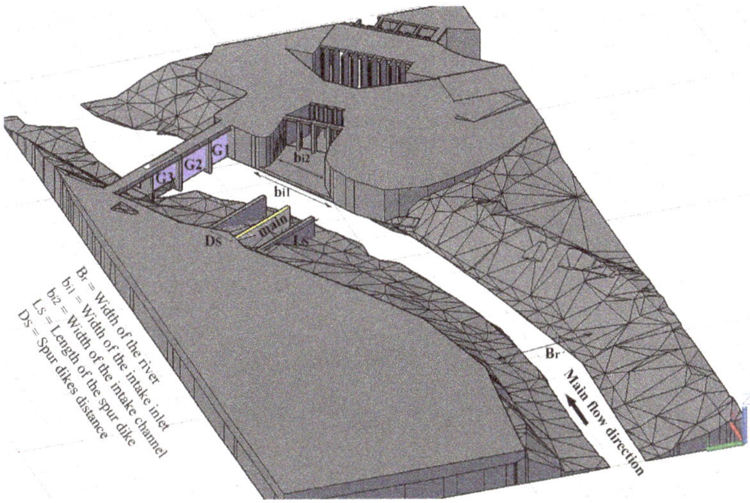

Figure 2. 3D view of the monitored area (and structures) and target parameters used in simulations. B_r: main river width; b_i1: intake inlet width; b_i2: intake channel width; L_S: spur dike length, D_S: distance of the spur dike base from the witness spur dike; G1, G2, and G3: dam gates.

2.2. Numerical Simulations

2.2.1. Notes on Flow-3D Model and Its Implementation

Flow-3D is a well-known and established computational fluid dynamics program. The model uses the volume of fluid (VOF) method on a gridded domain to solve the Navier–Stokes–Reynolds equations for three-dimensional analysis of incompressible flows. Flow-3D uses an advanced free surface flow tracking algorithm (TruVOF) developed by Hirt and Nichols [33], in which fluid configurations are defined in terms of a VOF function $F(x,y,z,t)$. Cells are represented by a cell-fill variable (i.e., fraction function) value representing the ratio of the fluid volume to the cell volume; the empty cells have a value of zero, whole cells a value of one, and cells that contain the free surface a value in the range 0–1. The water surface is then tracked in space and time as a first-order approximation according to the fluid-to-cell volume ratio and the location of the fluid in the surrounding cells. The TruVOF method considers only the fluid's value, not the air's; gas cells are considered empty. Previous studies used this method to reduce the time cost and graphically describe the free surface shape [34–36]. Considering that the main concern of this research is to increase the water diversion efficiency of the Hemmat Water Intake when the river has the lowest inflow, an initial flow of 12 m^3/s was considered in all the implementations and simulations, according to the actual conditions. The spur dike was alternatively located in three different places along the bank, in front of the intake inlet with a distance D_S of 7 m in comparison with the witness spur dike. This implies a relative distance D_S/b_{i1} equal to 1/3. The reason for choosing this distance is that placing a spur dike at a relative distance greater than 1/3 would either not conveniently address the flow towards the intake inlet or require a spur dike too close to the dam. As mentioned above, four spur dike lengths L_S were selected for this study, with values of the ratio L_S/B_r, respectively, equal to 24%, 32%, 40%, and 46%. Moreover, the following spur dike deviation angles were investigated: 75, 90, 105, 120, and 135 degrees. As mentioned earlier, the deviation angle is defined as the angle between the spur dike axis and the original river-bank line from which the spur dike extends. A spur dike pointing downstream has a deviation angle greater than 90 degrees.

2.2.2. Meshing and Boundary Conditions

The simulations were performed for different scenarios (Table 2) to investigate the effect of the spur dike location, length, and angle on the hydraulic efficiency of the lateral intake.

Table 2. Geometric conditions for the scenarios investigated in this study. D_S/b_{i1} is the ratio of the distance D_S between the base of the spur dike and the witness spur dike (i.e., S main) to the intake inlet width b_{i1}; L_S/B_r is the ratio of the spur dike effective length L_S to the river width B_r; and θ is the spur dike deflection angle.

Scenario	D_S/b_{i1}	L_S/B_r	θ (Degrees)
1	S main	0.24	90
2	S main	0.32	90
3	S main	0.40	90
4	S main	0.46	90
5	−1/3	0.32	90
6	+1/3	0.32	90
7	S main	0.32	75
8	S main	0.32	105
9	S main	0.32	120
10	S main	0.32	135
11	S main (all gates open)	0.32	90
12	No spur dike (all gates open)	-	-
13	No spur dike (3rd gate open)	-	-

The time duration for each run was considered to be 120 s after a trial-and-error analysis. Meshing around the spur dike and the intake structure included two parts: a first part from upstream of the spur dike to the outlet border (Mesh block 1) with smaller mesh elements including 2,804,660 cells; a second part from (nearly) the upstream face of the spur dike to the river approach section (Mesh block 2) with larger mesh elements including 1,000,000 cells. As the flow entered the next mesh block without changing, the boundary conditions between these two parts of the mesh cube were considered symmetric. Figure 3 shows two views of the meshing blocks.

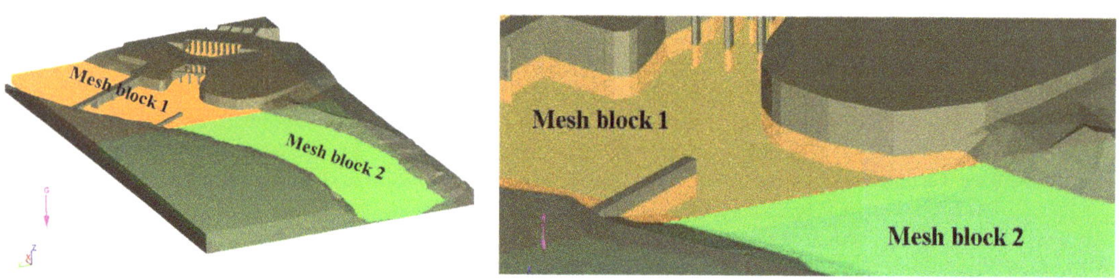

(**a**) (**b**)

Figure 3. Meshing around the intake structure and spur dike. (**a**) General view of the two mesh blocks; (**b**) magnification of the two mesh blocks around the intake inlet and the spur dike.

In each test, the number of mesh cells was determined and executed after investigating different meshes to find the optimal mesh.

The more mesh cells there were, the smaller their size as well as the more complete and the more accurate were the details, such as those for the spur dike and the entrance edge of the intake structure introduced to the software. Where the size of mesh cells was larger, or their amount was smaller, the edge of the spur dike was not well-introduced to the software. Table 3 presents some tests performed until the optimal mesh was found.

Table 3. Details of some tests to find the optimal mesh. The aspect ratio is the ratio of a cell's longest length to the shortest length. The ideal aspect ratio would be 1. X, Y, and Z directions are defined in Figure 2.

Mesh	1	2	3	4
Total number of real cells	1,348,763	3,445,038	1,975,560	2,804,660
Number of real cells (X direction)	289	393	326	365
Number of real cells (Y direction)	359	487	404	452
Number of real cells (Z direction)	13	18	15	17
Maximum aspect ratios (X-Y direction)	1.001	1.000	1.000	1.001
Maximum aspect ratios (Y-Z direction)	1.036	1.015	1.010	1.002
Maximum aspect ratios (Z-X direction)	1.034	1.016	1.011	1.000

The number of mesh cells was determined separately in each X, Y, and Z direction. According to Table 3, the maximum aspect ratio values were slightly higher than 1, which is a perfect value without errors. Finally, the optimal number of mesh cells (Mesh No. 4) without errors with a good finish of all parts of the structures was achieved. The optimal mesh had a maximum aspect ratio of around 1 for 2,804,660 mesh cells in all three directions. Then, the RNG turbulence model was applied, as recommended by Karami et al. [28] and Babagoli Sefidkoohi et al. [14]. The boundary conditions were an inflow discharge of 12 m^3/s for the river, two outflows including the lateral intake and the river reach downstream of the intake structure, and fixed riverbanks. Then, the border on both sides and the bottom of the river were considered as a wall (impermeable contours). The downstream boundary condition of the river was defined as outflow. Numerical simulations were carried out under clear-water conditions by considering the equivalent roughness of bed sediments. The bed roughness was calculated based on the particle size to apply the effect of the river bed on the flow velocity. Figure 4 and Table 4 provide some information on the numerical boundary conditions. Calibration was also done by comparing flow velocity and depth outputs to the measured data.

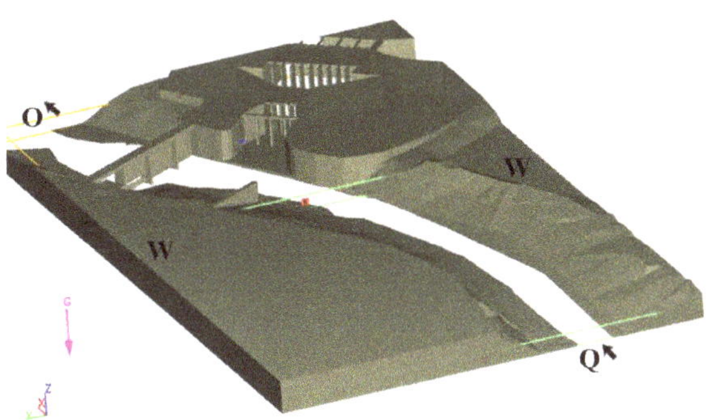

Figure 4. Overview on the upstream (Q), downstream (O), and lateral (W) boundary conditions.

Table 4. Details on the upstream, downstream, and lateral boundary conditions considered in this study.

Boundaries	Boundary Conditions
X Min	VFR (Volume Flow Rate)
X Max	O (Outflow)
Y Min	W (Wall)
Y Max	W (Wall)
Z Min	W (Wall)
Z Max	S (Symmetry)

Finally, in regard to the shear stresses distribution on the river bed, the following equations have been considered, according to Knight et al. [37]:

$$\begin{aligned} \tau_{bx} &= \frac{\rho g}{c^2}\overline{u}\sqrt{\overline{u}^2+\overline{v}^2} \\ \tau_{by} &= \frac{\rho g}{c^2}\overline{v}\sqrt{\overline{u}^2+\overline{v}^2} \\ \tau_b &= \sqrt{(\tau_{bx}^2+\tau_{by}^2)} \end{aligned} \tag{1}$$

where τ_{bx}, τ_{by}, and τ_b are the bed shear stresses in the longitudinal and transverse directions and the total shear stress, respectively. Also, $\overline{u}$ and $\overline{v}$ are the average velocities in the longitudinal and transverse directions at each section. Moreover, g, ρ, and c are the acceleration of gravity, the fluid density, and the Chézy coefficient, respectively.

2.2.3. Calibration of Flow-3D Model

The calibration of the Flow-3D model was done for the real flow of 12 m^3/s according to the velocity data at three points with distances of 100, 85, and 70 m upstream of the dam and in the central line of the river (Ahadiyan et al. [32]). The field data were compared with the numerical results for four distinct opening configurations of the gates. Namely, configuration#1 in which only gate No. 1 was open, configuration#2 in which only gate No. 2 was open, configuration#3 in which only gate No. 3 was open, and configuration#4 in which all gates were open. Ahadiyan et al. [32] showed a suitable agreement between the real and numerical data, with errors between 3.3% and 8.6%.

3. Results and Discussion

Several scenarios were investigated, including those aimed at increasing the efficiency of the Hemmat Water Intake and reducing the sediment deposition in conditions of water scarcity during dry seasons. Therefore, several simulations have been performed to detect the optimal configurations in terms of the best allocation, length, and deviation angle for a spur dike along the front of the intake inlet. Hence, different scenarios were analyzed through field investigation and numerical modelling.

In general, some considerations might be anticipated: (i) A spur dike causes a contraction of the flow path and, as a result, an increase in the flow velocity in the surroundings of its end and an increase in the average velocity in the contracted section. For a given river width and approaching flow discharge, these flow velocity amplifications become larger by increasing amounts as the spur dike length increases. Therefore, it is expected that the placement of a spur dike on the opposite side of a lateral intake would increase the flow velocities at the intake inlet as well as the shear stresses, thus reducing the sedimentation phenomena. The placement of the spur dike in comparison to the intake inlet would impact the position of the contracted flow region. It is understandable that, when the spur dike is aligned with the upstream edge of the intake inlet, the contracted flow region will most likely tend to extend along the entire intake inlet. Conversely, the placement of the spur dike upstream or downstream of the upstream edge of the intake inlet would only lead to partial overlapping between the contracted flow region and the intake inlet area. (ii) At a given river width and approaching discharge, the deviation angle for spur dikes pointing downstream would reduce the contracted flow region as the deviation angle increases. However, for deviation angles greater than 90 degrees (i.e., attracting spur dike), the scouring phenomena are less intense in comparison to a spur dike with a deviation angle of 90 degrees (i.e., deflecting spur dike) or deviation angle less than 90 degrees (i.e., repelling spur dike). Therefore, although an attracting spur dike would (slightly) reduce flow depths and flow velocities, scouring phenomena are alleviated and the spur dike stability is better preserved.

More specifically, in this study, when the effect of the gates is considered, the condition of only gate No. 3 being open is focused on. Indeed, gate No. 3 would determine the most critical state in terms of the minimum diverted discharge through the intake inlet due to its distance from the pumping station (greater than that of the other two gates). Moreover, in the case that the gates are open, the effect and efficiency of the spur dike become more meaningful and clearer when only gate No. 3 is open. Opening the other two gates would make the effect of the spur dike less discernible. The results focusing on different arrangements of open and closed gates, with the spur dike or not, can be found in Ahadiyan et al. [32]. Moreover, it may be important to highlight that, in the following analyses, the sedimentation processes are only presumed to be closely connected with the kinematic fields and/or shear stresses.

3.1. Changes in Flow Patterns at the Intake Inlet as the Position of the Spur Dike Changes

Figure 5 shows the flow pattern variations at the intake inlet affected by the spur dike position for steady flow and in the case of gate No. 3 being open. The spur dike is perpendicular to the approach flow and its alternative placements are: in front of the upstream edge of the intake structure with the base in the opposite bank (witness spur dike, Figure 5a, scenario 2), −7 m upstream of the position of the witness spur dike (Figure 5b, scenario 5), and +7 m downstream of the position of the witness spur dike (Figure 5c, scenario 6). The models were run with the river base discharge during a water shortage scenario and considering the three locations for the spur dike. However, the results on the kinematic fields were similar. Hence, the flow velocities around the intake structure and the spur dike did not change much in these three positions.

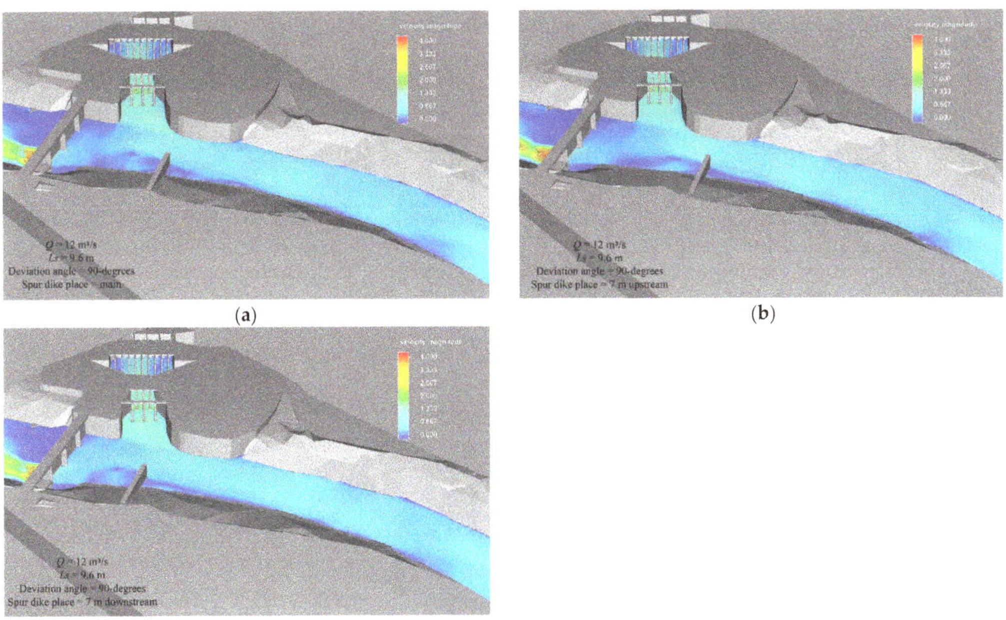

(**a**) (**b**) (**c**)

Figure 5. Flow velocity distribution around the dam, intake structure, and spur dike for the following positions of the latter: (**a**) in front of the upstream edge of the intake structure (witness spur dike); (**b**) 7 m upstream of the position of the witness spur dike; (**c**) 7 m downstream of the position of the witness spur dike. In the legend, flow velocities are in m/s and range from 0.0 to 4.0 m/s with steps of 0.667 m/s.

For example, the average flow velocity in the river (100 m upstream of the dam at the river centerline) was 1.5 m/s (V/V_{max} = 0.94, with V_{max} = maximum flow velocity in all runs) at the intake inlet (point 2 in Figure 6) for the witness spur dike. Meanwhile, in the case of the other positions of the spur dike on the upstream and downstream regions, this value reached 1.47 and 1.48 m/s (V/V_{max} = 0.93), respectively. In all three displacement modes, and especially in the case of the witness spur dike (Figure 5a), a large part of the main flow was directed toward the intake inlet due to the presence of the spur dike and its influential role in directing the flow.

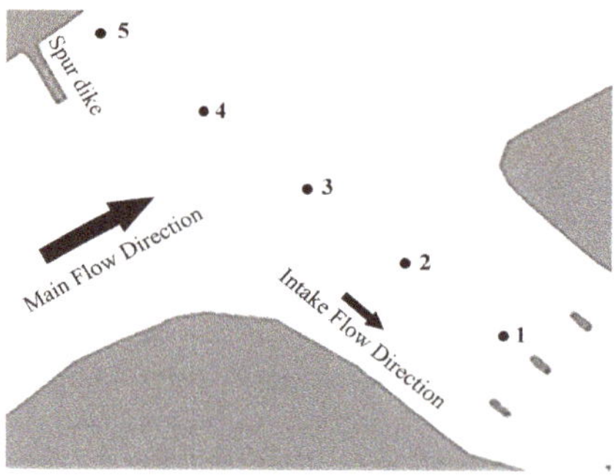

Figure 6. Position of the five control points for local velocity comparisons.

Figure 6 shows the position of the five control points for comparison of the local velocities for the different scenarios.

Figure 7 shows the average flow velocities from the centerline of the intake channel to the opposite riverbank (point 5), with an initial flow rate of 12 m^3/s, gate No. 3 being open, and three different positions of the spur dike. The results show that the flow velocities were very close at points 1, 2, and 3 for all three spur dike locations. On the other hand, a flow velocity gradient occurs at points 1 and 2, in comparison to the other points, due to the more significant impact of the spur dike on this area and their close proximity to the intake inlet. However, the local velocities at points 4 and 5 for $D_S/b_{i1} = -1/3$ are 0.55 m/s (V/V_{max} = 0.34) and 0.47 m/s (V/V_{max} = 0.29), respectively. Meanwhile, in the case of $D_S/b_{i1} = +1/3$, they were 0.29 m/s (V/V_{max} = 0.18) and 0.10 m/s (V/V_{max} = 0.06). The reasons for these discrepancies are the changes in the velocity gradient due to the spur dike base being oriented into the river flow, which causes the development of low-velocity areas and eddies. These low-velocity areas were also transferred (as shown in Figure 5) by changing the location of the spur dike, but they did not affect the average flow velocities entering the intake channel (i.e., points 1 and 2).

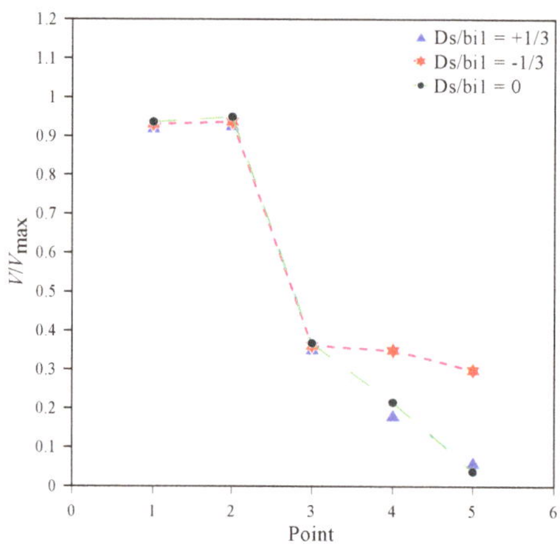

Figure 7. Flow velocities along the transect shown in Figure 6 for different positions of the spur dike.

According to Figures 8–10, four main vortices are formed by comparing the streamlines in the three positions of the spur dikes. They include: the small eddy upstream of the spur dike (Vortex 1) that is formed under the riverside shelter near the structure and is known as the primary eddy or horseshoe vortex; the larger eddy downstream of the spur dike (Vortex 2) that is known as the updraft eddy [38–40]; and two eddies in the intake basin (Vortex 3 and Vortex 4). The power of these eddies decreases or increases when changing the spur dike position. Vortex 1 becomes more powerful when the spur dike positions are downstream or upstream of the main spur dike (Figures 8–10). Meanwhile, this vortex becomes weaker in the case of the witness spur dike being located in front of the upstream edge of the intake inlet (Figure 9). Vortex 2 has an opposite behaviour. This vortex becomes smaller and weaker when moving the spur dike downstream or upstream with respect to the position of the witness spur dike (Figures 8 and 10 under scenarios 5 and 6, respectively, compared to Figure 9 with scenario 2). Moreover, Vortex 2 has larger dimensions and is more power compared to the other vortices, especially when the spur dike is positioned downstream of the main spur dike (Figure 8 with scenario 5). Interestingly, this result is consistent with those obtained by Koken and Constantinescu [41].

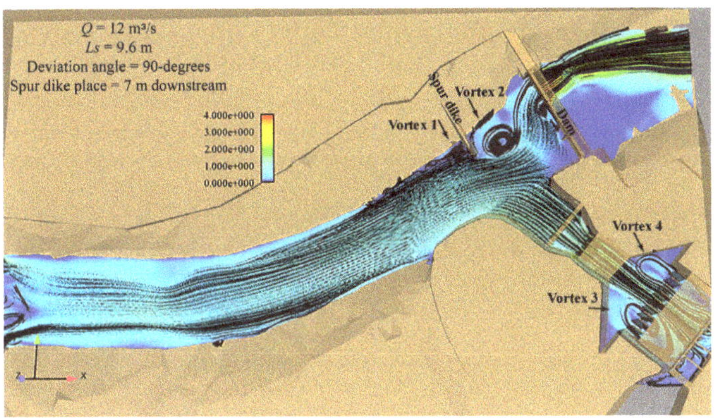

Figure 8. Simulation of the vortices at and around the intake structure in the case of a spur dike placed downstream of the witness spur dike. In the legend, flow velocities are in m/s and range from 0.0 to 4.0 m/s with steps of 1.0 m/s.

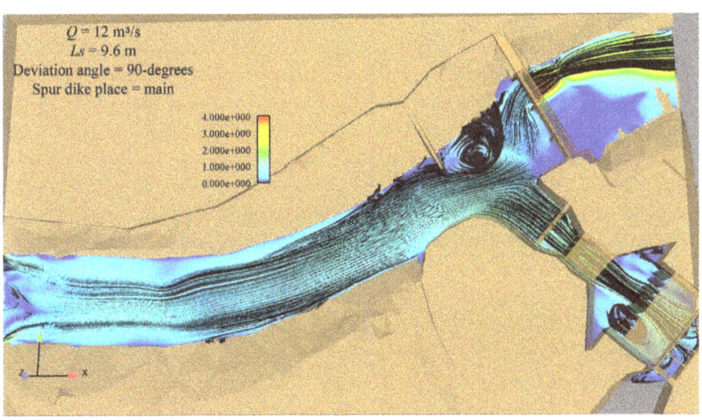

Figure 9. Simulation of the vortices at and around the intake structure in the case of the witness spur dike. In the legend, flow velocities are in m/s and range from 0.0 to 4.0 m/s with steps of 1.0 m/s.

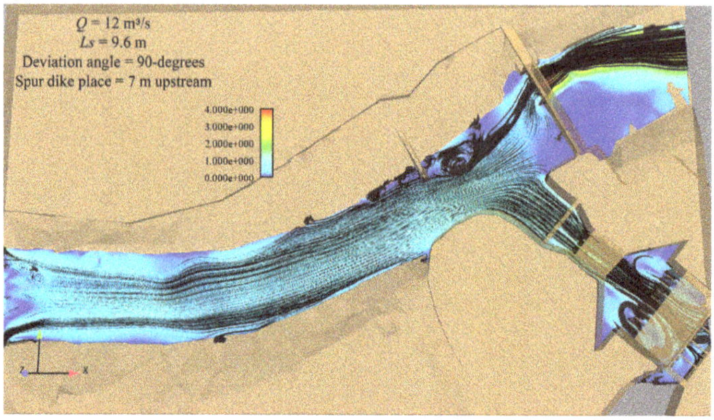

Figure 10. Simulation of the vortices at and around the intake structure in the case of a spur dike placed upstream of the witness spur dike. In the legend, flow velocities are in m/s and range from 0.0 to 4.0 m/s with steps of 1.0 m/s.

3.2. Changes in Flow Patterns at the Intake Inlet as the Length of the Spur Dike Changes

Figure 11a–d shows the effects of the spur dike length on the characteristics of the flow around the intake inlet (scenarios 1–4). The length of the spur dike was either 7.2, 9.6, 12, or 14 m. The spur dike was on the opposite side of the intake inlet, aligned with the upstream edge of the intake inlet (i.e., witness spur dike) and perpendicular to the original riverbank (90 degrees). Only gate No. 3 was open. It was found that the flow velocity at the intake inlet was directly related to the spur dike length. As the length of the spur dike increased, the river flow was better directed towards the intake structure, with an increase in discharge and flow velocity. The results in terms of flow velocity in Figure 12 show that the maximum spur dike length (L_S/B_r = 0.46, scenario 4) compared to the minimum spur dike length (L_S/B_r = 0.24, scenario 1) caused an increase of 11% for the inflow velocity at the intake inlet. As mentioned above, L_S is the spur dike length and B_r is the river width. Specifically, for an L_S equal to 14 m, the velocity was 1.58 m/s ($V/V_{\max}$ = 1.00) while, for an L_S equal to 7.2 m, the velocity was 1.42 m/s ($V/V_{\max}$ = 0.89). The effect of the spur dike length on the generation of eddies is stronger the longer the spur dike length is. There is no formation of vortices for the shorter spur dike with L_S/B_r = 0.24. In this case, the flow is uniformly directed toward gate No. 3.

Figure 12 compares the effects of the spur dike length on the flow velocities at the five points previously considered in Figure 6. The results show that the highest flow velocities in all cases are found at points 1 and 2, near the entrance of the intake channel. The highest flow velocities are found for the spur dike with an L_S/B_r = 0.46 (scenario 4) and the maximum value is found at point 2. At this point, the velocity is 1.58 m/s ($V/V_{\max}$ = 1.00) while, for L_S/B_r = 0.24 (scenario 1), the velocity is equal to 1.42 m/s ($V/V_{\max}$ = 0.89), which shows an increase of 11% when passing from the shortest spur dike to the longest one. The greatest change in flow velocities occurs at point 3, with a drastic decrease when moving from the longer to the shorter spur dike. Specifically, the average velocity at point 3 for the spur dike with an L_S/B_r = 0.24 is 0.52 m/s ($V/V_{\max}$ = 0.33), while at the same point the velocity is 1.55 m/s ($V/V_{\max}$ = 0.98) with an increase of 198% when the L_S/B_r = 0.46. This is due to the fact that point 3 is placed in a high-velocity area around the nose of the longest spur dike.

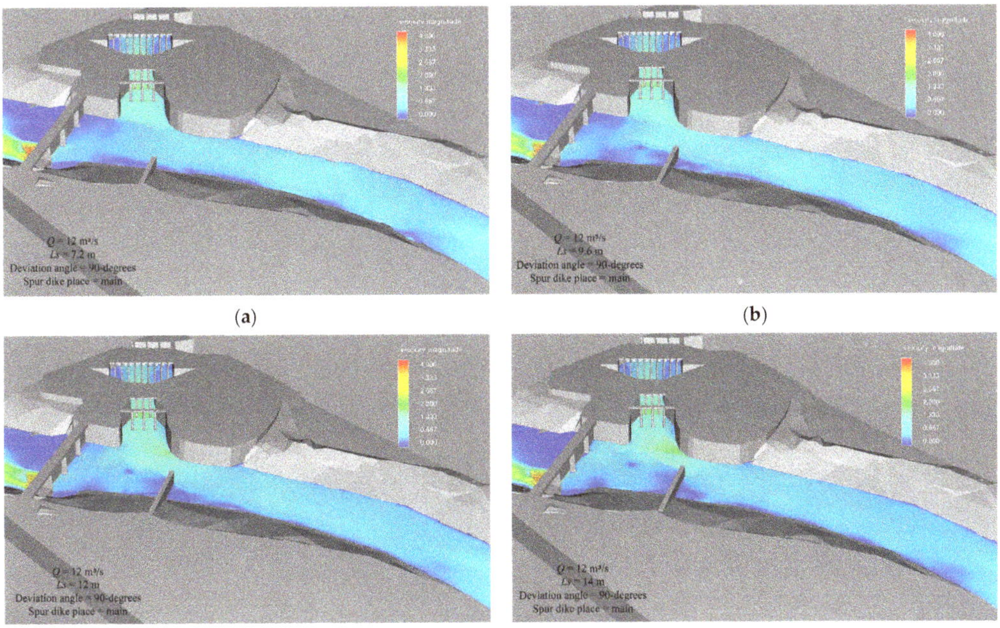

(a) (b) (c) (d)

Figure 11. Simulation of the kinematic field around the intake inlet in the case of a spur dike with: (**a**) L_S/B_r = 0.24; (**b**) L_S/B_r = 0.32; (**c**) L_S/B_r = 0.40; and (**d**) L_S/B_r = 0.46. The approach discharge Q is equal to 12 m^3/s and velocities are given in m/s. In the legend, flow velocities are in m/s and range from 0.0 to 4.0 m/s with steps of 0.667 m/s.

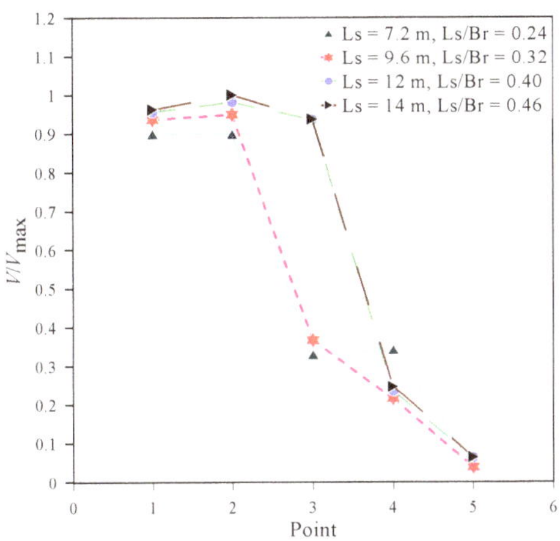

Figure 12. Flow velocities along the transect shown in Figure 6 for different lengths of the spur dike.

3.3. Changes in Flow Patterns at the Intake Inlet as the Deflection Angle of the Spur Dike Changes

Figure 13a–e shows the effects of the spur dike deflection angle on the flow patterns and characteristics around the intake inlet. Five deflection angles were considered, namely 75, 90, 105, 120, and 135 degrees, to which corresponded the scenarios 7, 2, 8, 9, and 10, respectively. As previously said, in this study, the deviation angle is defined as the angle between the spur dike axis and the original riverbank line from which the spur dike extends.

A spur dike pointing downstream has a deviation angle greater than 90 degrees, by way of example. The spur dike length was kept constant such that the L_S/B_r = 0.32. The numerical simulations revealed that no appreciable differences were noted in the kinematic field as the deflection angle varied.

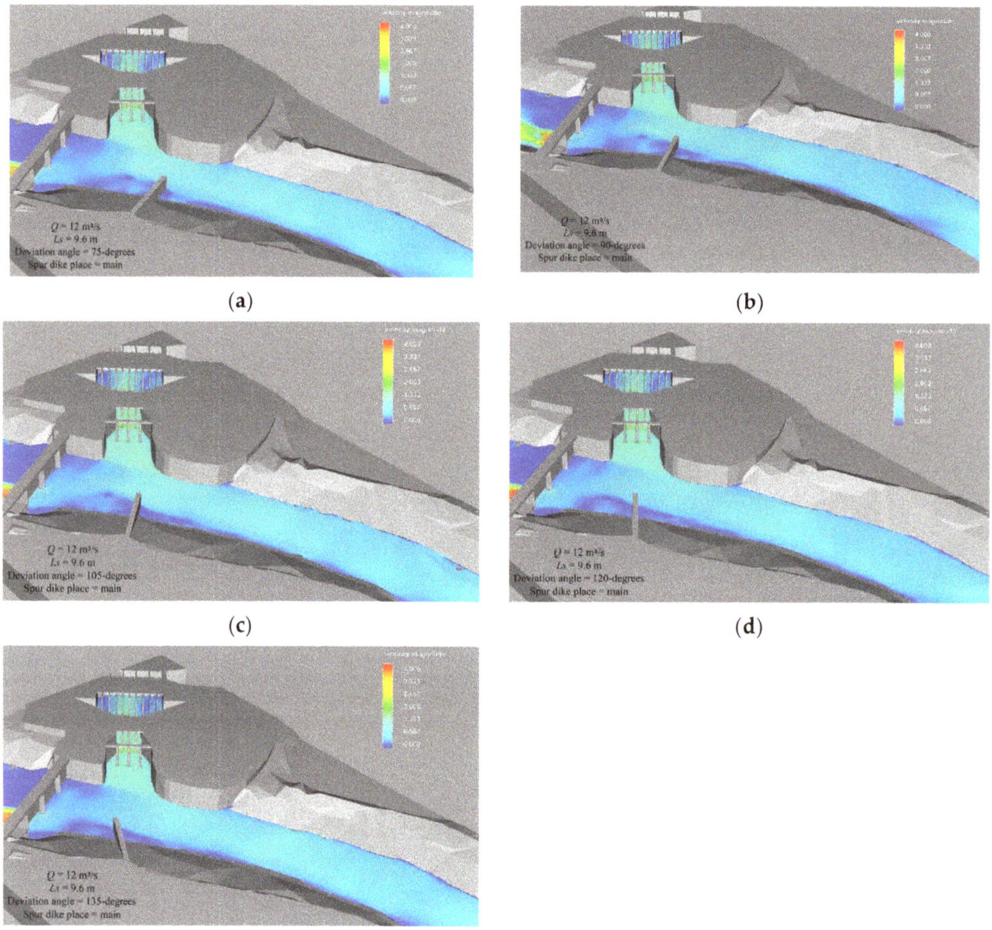

Figure 13. Simulation of the kinematic field around the intake inlet in the case of a spur dike with deflection angle of: (**a**) 75; (**b**) 90; (**c**) 105; (**d**) 120; and (**e**) 135 degrees. The approach discharge Q is equal to 12 m^3/s. In the legend, flow velocities are in m/s and range from 0.0 to 4.0 m/s with steps of 0.667 m/s.

Figure 14 compares the effects of the spur dike deflection angle on the flow velocities at the five points previously considered in Figure 6. It was found that no significant differences in the velocities occurred, especially at points 1 and 2 where the differences were less than 2%. This would imply that the flow velocities at the intake inlet are not much influenced by variations in the deflection angle between 75 and 135 degrees.

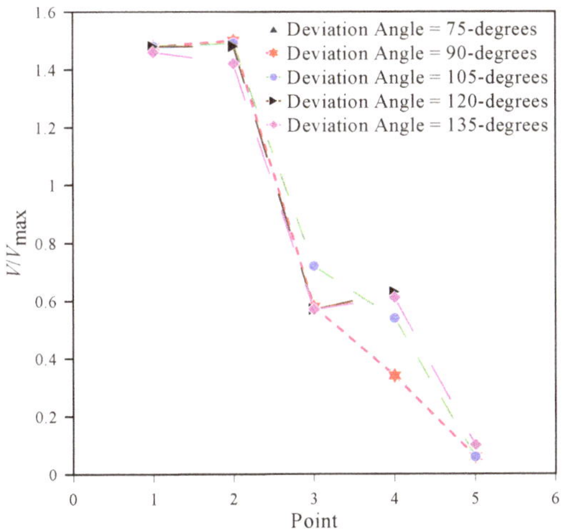

Figure 14. Flow velocities along the transect shown in Figure 6 for different deflection angles of the spur dike.

It should be noted that the deflection angle did not even affect the flow depths. The effects of changes in the spur dike deflection angle on the flow depth, with reference to 12 points along the transect (Figure 15), are shown in Figure 16. The results show that, for a spur dike deviation angle of 135 degrees, the flow depths at the intake inlet are the highest in comparison to the other configurations with different deviation angles. Specifically, the flow depth at the intake inlet increases by about 9% for the spur dike with a deflection angle of 135 degrees in comparison to the one with an angle of 75 degrees. In summary, the results showed that the flow depths increased, albeit in a limited way, with the spur dike deviation angle (from 75 to 135 degrees). Moreover, for all the deviation angles, the highest flow depths occurred at points 5, 6, 7, and 8 with a maximum value of the ratio Z/Z_{max} of about 0.35. Here, Z is the flow depth value and Z_{max} is the maximum flow depth value.

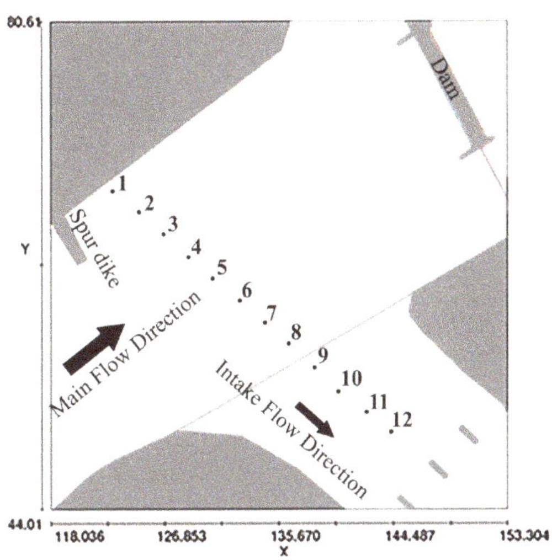

Figure 15. Position of the 12 control points for local flow depth comparisons.

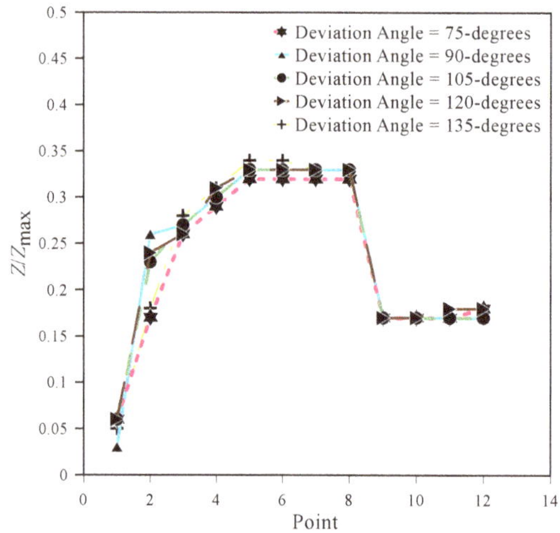

Figure 16. Flow depths Z along the transect shown in Figure 15 for different deflection angles of the spur dike. Only gate No. 3 is open and the witness spur dike is placed.

3.4. *Bed Shear Stresses Comparison for Different Scenarios*

Figure 17 shows three transects with control points for the comparison of shear stresses.

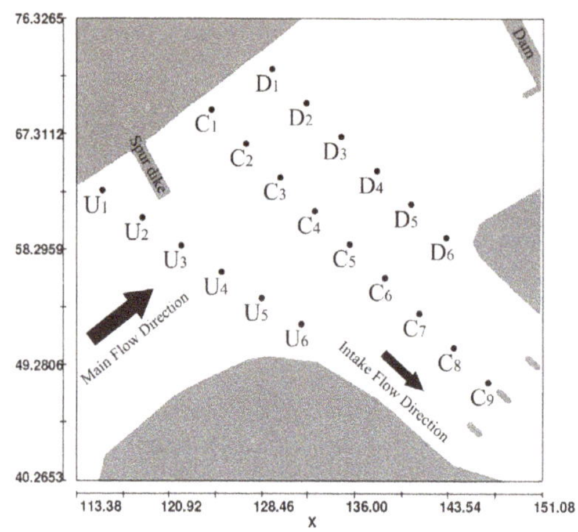

Figure 17. Indication of the three transects with the related control point positions for the comparison of shear stresses for different scenarios.

In Figures 18–20, the bed shear stresses along three transects, including the transect along the intake channel centerline including the points from C1 to C9, the transect upstream of the central one including the points from U1 to U6, and the transect downstream of the central one including the points from D1 to D6, are shown. The central transect with point C extends from the intake channel (along the channel centerline) until nearly the spur dike sidewall. Due to the steepness of the river sidewall and the entrance ramp of the intake channel, some points are located in the solid part of the structure (i.e., points C1, C7, C8, and C9), and thus have no value.

According to Figure 18, the results show that the highest values of the shear stresses, with significant differences from the other cases, occur when the spur dike is used and when all gates or gate No. 3 only are open. Moreover, the highest values of the shear stresses are related to the points C5, C4, and C3 regardless of the scenario because these points are located in the main river flow. For example, in the case of point C5 without a spur dike and when all gates are open, the shear stress τ_b is equal to 2.28 kPa. Similarly, τ_b is equal to 3.01 kPa when gate No. 3 only is open. Hence, the single opening of gate No. 3 increases the shear stress at point C5 (close to the intake inlet) by 32% in comparison to the configuration in which all gates are open. In this way, the main flow is more strongly diverted towards the intake channel. Therefore, opening gate No. 3 would reduce the possibility of sedimentation in comparison to the opening of all the gates.

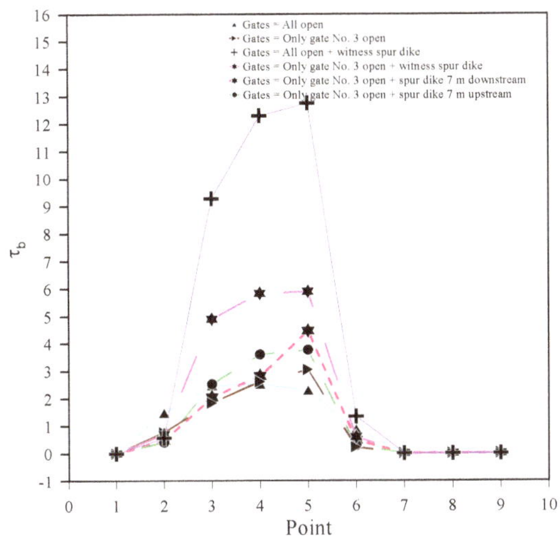

Figure 18. Shear stresses τ_b along the central transect shown in Figure 17. The values are in kPa.

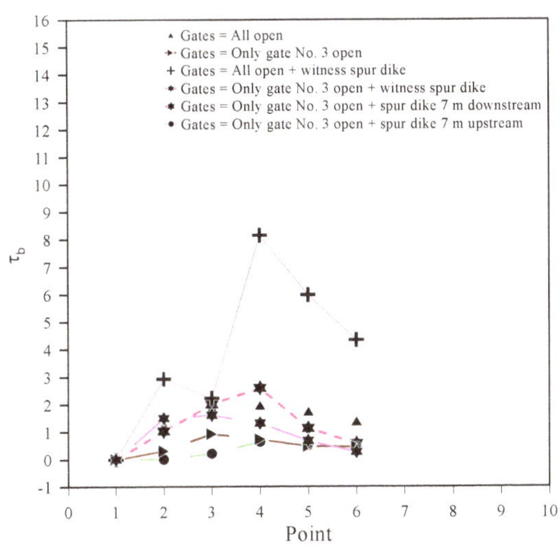

Figure 19. Shear stresses along the transect downstream of the central one shown in Figure 17. The values are in kPa.

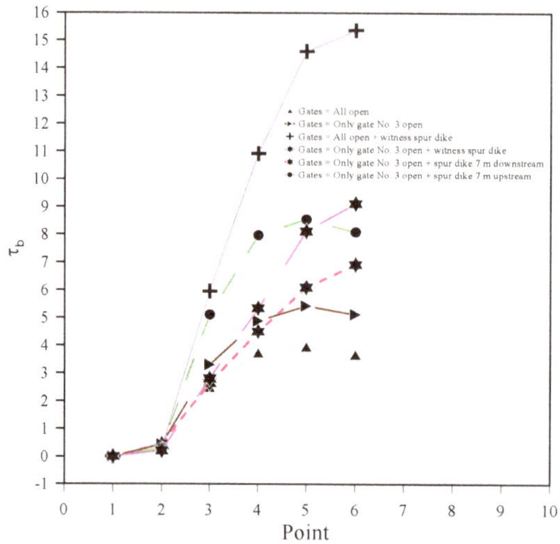

Figure 20. Shear stresses along the transect upstream of the central one shown in Figure 17. The values are in kPa.

On the other hand, the shear stress at point C5 increases up to 12.63 kPa when a spur dike is placed. Then, the shear stress increases more than five times in comparison to the case of the absence of a spur dike when the shear stress is equal to 2.28 kPa. Moreover, looking at point C6 (even closer to the intake inlet) in the case of the presence of a spur dike, the shear stress is equal to 1.31 kPa with an increase of 50% in comparison to the case without a spur dike, in which the shear stress value drops to 0.87 kPa. Hence, the spur dike would increase the flow strength at the intake inlet. Likewise, the shear stress decreased at point C2 (because the spur dike armors this point) from 1.47 to 0.51 kPa (Figure 18), with a reduction by a factor of three. However, the possibility of sedimentation increases in this area due to the lower shear stresses. This process causes the gradual modification of the riverbank on the left. In addition, the shear stresses are 9.11 and 12.27 kPa at points C3 and C4, respectively, when the spur dike is placed. This shows increases of 270% and 386% in comparison to the condition without the spur dike. Such a situation occurs due to the presence of an area with maximum velocities near the nose of the spur dike and the flow separation from the structure. In these conditions, scouring occurs around the nose of the spur dike and the eroded area would develop downstream of the structure, which is a result also obtained by Jafari and Sui [42].

At the downstream points, and in particular at points D5 and D6 (Figure 19), the shear stresses are, in presence of a spur dike, 6.34 and 4.66 kPa, respectively. In the case of the absence of the spur dike, the shear stresses at the above points decrease by 161% and 86%. Indeed, when a spur dike is placed, a more significant part of the main flow is directed to the intake, which faces the high-velocity flow in the river. Points D5 and D6 are located along the downstream wall of the intake. This area is characterized by lower shear stresses in comparison to the central and upstream transects due to the direction of the flow toward the intake channel. However, the shear stress at point D3 is almost equal in both cases with and without a spur dike, being about 2 kPa (Figure 19). Meanwhile, this point is located in the river center and downstream of the intake inlet. In other words, the surrounding area is not much affected by the spur dike and the lateral intake, which would explain the almost equal values of the shear stress in both cases.

At the upstream points, in particular points U6 and U5, the values of the shear stress, in the case of the presence of the spur dike, are 15.39 and 14.70 kPa, respectively (Figure 20), whereas, in the case of the absence of the spur dike, both values are about 3.5 kPa, which would show a notable decrease. However, these results are in contrast with those for points

U1 and U2, where the shear stresses in the presence of a spur dike are even smaller than those found in the case without a spur dike. Actually, the presence of the spur dike would cause the surrounding area of points U1 and U2 to be included in the region where the main flow impacts the spur dike, with a significant reduction in the flow velocities and shear stresses.

3.5. *Comparison of Results with Previous Similar Literature Studies*

Patel et al. [43] found, experimentally, that the maximum local scour occurs at the leading edge of the spur dike; the present research also achieved the same results, though in terms of shear stresses. Karami et al. [28] showed, through a 3D Computational Fluid Dynamic (CFD) code, that placing the spur dike and submerged vanes in the main channel and in front of the intake inlet would increase shear stresses near the base of the vanes and spur dikes themselves. This outcome also complies with the results of the present study. In addition, by comparing the effects of changing the relative spur dike length (i.e., L_S/W_c = 0.20, 0.25, and 0.30, with L_S being spur dike length and W_c the channel width) on the intake efficiency, Karami et al. [28] found that the spur dike with an L_S/W_c = 0.20 allowed a diversion ratio from the initial value of 11% (without dikes and vanes) to 22.2%, with a simultaneous reduction in sedimentation. Iqbal and Tanaka's [40] experimental findings revealed that elongating the spur dike wing length reciprocally affected the depth-averaged velocity (at the spur dike head and near the adjacent spur dike bank), concurrently impacting the flow deflection and backwater rising. Therefore, the general results of these studies are similar, and the presence of a spur dike on the bank opposite the intake inlet would increase the efficiency of the intake structures. They further highlighted that increasing the spur dike length would cause a decrease in the diverted discharge. Conversely, the results of the present study showed that, when placing the longest spur dike, the inflow velocity at the intake inlet increases by 11% in comparison to the shortest spur dike. Actually, Karami et al. [28] considered a spur dike just upstream of the intake inlet and placed perpendicular to the flow. In such a condition, increasing the spur dike length would direct the main flow towards the upstream intake wall, so less flow enters the intake structure. Further, the results of this research showed that the breakwater induced by the spur dike increases the shear stresses for the flow entering the intake structure. However, these findings are in contrast with those obtained by Karami et al. [28], who used parallel and zig-zag submerged vanes at angles of 10 and 30 degrees. Submerged vanes reduce the incoming shear stresses because these plates are placed before the water intake opening. In the present research, the spur dike can divert the main flow and the high-speed areas towards the intake structure. Regarding the development of vortices and their positioning, as well as the generation of low- and high-velocity regions, the results of this study were consistent with the results of earlier research such as that by Tripathi and Pandey [39] (experimental and numerical study), Abbasi et al. [38] (numerical study), and Koken and Constantinescu [41] (experimental and numerical study).

Interestingly, the streamlines around the spur dikes in Flow-3D numerical modelling appear to be in harmony with the experimental evidence reported in some experimental studies on local scour at spur dikes (e.g., Pandey et al. [44] and Aung et al. [45]). As mentioned earlier, in this study, the erosion and sedimentation processes are only presumed to be closely connected with the kinematic fields and/or shear stresses. However, in-depth analysis and comparison between the numerical results of this research and the experimental outcomes regarding the bed morphological changes around spur dikes could lead to further validations.

4. Conclusions

This study was aimed to increase the efficiency and inflow at the lateral intake as well as to reduce the sedimentation area at the intake inlet. To this purpose, the effect on the flow patterns of a spur dike is verified by performing numerical simulations. Three spur dike locations, four spur dike lengths, and five spur dike deflection angles were investigated.

The results generally showed that there are no noticeable changes in the kinematic field of the flow entering the intake when changing the spur dike position. According to the findings, the flow velocities at the intake inlet are directly related to the spur dike length. In particular, the inlet flow velocity increases with an increase in the L_S/B_r, where L_S is the effective spur dike length and B_r is the river width. The longest spur dike, in comparison to the shortest one, would increase the flow velocity at the intake inlet by 11%. In the comparison of four different lengths, the highest inlet velocities were found for the values of L_S/B_r equal to 0.40 and 0.46 (scenarios 3 and 4), with values of 1.55 m/s (V/V_{max} = 0.98) and 1.58 m/s (V/V_{max} = 1.00). The average velocity at point 3 and for the case with L_S/B_r = 0.24 was found to be equal to 0.52 m/s (V/V_{max} = 0.33). In the case of a longer spur dike with L_S/B_r = 0.46, the velocity at the same point was found to be equal to 1.55 m/s (V/V_{max} = 0.98), with an increase of 198%. On the other hand, the results showed that the inlet flow velocity did not change much with the changing of the spur dike deflection angle between 75 and 135 degrees. However, the flow depth at the intake inlet increased by 9% when the spur dike was placed with a deflection angle of 135 degrees in comparison to 75 degrees. Based on the shear stress results, it was found that, in all scenarios, the highest shear stresses were related to the centerline of the river. Also, the shear stress in front of the intake inlet (point C5), in the case of the presence of a spur dike, was found to be equal to 12.63 kPa and, in comparison to the case without a spur dike, increased by 454%. Similarly, at the intake inlet (point C6), the shear stress increased by 50% in the case of the presence of a spur dike. Conversely, in the case of point C2, the spur dike reduced the shear stress from 1.47 to 0.51, with a decrease of 188%. Finally, for the longest spur dike (L_S/B_r = 0.46) and for the deflection angle of 135 degrees, the main flow of the river is directed towards the intake structure with the highest velocity and depth and in the best way. Also, the risk of sediment settling is reduced due to the increased flow velocity towards the intake structure. Hence, the flow intensity in the intake channel increases, and the required flow at the pumping station is supplied. However, the latter configuration could not be the optimal one among all the possible combinations (not all examined in this study) for the placement, length, and deviation angle of the spur dike. Moreover, there are further design configurations (for instance: spur dike of different shape, permeable spur dike, submerged spur dikes or vanes, spur dikes in cascade) which can be considered to further narrow down the search for optimal spur dike design. On the other hand, in the case of the placement of the main spur dike, the shear stress at the intake inlet would increase by 50% in comparison to the case without a spur dike, and this indicates the effect of the spur dike in increasing the flow entering the intake inlet. All the results showed that the placement of a simple spur dike in front of the intake opening would increase the velocities and shear stresses of the flow entering the lateral intake. This is very practical in increasing the water diversion ratio and reducing the possibility of sedimentation at the intake inlet and inside the intake channel. All of these results are possible through the use of hydraulic works of moderate size and cost (in this specific case the use of spur dikes) to enhance the hydraulic efficiency of lateral intakes without (or with little) maintenance over time. An accurate numerical simulation could reveal attractive details for a good functionality even in periods of water scarcity. However, only low ordinary flow conditions were considered here. Future in-depth analyses could consider the impact of the spur dike on the bed morphological changes (e.g., scouring phenomena and aggradation patterns) when floods occur. The design proposals resulting from this study will see their realization in the near future. Therefore, a comparison between numerical results and empirical evidence can only be discussed in a subsequent paper. In this study, only a qualitative and partial comparison with the literature was made possible. On the other hand, the results here presented appear to have the potential of possible extension to other real-word river engineering projects. They are mainly based on numerical simulations which can be applied in similar contexts, but, in any case, would require selected field data for calibration and validation. When 3D processes show potential to become significant, experiments on physical models could be of important help.

Author Contributions: Conceptualization: All authors; Methodology: All authors; Software: S.A.C. and J.A.; Validation: All authors; Formal Analysis: All authors; Investigation: All authors; Resources: All authors; Data Curation: All authors; Writing—Original Draft Preparation: S.A.C., H.A.N. and J.A.; Writing—Review and Editing: All authors; Visualization: All authors; Supervision: G.O., M.V. and J.A.; Project Administration: G.O., M.V. and J.A.; Funding Acquisition: J.A. and S.M.S. All authors have read and agreed to the published version of the manuscript.

Funding: This work was supported by the Shahid Chamran University of Ahvaz and Khuzestan Water and Power Authority.

Data Availability Statement: All data generated or analysed during this study are available from the corresponding authors on request.

Acknowledgments: The authors are grateful to the Research Council of the Shahid Chamran University of Ahvaz, the Khuzestan Water and Power Authority (KWPA), and the Center of Excellence of the Irrigation and Drainage Networks Improvement and Maintenance (Ahvaz, Iran) for their valuable support.

Conflicts of Interest: The authors declare no conflicts of interest.

References

1. Tadda, M.A.; Ahsan, A.; Imteaz, M.; Shitu, A.; Danhassan, U.A.; Muhammad, A.I. Operation and Maintenance of Hydraulic Structures. In *Hydraulic Structures—Theory and Applications*; Ahsan, A., Ed.; IntechOpen: London, UK, 2020. [CrossRef]
2. Montaseri, H.; Tavakoli, K.; Evangelista, S.; Omidvar, P. Sediment transport and bed evolution in a 180° curved channel with lateral intake: Numerical simulations using Eulerian and Discrete Phase models. *Int. J. Mod. Phys. C* **2020**, *31*, 2050113. [CrossRef]
3. Moghadam, M.K.; Keshavarzi, A.R. An optimised water intake with the presence of submerged vanes in irrigation canals. *Irrig. Drain.* **2010**, *59*, 432–441. [CrossRef]
4. Alomari, N.K.; Yusuf, B.; Mohammad, T.A.; Ghazali, A.H. Influence of diversion angle on water and sediment flow into diversion channel. *Int. J. Sediment Res.* **2020**, *35*, 600–608. [CrossRef]
5. Masjedi, A.; Foroushani, E.P. Experimental effect of flow depth on ratio discharge in lateral intakes in river bend. *IOP Conf. Ser. Earth Environ. Sci.* **2012**, *15*, 062054. [CrossRef]
6. Farshad, R.; Kashefipour, S.M.; Ghomeshi, M.; Oliveto, G. Temporal scour variations at permeable and angled spur dikes under steady and unsteady flows. *Water* **2022**, *14*, 3310. [CrossRef]
7. Montaseri, H.; Asiaei, H.; Baghlani, A.; Omidvar, P. Numerical study of flow pattern around lateral intake in a curved channel. *Int. J. Mod. Phys. C* **2019**, *30*, 1950083. [CrossRef]
8. Neary, V.S.; Odgaard, A.J. Three-dimensional flow structure at open-channel diversions. *J. Hydraul. Eng.* **1993**, *119*, 1223–1230. [CrossRef]
9. Jalili, H.R.; Dalir, A.H.; Farsadizadeh, D. Effect of mouth shape on the flow pattern and sedimentation in the vertical lateral diversion. *Water Soil* **2012**, *25*, 1321–1331.
10. Al-Zubaidy, R.A.; Ismaeil, R.H. Sediment control at the lateral channel inlet. *IOP Conf. Ser. Earth Environ. Sci.* **2022**, *961*, 012096. [CrossRef]
11. Abolghasemi, M.; Ghodsian, M.; Sanginabadi, H. The effect of flow diversion on sediment entering intake located at section 52° in sinuous rivers. *J. Hydraul.* **2011**, *6*, 41–52.
12. Kashyap, S.; Constantinescu, G.; Rennie, C.D.; Post, G.; Townsend, R. Influence of channel aspect ratio and curvature on flow, secondary circulation, and bed shear stress in a rectangular channel bend. *J. Hydraul. Eng.* **2012**, *138*, 1045–1059. [CrossRef]
13. Azimi, H.; Shabanlou, S.; Kardar, S. Flow field within rectangular lateral intakes in the subcritical flow regimes. *Model. Earth Syst. Environ.* **2019**, *5*, 421–430. [CrossRef]
14. Babagoli Sefidkoohi, R.; Shahidi, A.; Ramezani, Y.; Kahe, M. Simulation of flow pattern in intake by using a numerical model. *Water Harvest. Res.* **2017**, *2*, 24–36.
15. Safarzadeh, A.; Khaiatrostami, B. Mean flow characteristics, vertical structures and bed shear stress at open channel bifurcation. *J. Appl. Res. Water Wastewater* **2017**, *4*, 299–304.
16. Tavakoli, K.; Montaseri, H.; Omidvar, P.; Evangelista, S. Numerical simulation of sediment transport in a U-shaped channel with lateral intake: Effects of intake position and diversion angle. *Int. J. Mod. Phys. C* **2019**, *3*, 1950071. [CrossRef]
17. Sayed, T. An experimental study of branching flow in open channels. *Limnol. Rev.* **2019**, *19*, 93–101. [CrossRef]
18. Heidari Rad, P.; Kamanbedast, A.; Heidarnejad, M.; Masjedi, A.; Hasonizadeh, H. The effect of convergence and divergence on flow pattern and sediment transport in lateral intakes using physical and numerical models. *Ain Shams Eng. J.* **2020**, *11*, 445–454. [CrossRef]
19. Meshkati, S.M.; Salehi, S. Numerical simulation of 3D flow pattern at lateral intake in 180-degree bend. *Iran. J. Soil Water Res.* **2021**, *52*, 37–51.
20. Niknezhad, F.; Mahmoudi, A.; Vaghefi, M. A Study of the effects of a lateral intake's positions on the flow pattern around a bridge pier in a 180° bend using SSIIM2. *Arab. J. Sci. Eng.* **2023**, *48*, 12903–12929. [CrossRef]

21. Davoodi, L.; Bejestan, M.S. Control of sediment entry to intake on a trapezoidal channel by submerged vane. *Ecol. Environ. Conserv.* **2012**, *18*, 165–169.
22. Sarhadi, A.; Jabbari, E. Investigating effect of different parameters of the submerged vanes on the lateral intake discharge located in the 180 degree bend using the numerical model. *J. Civ. Eng.* **2017**, *3*, 1176–1187. [CrossRef]
23. Firozjaei, R.M.; Salehi Neyshabouri, S.A.A.; Amini Sola, S.; Mohajeri, S.H. Numerical simulation on the performance improvement of a lateral intake using submerged vanes. *IJST Trans. Civ. Eng.* **2019**, *43*, 167–177.
24. Baltazar, J.; Alves, E.; Bombar, G.; Cardoso, H.A. Effect of a submerged vane-field on the flow pattern of a movable bed channel with a 90° lateral diversion. *Water* **2021**, *13*, 828. [CrossRef]
25. Moghadam, M.K.; Amini, A.; Keshavarzi, A. Intake design attributes and submerged vanes effects on sedimentation and shear stress. *Water Environ. J.* **2019**, *34*, 374–380. [CrossRef]
26. Shamloo, H.; Asnaashari, A.; Pirzadeh, B. Numerical investigation of the effects of submerged vanes and spur dikes on sediment control at lateral intakes. In Proceedings of the 12th Annual Conference of Thai Society of Agricultural Engineering "International Conference in Agricultural Engineering" (Novelty, Clean and Sustainable) Chon-Chan Pattaya Resort, Chonburi, Thailand, 31 March–1 April 2011.
27. Daneshfaraz, R.; Bagheri, Y.; Ghaderi, A.; Ghahremanzadeh, A. An analysis of flowing pattern around T-shaped spur dike at 90° arc, based on Fluent and Flow-3D models. *Int. J. Water Resour. Dev.* **2015**, *3*, 1–9.
28. Karami, H.; Farzin, S.; Sadrabadi, M.T.; Moazeni, H. Simulation of flow pattern at rectangular lateral intake with different dike and submerged vane scenarios. *Water Sci. Eng.* **2017**, *10*, 246–255. [CrossRef]
29. Geravandi, E.; Kamanbedast, A.A.; Masjedi, A.R.; Heidarnejad, M.; Bordbar, A. Laboratory investigation of the impact of armor dike simple and L-shaped in upstream and downstream intake of the hydraulic flow of the Kheirabad River to help physical model. *Fresenius Environ. Bull.* **2018**, *27*, 263–276.
30. Zamani, M.; Rabiefar, H.; Rostami, M. Experimental evaluation of spur dikes placement position effect on the hydraulic and erosion conditions of intakes. *Water Supply* **2020**, *20*, 900–908. [CrossRef]
31. Moradinejad, A.; Saneie, M.; Ghaderi, A.; Zamanieh Shahri, S.M. Experimental study of flow pattern and sediment behavior near the intake structures using the spur dike and skimming wall. *Appl. Water Sci.* **2019**, *9*, 195. [CrossRef]
32. Ahadiyan, J.; Chenari, S.A.; Nadian, H.A.; Katopodis, C.; Valipour, M.; Sajjadi, S.M.; Omidvarinia, M. Sustainable systems engineering by CFD modeling of lateral intake flow with flexible gate operations to improve efficient water supply. *Int. J. Sediment Res.* **2024**, *in press*. [CrossRef]
33. Hirt, C.W.; Nichols, B.D. Volume of Fluid (VOF) method for the dynamics of free boundaries. *J. Comput. Phys.* **1981**, *39*, 201–225. [CrossRef]
34. Di Francesco, S.; Biscarini, C.; Manciola, P. Numerical simulation of water free-surface flows through a front-tracking lattice Boltzmann approach. *J. Hydroinform.* **2015**, *17*, 1–6. [CrossRef]
35. Daneshfaraz, R.; Aminvash, E.; Bagherzadeh, M.; Ghaderi, A.; Kuriqi, A.; Najibi, A.; Ricardo, A.M. Laboratory investigation of hydraulic parameters on inclined drop equipped with fishway elements. *Symmetry* **2021**, *13*, 1643. [CrossRef]
36. Ghaderi, A.; Abbasi, S.; Di Francesco, S. Numerical study on the hydraulic properties of flow over different pooled stepped spillways. *Water* **2021**, *13*, 710. [CrossRef]
37. Knight, D.W.; Omran, M.; Tang, X. Modeling depth-averaged velocity and boundary shear in trapezoidal channels with secondary flows. *J. Hydraul. Eng.* **2007**, *133*, 39–47. [CrossRef]
38. Abbasi, S.; Kamanbedast, A.; Ahadian, J. Numerical investigation of angle and geometric of L-shape groin on the flow and erosion regime at river bends. *World Appl. Sci. J.* **2011**, *15*, 279–284.
39. Tripathi, R.P.; Pandey, K.K. Scour around spur dike in curved channel: A review. *Acta Geophys.* **2022**, *70*, 2469–2485. [CrossRef]
40. Iqbal, S.; Tanaka, N. Hydraulic performance assessment of various submerged pile designs around an emerged dike. *Water Sci. Eng.* **2024**, *in press*. [CrossRef]
41. Koken, M.; Constantinescu, G. An investigation of the flow and scour mechanisms around isolated spur dikes in a shallow open channel: 1. Conditions corresponding to the initiation of the erosion and deposition process. *Water Resour. Res.* **2008**, *44*, W08406. [CrossRef]
42. Jafari, R.; Sui, J. Velocity field and turbulence structure around spur dikes with different angles of orientation under ice covered flow conditions. *Water* **2021**, *13*, 1844. [CrossRef]
43. Patel, H.K.; Qi, M.; Kumar, B. Downward seepage effects on flow near a L-shape spur dike and bed morphology. *Int. J. Sediment Res.* **2024**, *39*, 194–208. [CrossRef]
44. Pandey, M.; Valyrakis, M.; Qi, M.; Sharma, A.; Lodhi, A.S. Experimental assessment and prediction of temporal scour depth around a spur dike. *Int. J. Sediment Res.* **2021**, *36*, 17–28. [CrossRef]
45. Aung, H.; Onorati, B.; Oliveto, G.; Yu, G. Riverbed morphologies induced by local scour processes at single spur dike and spur dikes in cascade. *Water* **2023**, *15*, 1746. [CrossRef]

water

MDPI

Technical Note

Experimental Study on Upstream Water Level Rise of Submerged Rock Weirs

Wen Zhang [1,2], Xingnian Liu [3] and Binrui Gan [1,2,*]

[1] Key Laboratory of Hydraulic and Waterway Engineering of the Ministry of Education, Chongqing Jiaotong University, Chongqing 400074, China; zhangwen@cqjtu.edu.cn
[2] College of River and Ocean Engineering, Chongqing Jiaotong University, Chongqing 400074, China
[3] State Key Laboratory of Hydraulics and Mountain River Engineering, Sichuan University, Chengdu 610065, China; liuxingnian@126.com
* Correspondence: binruigan@cqjtu.edu.cn

Abstract: Rock weirs, typically created by the placement of loose rocks, are eco-friendly hydraulic structures used for raising the upstream water level, which has benefits for irrigation, shipping, and grade control. Although rock weirs are frequently submerged in rivers, few studies have systematically investigated their impacts on the upstream water level under submerged conditions. A series of flume experiments regarding this topic were conducted. Different flow discharges, tail-water depths, and void ratios were adopted in the experiments. The results show that (1) the submerged rock weirs primarily function to raise the upstream water level, while having a limited impact on the tail-water level; (2) for a given tail-water depth and void ratio, the upstream water level rise increases with increased discharge, although this response becomes insignificant as tail-water depth increases; (3) as void ratio increases, the upstream water level rise is expected to decrease for a given tail-water depth and discharge; and (4) based on the data and observations, a predictor including the effects of Froude number, submergence, and void ratio is proposed for estimating the upstream water level rise of submerged rock weirs. These results contribute to further understanding the hydraulic properties of rock weirs and are important for river training practices using rock weirs.

Keywords: rock weir; upstream water level; flume experiment; submergence; void ratio

1. Introduction

Rock weirs are eco-friendly hydraulic structures typically created by the placement of loose rocks [1]. They usually span the full width of the channel to increase local flow resistance, thereby increasing the upstream water level, which is beneficial for irrigation, shipping, and grade control [2,3]. Also, rock weirs are more eco-friendly than solid weirs as they can improve river longitudinal connectivity [4], mimic heterogeneous habitats to support aquatic life [5], and promote hyporheic exchange to enhance water quality [6]. Consequently, rock weirs are widely used for training or restoring small- and medium-sized rivers [7]. Despite their popularity, the impact of rock weirs on the upstream water level is not yet well understood due to the complexity of structures formed by rocks (e.g., structural permeability [8,9]), limiting their application in river engineering.

In recent decades, many studies have been conducted to study the response of the upstream water level to the presence of rock weirs. In earlier works, many studies often simplified rock weirs as broad-crested weirs, using modified head–discharge relationships (e.g., modifying the discharge coefficient [10–15]) to estimate the upstream water level. Over the past decade, many studies specific to rock weirs were conducted. Holmquist-Johnson [16] experimentally studied the responses of water depth upstream of U-shaped rock weirs (in plan), proposing an empirical equation including the effects of weir geometry and normal water depth for predicting the upstream water depth. Thornton [17] investigated the impacts of W-, U-, and A-shaped rock weirs on the river water level using the experimental data from Meneghetti [18]; subsequently, an equation calculating the water level upstream for each type of rock weirs was proposed. Baki [19] numerically studied the flow regimes and hydraulics of I- and V-shaped rock weirs; subsequently, they proposed an empirical equation to predict the weir flow depth including the effects of submergence, weir geometry, and bed slope. Kupferschmidt and Zhu [20] experimentally studied the impacts of weir geometry, discharge, and bed slope on the water level along a flume with I- or V-shaped rock weirs; subsequently, they proposed a predictor for the water level along the flume center line. Zeng [21] numerically studied the flow behavior around I-shaped rock weirs and proposed a head–discharge relationship including the effect of weir permeability. Although existing studies provide significant knowledge regarding the impact of rock weirs on the upstream water level, these studies are usually based on non-submerged or partially submerged conditions. In practice, rock weirs are frequently submerged in rivers as they are low-head hydraulic structures [1]. Given that the behavior of flow over rock weirs can be affected by the tail-water [22,23] and the weir permeability [24,25] under fully submerged conditions, investigations on the impact of rock weirs on the upstream water level are far from complete.

The reported study conducted a series of flume experiments (using a straight flume with a fixed bed) to investigate the upstream water level rise of submerged rock weirs. Section 2 presents the details of the experimental setup. Section 3 addresses and discusses the impacts of the Froude number, weir submergence, and weir void ratio on the upstream water level rise of submerged rock weirs, proposing an equation for predicting the upstream water level rise. These results present new insights in hydraulic characteristics of rock weirs, which is important for weir design and river engineering.

2. Methods

2.1. Experimental Setup

All tests were conducted in a glass-walled straight flume measuring 20 m (length) × 0.8 m (depth) × 0.5 m (width) in the Key Laboratory of Hydraulic and Waterway Engineering of the Ministry of Education at Chongqing Jiaotong University, China (Figure 1). This flume was equipped by a water pump that could circulate the water in it. The flume discharge was varied with the speed of the water pump, which could be adjusted by a variable electronic speed control unit (the maximum discharge was about 60 L/s). A flow straightener was set at the flume inlet to ensure unidirectional flows, and a tail-water gate was used at the downstream end of the flume to adjust the tail-water depth. The flume bed was flat. The water level along the flume could be measured using cameras focused through transparent-grid sheets attached to both sides of the flume (the grid scale was 1 cm × 1 cm, and the accuracy of the measurement method was ±2 mm). No sediment was used in this study, indicating that all tests were conducted on fixed bed conditions.

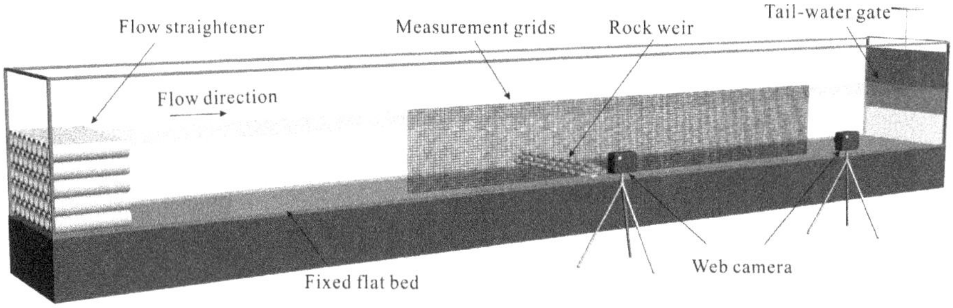

Figure 1. Illustration of experimental flume and measurement devices.

The rock weir used in this study was composed of natural loose rocks. It was placed 10 m downstream of the flume inlet with a 50 mm protrusion (i.e., weir height z) above the flume bed and spanned the full width of the flume. The rock weir was I-shaped in plan and had a triangular cross-section. The upstream and downstream weir slope were equal to the submerged angle of repose of rocks (about 40°). Three types of rocks were used to place the weir (i.e., D = 12 mm, 24 mm and 33 mm). Each type of rock was approximately uniform in size with a coefficient of uniformity ($CU = D_{60}/D_{10}$, where D_{60} and D_{10} are the diameters for which 60% and 10% of rocks are finer, respectively) less than two (Bell [26]). By using these three types of rocks, three weir void ratios (i.e., e = 0.13, 0.21, and 0.35, indicating the ratio of the weir voids volume to all the solids volume) were obtained. The volume of voids was determined by subtracting the volume of solids from the total volume of the weir, and the volume of solids was obtained after measuring the total rock weight and dividing it by the rock density (2650 kg/m^3 [27]).

Table 1 summarized the experimental conditions of this study. Figure 2 plots the main nomenclatures used in this study and their meanings are as follows: h_t = tail-water depth; h_0 = approach flow depth; H_d = water level difference across the weir (i.e., $H_d = h_0 - h_t$); U_0 = approach flow velocity; and φ = weir slope. Figure 2 also plots the coordinate system (x-O-y) for measurement in this study. The coordinate origin (O) was at the center of the weir bottom in the cross-section view. Three tail-water depths (i.e., h_t = 100 mm, 120 mm, and 150 mm) were designed to form different weir submergence (h_t/z). For each h_t/z, a range of discharge (Q) was designed. The minimum discharge corresponds to a small variation of the upstream water level, and the maximum one should be below the rock weir failure threshold specified by Zhang [23]. In total, 36 test trials were conducted in this study.

Table 1. Summary of the experimental conditions.

Q (L/s)	z (mm)	D (mm)	e	h_t (mm)	H_d (mm)	h_0 (mm)	Fu
9~18	50	12	0.13	100	5~15	105~115	0.17~0.29
11~21	50	12	0.13	120	4~11	124~131	0.16~0.29
18~26	50	12	0.13	150	2~6	152~156	0.19~0.27
5~23	50	24	0.21	100	1~26	101~126	0.11~0.33
11~30	50	24	0.21	120	3~18	123~138	0.17~0.37
16~39	50	24	0.21	150	1~14	151~164	0.19~0.40
9~23	50	33	0.35	100	3~27	103~127	0.17~0.33
14~29	50	33	0.35	120	3~21	123~141	0.20~0.35
24~39	50	33	0.35	150	4~15	154~165	0.25~0.37

Notes: Fu is the Froude number of the approach flow, $Fu = U_0/(gh_0)^{0.5}$.

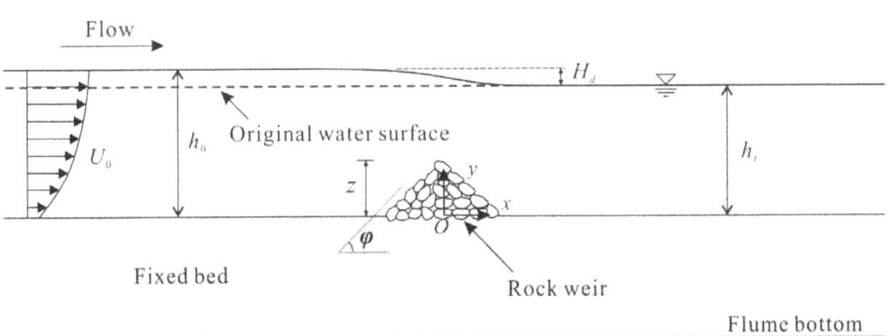

Figure 2. Definition of main test nomenclatures and the coordinate system (x-O-y) for measurement.

Prior to each test trial, water was added slowly into the flume to the designed h_t (i.e., the initial water depth in the flume). Subsequently, the discharge in the flume was increased to the designed value, with h_t fixed through adjusting the tail-water gate. After the flow in the flume became stable, the water level along the flume was recorded by the camera. The water surface profile in the flume was expected to be quasi-two-dimensional along the longitudinal direction due to the aspect ratio of the flow = 3.3~5.0 (i.e., the ratio of flow width to flow depth in the cross-section of the flume [28]). The longitudinal range for water level records was $x = -1.5$ m~1.5 m. A similar approach was applied to the subsequent next trial until all test trials were performed.

2.2. Framework of Analysis

Under submerged conditions, the water level rise upstream of a rock weir is approximately equal to the water level difference across the weir (i.e., H_d depicted in Figure 2), which can be determined by the following parameters:

$$H_d = f(z, D, b, \Phi, U, h, \nu, g, B, e) \quad (1)$$

where b = weir width; g = acceleration of gravity; U = flow velocity in flume without weir; h = flow depth in flume without weir; B = flume width; ν = fluid kinematic viscosity; and Φ = rock shape factor. Equation (1) can be expressed non-dimensionally with g and D chosen as the repeating variables:

$$H_d/z = f(D/z, b/z, \Phi, U/g^{0.5}h^{0.5}, h/z, Uh/\nu, B/z, e) \quad (2)$$

In this study, the rock materials are uniform in shape (the effect of Φ is negligible); the values for b/z and B/z are constant; the D/z parameter is related to e for a given flume scale [29]; Uh/ν can be neglected for a fully turbulent flow, as envisaged; and for a given Q, U and h can be approximated as the flow velocity (U_t) and water depth (h_t) in the tail-water region of a submerged weir, respectively [3]. Hence, Equation (2) can be simplified as:

$$H_d/z = f(U_t/g^{0.5}h_t^{0.5}, h_t/z, e) \quad (3)$$

where U_t can be calculated as $Q/(B \cdot h_t)$. Equation (3) assumes that the water level rise upstream of submerged rock weirs is driven by submergence h_t/z, the void ratio e, and the Froude number before building the weir $U_t/g^{0.5}h_t^{0.5}$.

3. Results and Discussion

3.1. Effect of Tail-Water Depth on Upstream Water Level Rise

Figure 3 plots the water level rise occurring upstream and downstream of the rock weir ($e = 0.21$) for different Q and h_t values. Δh denotes the differences between the actual and the initial water depth (i.e., h_t) along the weir for each test trial. Figure 3 shows that an obvious Δh occurs upstream of the weir ($x < 6$ cm). Near the weir crest (-6 cm $< x < 6$ cm), the increased flow velocity decreases the flow potential energy. Subsequently, the water level begins to decrease to connect with the tail-water level ($x > 6$ cm). Figure 3 summarizes three characteristics for Δh along the submerged weir. (1) The tail-water level is nearly unchanged in this study, indicating that the submerged rock weirs mainly affect the water level upstream of the weir (see each subplot in Figure 3a). (2) With increasing Q, a larger Δh upstream of the weir is expected for each h_t (see each subplot in Figure 3a). As the flow inertia is positively related to the flow intensity [3], the increased discharge enhances the weir obstruction on the approaching flow; according to Bernoulli's continuity equation [30], the velocity of approaching flow reduces greatly, resulting in a significant increase in Δh. (3) The larger the h_t, the smaller the Δh upstream of the weir. This phenomenon is found by comparing the maximum Δh value for different h_t values in Figure 3. For example, the maximum Δh value in Figure 3c is smaller than that in Figure 3a, although the corresponding discharge is larger in Figure 3c, which suggests that the effect of rock weirs on raising the upstream water level decreases with increasing h_t.

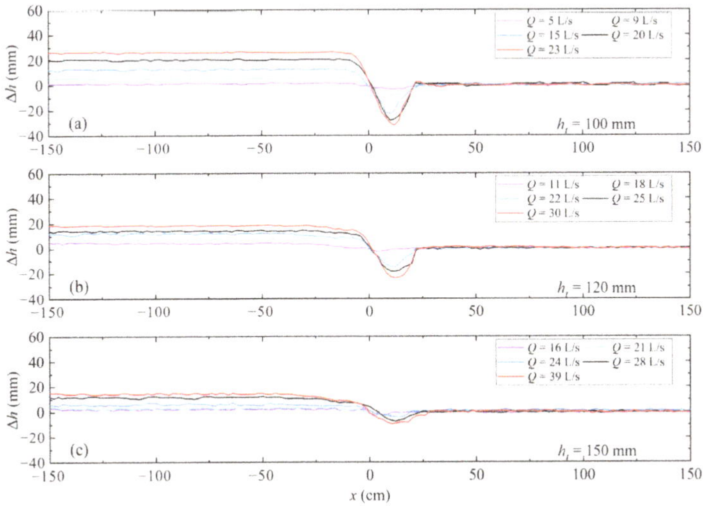

Figure 3. Water level rise along the submerged rock weirs for various h_t values at e = 0.21: (**a**) h_t = 100 mm; (**b**) h_t = 120 mm; and (**c**) h_t = 150 mm.

3.2. Effect of Weir Void Ratio on Upstream Water Level Rise

Figure 4 shows the water level rise occurring upstream and downstream of the rock weir for different e and h_t values. In each subplot, the discharge Q is fixed. Figure 4 shows that the submerged rock weirs with different e remain only affecting the upstream water level. Namely, the tail-water level is insensitive to the change in e. Furthermore, for the same Q, with larger e, a smaller upstream Δh is expected. This phenomenon is related to the characteristics of the flow passing the weir. Compared with a solid weir, the flow passing a rock weir consists of two components, namely, the flow over the weir (weir overflow) and the flow through the weir (interstitial flow) [21]. Since the increase in e can lead to more flow through the weir, the weir overflow and the corresponding upstream Δh are reduced. This finding agrees well with that in Mohamed [25] based on gabion weirs.

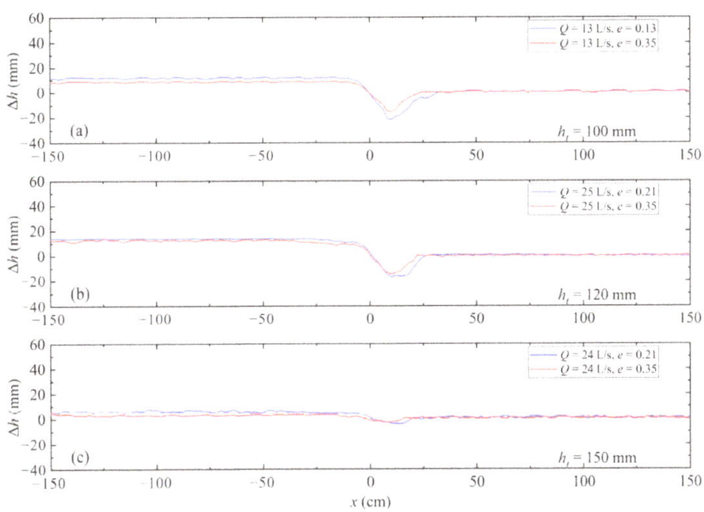

Figure 4. Water level rise along the submerged rock weirs for various e values: (**a**) h_t = 100 mm; (**b**) h_t = 120 mm; and (**c**) h_t = 150 mm.

3.3. Estimation of Upstream Water Level Rise

Based on Equation (3), Figure 5 shows the sensitivity of the dimensionless upstream water level rise (i.e., H_d/z) to $U_t/g^{0.5}h_t^{0.5}$, h_t/z, and e. Figure 5a–c indicate that for a given e and h_t/z, H_d/z increases with increasing $U_t/g^{0.5}h_t^{0.5}$. The increased Froude number raises the kinetic energy of the flowing water, resulting in a more pronounced conversion of kinetic energy to potential energy as the water approaches the weir, thereby raising the water level. Figure 5d–f suggest that for a given e and $U_t/g^{0.5}h_t^{0.5}$, H_d/z decreases with increasing h_t/z. This is because with increasing submergence, a decrease in weir obstruction on the approaching flow is expected, allowing the flows to pass the weir more easily and reducing upstream water level rise accordingly. Figure 5g–i indicate that for a given h_t/z and $U_t/g^{0.5}h_t^{0.5}$, H_d/z decreases with increasing e. As mentioned in Section 3.2, an increase in e leads to more flow passing through the weir, reducing the weir obstruction on the approaching flow to decrease the upstream water level rise.

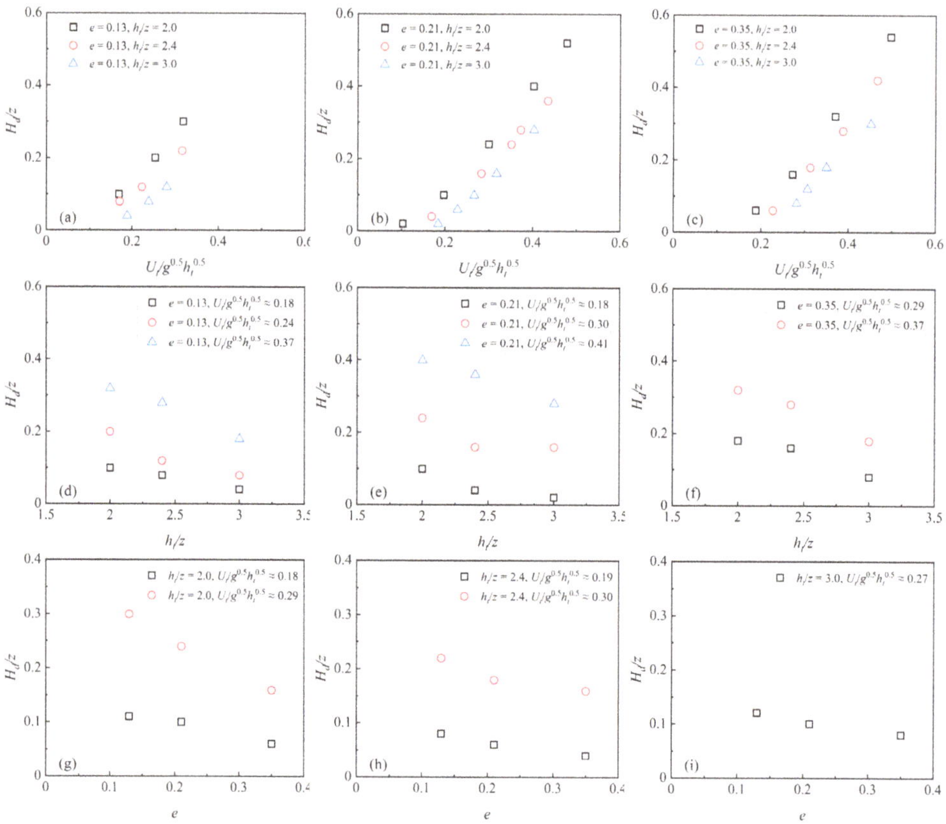

Figure 5. Sensitivity of H_d/z to $U_t/g^{0.5}h_t^{0.5}$ (**a–c**), h_t/z (**d–f**), and e (**g–i**).

Based on the above analysis and Equation (3), Equation (4) is proposed for predicting the water level rise upstream of I-shaped submerged rock weirs using all data in this study (in each test trial, the h_0 value used for calculating H_d is the mean value of h_0 measured from -1.5 m < x < -0.5 m):

$$H_d/z = 2.992 \cdot (U_t/g^{0.5}h_t^{0.5})^{1.915} \cdot (e)^{-0.229} \cdot (h_t/z)^{-0.939} \quad (4)$$

Equation (4) has a multiple correlation $R^2 = 0.97$. Figure 6 compares the normalized upstream water level rise H_d/z predicted using Equation (4) with the measured values, and shows good agreement with a Mean Relative Deviation (MRD [31] = 0.12). In this study, the MRD is calculated as:

$$\text{MRD} = \frac{1}{n}\sum_{i=1}^{n}\left|\frac{x_i - y_i}{y_i}\right| \tag{5}$$

where n = the number of data points; x_i = the predicted value of the i-th data point; and y_i = the measured value of the i-th data point.

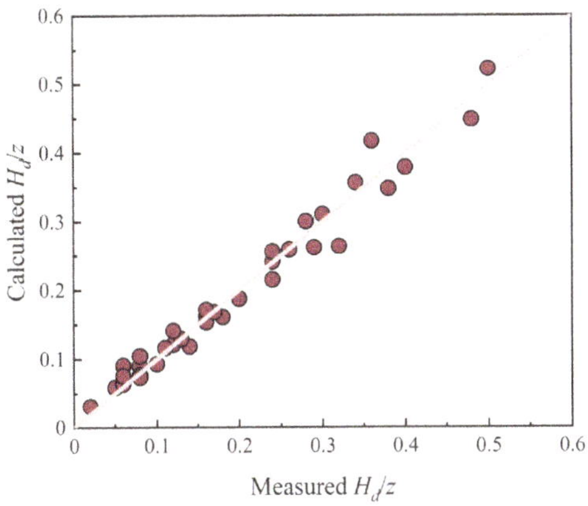

Figure 6. Comparison of measured and calculated H_d/z.

The estimated water level rise using Equation (4) may also be used for determining the head–discharge relationship of I-shaped rock weirs under submerged conditions. In previous studies, the Poleni formula (Equation (6)) was widely used for calculating the head–discharge relationship of rock weirs after the discharge coefficient is determined [21]:

$$Q = \frac{2}{3}C_d B\sqrt{2g}H^{1.5} \tag{6}$$

where C_d = the dimensionless discharge coefficient, and H = the head of water above the weir crest. Since the term H relates to the water level rise upstream of submerged weirs (i.e., $H = H_d + h_t - z$), Equation (4) may be used for calibrating the discharge coefficient C_d in Equation (5) when the terms Q and B are known. Then, the head–discharge relationship of I-shaped rock weirs under submerged conditions can be determined, which facilitates the design of river training practices.

It should be noted that, as this study is based on flume experiments and can be inevitably affected by scale effects, and Equation (4) is valid for $2 < h_t/z < 3$, $0.13 < e < 0.35$, and $0.10 < U_t/g^{0.5}h_t^{0.5} < 0.47$, more field measurements or experimental data using flumes at different scales are required to examine the reliability and practicability of the findings and predictors of this study.

4. Conclusions

Flume experiments were conducted to study the upstream water level rise of submerged rock weirs. The main findings are as follows.

The recorded water level along the flume showed that under submerged conditions, the rock weirs primarily raise the upstream water level, while exerting minimal influence on changing the tail-water level. For a given tail-water depth and void ratio of rock weirs, the upstream water level rise increases with increasing discharge. However, this response becomes insignificant as tail-water depth increases. Furthermore, as the weir void ratio increases, the water level rise upstream of the weir is expected to decrease for a given discharge and tail-water depth.

Based on the experimental data and observations, a predictor including the effects of the Froude number before building the weir, weir submergence, and the weir void ratio was proposed for estimating the water level rise upstream of I-shaped rock weirs under submerged conductions. The computed water level rise agreed well with the measured data. The estimated upstream water level rise may also be useful for determining the head–discharge relationships of submerged rock weirs, which can assist in the design of rock weirs for river engineering.

Author Contributions: Conceptualization, W.Z.; methodology, W.Z. and B.G.; supervision, X.L.; formal analysis, W.Z.; visualization, W.Z.; data curation, W.Z.; funding acquisition, W.Z. and B.G.; writing—original draft preparation, W.Z.; writing—review and editing, W.Z., B.G. and X.L. All authors have read and agreed to the published version of the manuscript.

Funding: This research was funded by the National Key R&D Program of China (2023YFC3208800), the National Natural Science Foundation of China (52309078), the Chongqing Natural Science Foundation Postdoctoral Science Fund (CSTB2023NSCQ-BHX0201), the Science and Technology Research Program of Chongqing Municipal Education Commission (KJQN202300727), the Open Fund of Key Laboratory of Hydraulic and Waterway Engineering of the Ministry of Education, Chongqing Jiaotong University (SLK2023B12), and the Open Fund of State Key Laboratory of Hydraulics and Mountain River Engineering (SKHL2320).

Data Availability Statement: Data related to this study are available from the authors upon reasonable request.

Acknowledgments: The first author would like to thank Shengxian Xu for assisting in analyzing part of results of this study.

Conflicts of Interest: The authors declare no conflicts of interest.

References

1. USBR (United States Bureau of Reclamation). *Rock Weir Design Guidance*; USBR Technical Service Center, Sedimentation and River Hydraulics Group: Denver, CO, USA, 2016.
2. Nichols, M.H.; Polyakov, V.O. The impacts of porous rock check dams on a semiarid alluvial fan. *Sci. Total Environ.* **2019**, *664*, 576–582. [CrossRef]
3. Chen, Z.C.; Shao, X.J.; Zhang, J.W. Experimental study on the upstream water level rise and downstream scour length of a submerged dam. *J. Hydraul. Res.* **2005**, *43*, 703–709. [CrossRef]
4. Martens, K.D.; Connolly, P.J. Effectiveness of a redesigned water diversion using rock vortex weirs to enhance longitudinal connectivity for small salmonids. *N. Am. J. Fish Manage.* **2010**, *30*, 1544–1552. [CrossRef]
5. Baki, A.B.M.; Zhu, D.Z.; Harwood, A.; Lewis, A.; Healey, K. Hydraulic design aspects of rock-weir fishways with notch for habitat connectivity. *J. Ecohydraulics* **2019**, *5*, 94–109. [CrossRef]
6. Gordon, R.P.; Lautz, L.K.; Daniluk, T.L. Spatial patterns of hyporheic exchange and biogeochemical cycling around cross-vane restoration structures: Implications for stream restoration design. *Water Resour. Res.* **2013**, *49*, 2040–2055. [CrossRef]
7. Radspinner, R.R.; Diplas, P.; Lightbody, A.F.; Sotiropoulos, F. River training and ecological enhancement potential using in-stream structures. *J. Hydraul. Eng.* **2010**, *136*, 967–980. [CrossRef]
8. Wang, C.L.; Zhao, Y.; Ning, B.; Bi, J. Permeability evolution of coal subjected to triaxial compression based on in-situ nuclear magnetic resonance. Int. *J. Rock Mech. Min.* **2022**, *159*, 105213.
9. Wang, C.L.; Pan, L.H.; Zhao, Y.; Zhang, Y.F.; Shen, W.K. Analysis of the pressure-pulse propagation in rock: A new approach to simultaneously determine permeability, porosity, and adsorption capacity. *Rock Mech. Rock Eng.* **2019**, *52*, 4301–4317. [CrossRef]
10. Pasternack, G.B.; Ellis, C.R.; Marr, J.D. Jet and hydraulic jump near-bed stresses below a horseshoe waterfall. *Water Resour. Res.* **2007**, *43*, W07449. [CrossRef]
11. Wyrick, J.R.; Pasternack, G.B. Modeling energy dissipation and hydraulic jump regime responses to channel nonuniformity at river steps. *J. Geophys. Res.* **2008**, *113*, F03003. [CrossRef]
12. Horton, R.E. *Weir Experiments, Coefficients and Formulas*; USGS (United States Geological Survey): Washington, DC, USA, 1907.
13. Ramamurthy, A.S.; Tim, U.S.; Rao, M.V.J. Characteristics of square-edged and round-nosed broad-crested weirs. *J. Irrig. Drain. Eng.* **1988**, *114*, 61–73. [CrossRef]
14. Dust, D.; Wohl, E. Characterization of the hydraulics at natural step crests in step-pool streams via weir flow concepts. *Water Resour. Res.* **2012**, *48*, 1–14. [CrossRef]
15. FAO/DVWK (Food and Agriculture Organization of the United Nations with Deutscher Verband fur Wasserwirtschaft und Kulturbau). *Fish Passes: Design, Dimensions, and Monitoring*; DVWK: Rome, Italy, 2002.

16. Holmquist-Johnson, C.L. Numerical Analysis of River Spanning Rock U-Weirs: Evaluating Effects of Structure Geometry on Local Hydraulics. Ph.D. Thesis, Colorado State University, Fort Collins, CO, USA, 2011.
17. Thornton, C.I.; Meneghetti, A.M.; Collins, K.; Abt, S.R.; Scurlock, S.M. Stage-discharge relationships for U-, A-, and W-weirs in un-submerged flow conditions. *J. Am. Water Resour. Assoc.* **2011**, *47*, 169–178. [CrossRef]
18. Meneghetti, A.M. Stage-Discharge Relationships for U-, W-, and A-Weirs in Unsubmerged Conditions. Ph.D. Thesis, Colorado State University, Fort Collins, CO, USA, 2009.
19. Baki, A.B.M.; Zhu, D.Z.; Harwood, A.; Lewis, A.; Healey, K. Rock-weir fishway. I: Flow regimes and hydraulic characteristics. *J. Ecohydraul.* **2017**, *2*, 122–141. [CrossRef]
20. Kupferschmidt, C.; Zhu, D.Z. Physical modelling of pool and weir fishways with rock weirs. *River Res. Appl.* **2017**, *33*, 1130–1142. [CrossRef]
21. Zeng, Y.X.; Ismail, H.; Liu, X.F. Flow decomposition method based on computational fluid dynamics for rock weir head-discharge relationship. *J. Irrig. Drain. E.* **2021**, *147*, 04021030. [CrossRef]
22. Pagliara, S.; Palermo, M.; Kurdistania, S.M.; Hassanabadi, L.S. Erosive and hydrodynamic processes downstream of low-head control structures. *J. Appl. Water. Eng. Res.* **2015**, *3*, 122–131. [CrossRef]
23. Zhang, W.; Nie, R.H.; Melville, B.W.; Whittaker, C.N.; Shamseldin, A.Y.; Liu, X.N.; Wang, L. Failure of rock weirs due to rock dislodgements. *J. Hydraul. Res.* **2023**, *61*, 356–368. [CrossRef]
24. Pagliara, S.; Palermo, M. Rock grade control structures and stepped gabion weirs: Scour analysis and flow features. *Acta. Geophys.* **2013**, *61*, 126–150. [CrossRef]
25. Mohamed, H.I. Flow over gabion weir. *J. Irrig. Drain. E.* **2010**, *136*, 573–577. [CrossRef]
26. Bell, F.G. *Engineering Properties of Soils and Rocks*, 3rd ed.; Butterworth-Heinemann: Oxford, Boston, 1992.
27. Zhao, Y.; Liu, H.H. An elastic stress-strain relationship for porous rock under anisotropic stress conditions. *Rock Mech. Rock Eng.* **2012**, *45*, 389–399. [CrossRef]
28. Zhang, W.; Wang, L.; Melville, B.W.; Guan, D.W.; Whittaker, C.N.; Shamseldin, A.Y. Characteristics of the flow field within a developing scour hole at a submerged weir. *J. Hydraul. Res.* **2021**, *60*, 283–294. [CrossRef]
29. Ettema, R.; AuBuchon, J.; Holste, N.; Varyu, D.; Baird, D.; Padilla, R.; Posner, A.; Thornton, C. Large flume tests on flow dislodgment of rocks forming bendway weirs. *J. Hydraul. Eng.* **2020**, *146*, 04020008. [CrossRef]
30. Shatalov, A.; Hafez, M. Numerical solutions of incompressible Navier–Stokes equations using modified Bernoulli's law. *Int. J. Numer. Meth. Fl.* **2003**, *43*, 1107–1137. [CrossRef]
31. Guan, D.W.; Melville, B.W.; Friedrich, H. Live-bed scour at submerged weirs. *J. Hydraul. Eng.* **2015**, *141*, 04014071. [CrossRef]

water

MDPI

Article

An Empirical Relation for Estimating Sediment Particle Size in Meandering Gravel-Bed Rivers

Arman Nejat Dehkordi [1], Ahmad Sharafati [1,*], Mojtaba Mehraein [2] and Seyed Abbas Hosseini [1]

[1] Department of Civil Engineering, Science and Research Branch, Islamic Azad University, Tehran, Iran; dehkordiarman@yahoo.com (A.N.D.); abbas_hoseyni@srbiau.ac.ir (S.A.H.)
[2] Faculty of Engineering, Kharazmi University, Tehran, Iran; mehraein@khu.ac.ir
* Correspondence: asharafati@gmail.com

Abstract: This paper aims to obtain a relation for estimating the median size of bed sediment, d_{50}, at the bends of meandering rivers based on real data. To achieve such a purpose, field data, including topographic, sediment sampling, and flow measurements, were collected from various rivers in Iran at different times of the year. Then, the Buckingham Π-theorem was applied to identify the effective dimensionless numbers such as the Shields function, Reynolds particle number, Froude number, submerged specific gravity of sediment, and aspect and curvature ratios. A correlation analysis was conducted between such factors to eliminate those dependent on others. In the following, three regression techniques, containing the power function approach, the General Additive Model (GAM), and the Multivariate Adaptive Regression Spline (MARS), were chosen to achieve the best relation. The obtained results indicated that the developed MARS model produced a better result than the others and was much more satisfactory, with a coefficient of determination (R^2) of 0.96 and 0.95 and Root-Mean-Square Error (RMSE) of 140.64 and 140.47 for the training and testing phases, respectively. Furthermore, the MARS outputs were validated with an analytical method, which showed that MARS fitted with the field data much better. Consequently, the distinguished merit of this study is the development of a relation for determining d_{50} that shows which geometric and hydraulic parameters have the most effect on sediment size in the river bend.

Keywords: sediment particle size; river bend; field data; dimensional analysis; regression models

1. Introduction

Determining the sediment particle size in alluvial rivers is essential for many purposes, such as sediment transport mechanisms [1], prediction of river morphodynamics [2], estimation of bed roughness [3], and maintenance of ecological conditions [4]. One of the few studies conducted is related to Bridge's work in 1977, which established the median grain size of bed sediments in bends using the balance between the forces acting on grains. Although analytical relations are presented for both circular and non-circular bends separately in this study, the geometric characteristics of the bends are not directly presented in the relations [5]. In another study by Odgard in 1984, the grain size was studied in the bed's armor layer, which ended in an analytical method with a relation. Such a study was compared with the data collected by other scientists [6,7]. However, the main focus of the current research has been on the size distribution of sediment particles in the armor layer in non-meandering beds. Julien and Anthony used a three-dimensional approach to determine the direction of particles in motion with different sizes in natural meander bends [8]. Also, the work of Wright and Parker using the numerical finite difference method with field data established a one-dimensional mathematical formulation for downstream fining in sand-bed rivers [9]. Jang et al. studied the effects of lifting force on bed sediment size using a numerical model in another investigation. The outcome showed that the consequences of the mentioned force were significant [10]. Kuhnle et al. surveyed the impact of bed shear stress on the bed grain size distribution with a poorly sorted combination of sand and gravel. They proved the increasing of the bed median sediment size with bed shear stress [11]. In a series of experiments by McKie et al., the various distance effects of stabilized large sediment particles on the bed and transported sediment sizes in gravel-bed rivers were investigated. The results reflected that reducing the spaces between the particles made the transported and bed sediment sizes coarser [12]. White and Nelson used a sinusoidal flume with constant flow rates to study the mechanisms influencing sediment sorting patterns in a channel bed. They reported that interactions between channel shape, bed topography, and secondary currents led to fine and coarse sediment deposition on bars and pools, respectively [13].

Additionally, some research has used traditional regression approaches to predict the relation between the distribution of sediment particles and other hydraulic variables. Yen and Lee formulated transverse sediment sorting in a laboratory channel bend under unsteady flow conditions and determined a regression equation for the variation of sediment size [14]. Pitlick et al. expressed a regression relation for grain size in bed surface and substrate layers with flow properties in gravel-bed rivers [15]. Also, using the least squares linear regression method, Naito et al. set out a study about a total bed material relation for a fine sediment mixture to treat grain sorting in sand–silt bed rivers [16]. However, these techniques may need to be more efficient to fully explain the effect of channel geometry variables and flow parameters in an alluvial channel [17].

On the other hand, various advanced models like Generalized Additive Model (GAM), Multivariate Adaptive regression Spline (MARS), Support Vector Machine (SVM), and such have been used for different hydraulically and hydrologically matters [18–22]. Among these, GAM and MARS sound fairly comprehensive to expand a set of formulas for estimating the particle sediment size in bends. Recently, these techniques have been successfully used to present a variety of subjects related to flow and sediment dynamics [23, 24]. Each regression model has its own strengths and weaknesses; for example, although the power model errors may increase with multi-collinearity between the parameters, it is one of the best among the other traditional ones to predict the behavior of natural phenomena. Also, the GAM interpretation results may be challenging due to the complexity of their functions. However, it applies smooth functions to the predictor variables instead of describing a quantity as a linear or polynomial function [25]. Regarding MARS, it is a flexible method with optimal accuracy compared to the other techniques to clarify outputs using linear and nonlinear relations [26]. Not only does MARS create nonparametric modeling without using functional equations between inputs and outputs to recognize the effect of relevant parameters, but it also selects essential variables automatically to predict the results with a high analytical speed in comparison with the other techniques [27]. The superiority of the MARS model is that its structure is additive and interactive, creating fewer variable interactions and fewer deviations [28].

Despite conducting numerous studies on the sediment dynamics in natural meandering bends, it is impossible to predict with certainty what parameters of the bend and which elements of flow characteristics can affect the size of the bed sediment particles. Also, there are some questions, such as whether there is a relation between the size of the bed sediment particles as well as the large-scale components of the bend and flow characteristics, or whether it is possible to estimate the size of these particles under the influence of bend physical specifications. So, calculating sediment particle sizes, especially in meandering rivers, requires more attention and is a fundamental goal of sediment studies. For this reason, this study focused on estimating the median particle size of sediments in alluvial gravel beds, and attempts were made to find a reliable relation between the channel bend geometry and the flow characteristics. To achieve this aim, field measurements were made to collect data in the natural meandering rivers. In addition, dimensional analysis was applied to diminish the number of variables involved and to introduce a nondimensional framework based on the physical insight of the problem and the results of other previously conducted research. Then, three regression forms—the power model as a classical one, GAM, and

MARS as advanced techniques—were adopted to develop a mathematical relation between parameters obtained from the dimensional analysis to choose the best-fitted model using the R programming package v.4.3.2 [29]. Furthermore, the outputs were compared, and the results of the superior model were evaluated with Bridge's analytical relation. Finally, parameter sensitivity and uncertainty analysis were employed to understand the trend of the superior model estimations and ensure the robustness of this one.

Collecting the field data and identifying the effective parameters can strengthen this study and provide a reliable understanding of this issue. Additionally, developing an empirical relation to determining the median size of sediment particles using three different regression models and presenting the best model can help fill such a gap. Validation of outputs with the analytical model can establish acceptable trust in the model and its results.

2. Materials and Methods

2.1. Field Measurements

A number of 187 sediment samples were collected from different reaches of four meandering rivers named the Niakan, Zayandehrood, Karun, and Khersan in Chaharmahal Va Bakhtiari and Koohkelooyeh Va BoyerAhmad, the provinces of Iran. The sampling sites were selected based on availability, and the bed materials ranged from different gravel sizes. Moreover, no regulatory structures or human activities were accessible up to 10 times the bankfull width from upstream and downstream reaches. Accordingly, nine free bends with different curvature radii were considered, four of which were located in the Niakan River in Niakan valley (50°09′06″ E, 32°31′52″ N & 50°09′11″ E, 32°31′33″ N), two in the Zayandehrood in the vicinity of Owregan village (50°25′33″ E, 32°36′18″ N), two in the Karun near DoPolan region (50°35′50″ E, 31°54′31″ N & 50°37′21″ E, 31°52′31″ N), and one in the Khersan downstream of Kata village (51°15′30″ E, 31°11′05″ N). Figure 1 shows the locations of the study areas.

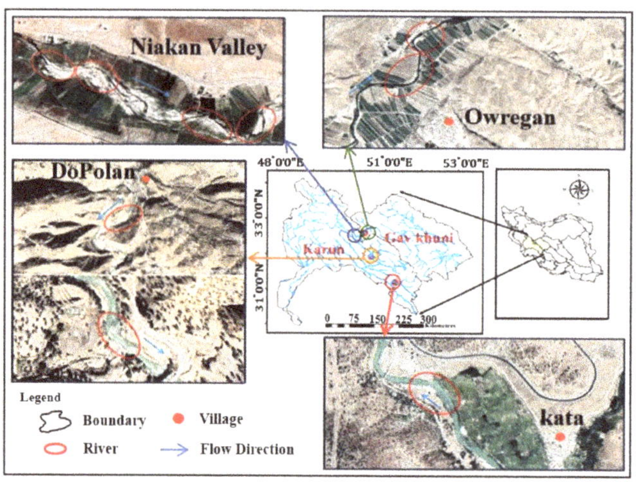

Figure 1. The location of the study areas in Zayandehrood, Karun, Khersan, and Niakan Rivers, Iran.

To gather bed-surface sediments and flow data, a minimum of three and a maximum of five cross-sections for each river were chosen based on the dimensions of the bends. However, one or two extra cross-sections were considered for accuracy in large bends. Cross-sections were surveyed by dividing them into several subsections, each ranging from 0.5 to 6 m. Following that, from the beginning to the end of each interval, bed samples were gathered into 200 pebbles lying to a depth of up to 1.97 m. Simultaneously, water depth and flow velocity (streamwise component) were recorded at each sampling point along the cross-sections. A current meter (Ab Andishan Zayandehrood, Isfahan, Iran) with the capability of flow velocity measurement in the 0.05 to 5 m/s range with an accuracy of $\pm0.1\%$ to $\pm5\%$ was used to gain the water velocity.

Also, surveying was undertaken using multi-frequency G.P.S. and total station instruments (Ruide Surveying Instrument Co. Ltd., Guangzhou, China) to obtain the topographic data of the river bends by two sets of measurements along the water surface and floodplain. The distance between the surrounding points was determined to be between 0.5 and 1 m to accurately record the geometry of the bends (about 2000 points for each bend on average). Field collection of data and sampling were conducted in the rivers in October, November 2019, and May 2020.

In the next step, after drying, all the sediment samples were sieved and weighed based on laboratory methods, using mechanical sieving, submerged, and direct shear tests to characterize median grain sizes (d_{50}), Specific Gravity (SG), and internal friction angle (φ). The point sample weights ranged from approximately 120 to 800 g, and all the samples were gravelly sediments based on Unified Soil Classification System (U.S.C.S.) [30].

Calculated flow parameters, including the average flow velocity and the bed shear stress, were determined regarding the mid-section method and consideration of the formula $\tau_b = \rho_w g R_h S$, respectively, where ρ_w shows the water density, g stands for the acceleration of gravity, R_h signifies the hydraulic radius, and S is the energy slope. In addition, after processing geometric field data, the radius curvatures of the bends, R_c, were obtained by adjusting a circle on the center line of the bend. Similarly, the observed data via fundamental relations distinguished other flow-relevant and channel geometry variables. Table 1 summarizes the range of both the observed and calculated data.

Table 1. Range of the observed data and calculated parameters.

Variables	Units	Minimum	Maximum
Flow depth, h	m	0.15	1.97
Channel top width, T	m	3.60	58.5
Flow velocity, u	m/s	0.10	1.44
Mean sediment size, d_{50}	mm	15	53
Specific Gravity, SG	-	2.66	2.74
Angle of integral friction, φ	°	24	32
Curvature radius, R_c	m	50	287
Longitudinal slope, γ	-	0.005	0.01
Transverse slope, α	-	0.0015	0.0075

Since the purpose is to calculate d_{50} along the bend, the sediment characteristics in each cross-section, identical to the values of the flow parameters, were averaged and used in the dimensional analysis.

2.2. Dimensional Analysis

The bed sediment particle size in a river bend with a radius curvature of R_c, the cross-sectional mean flow velocity of $\overline{u}$, and the shear stress of τ_b were expressed by the following functional form:

$$f(T, R_c, D_h, d_{50}, \overline{u}, \tau_b, g, \mu, \rho_w, \rho_s - \rho_w) = 0, \quad (1)$$

where T stands for channel top width, D_h signifies hydraulic depth defined via dividing the flow area by channel top width ($D_h = \frac{A}{T}$). d_{50} symbolizes median sediment grain size, ρ_s mirrors sediment density, μ shows water dynamic viscosity, and g and ρ_w are the same parameters previously described.

The functional relation was obtained using the Buckingham Π-theorem [31] with the number of variables and the base quantities (i.e., length, time, and mass). The number of seven Π_i obtained was generated by choosing ρ_w, τ_b, and d_{50} as repeated variables. The first dimensionless parameter (Π_1) was developed via dividing T by d_{50} ($\Pi_1 = \frac{T}{d_{50}}$). This parameter introduces the median sediment particle size in a given section of the bend with a width of T. Changes in sediment size in river bends depend on cross-circulation motion related to the curvature ratio [32], so $\frac{R_c}{d_{50}}$ was selected as the second-dimensional group ($\Pi_2 = \frac{R_c}{d_{50}}$), which can be changed to the curvature ratio via the combination with Π_1; $\Pi_2' = \Pi_2 \times \Pi_1^{-1} = \frac{R_c}{T}$. It is also reported that the aspect ratio and submerged specific gravity can affect the sediment particle size [33]; thus, $\Pi_3 = \frac{D_h}{d_{50}}$ that can be changed to $\Pi_3' = \Pi_3^{-1} \times \Pi_1 = \frac{T}{D_h}$ and $\Pi_4 = \frac{\rho_s - \rho_w}{\rho_w} = SG - 1$.

Another effective parameter produced by shear stress, $\Pi_5 = \frac{\tau_b}{\rho_w g d_{50}}$, can be converted to a new dimensionless group by division of Π_5 and Π_4 ($\Pi_5' = \frac{\tau_b}{\rho_w g (SG-1) d_{50}}$) to obtain a known parameter presented as the Shields number, $\theta_{shields}$. The following dimensionless developed parameter was $\Pi_6 = \frac{d_{50}\sqrt{\rho_w \tau_b}}{\mu}$, considering $\tau_b = \rho_w u_*^2$ and $\vartheta = \mu/\rho_w$, in which u_* is the shear velocity and ϑ is the water kinematic viscosity that equals 10^{-6} m^2/s, simplified as $\Pi_6' = \frac{u_* d_{50}}{\vartheta}$ named particle Reynolds number, R_{e*}.

The role of natural flow regimes introduced by another nondimensional group developed via $\Pi_7 = \frac{\overline{u}\sqrt{\rho_w}}{\sqrt{\tau_b}}$. This parameter, rewritten to another combined dimensionless group, $\Pi_7' = \left(\frac{\Pi_7^2 \times \Pi_5}{\Pi_3}\right)^{1/2}$, led to the Froude number, $F_r = \frac{\overline{u}}{\sqrt{g D_h}}$.

By considering the mentioned dimensionless groups, the following functional relation can be expressed between the sediment particle size and the variables:

$$\frac{T}{d_{50}} = f'\left(\frac{R_c}{T}, \frac{T}{D_h}, SG - 1, \theta_{shields}, Re_*, Fr\right) \tag{2}$$

where the values R_c/T, T/D_h, $SG - 1$, $\theta_{shields}$, Re_*, and Fr are taken as input variables to determine T/d_{50}.

Since some parameters may depend on others, it is essential to figure out and omit them to ensure all the input variables are independent. So, the correlation analysis method was used to evaluate the data dependency.

2.3. Correlation Analysis between Variables

Multicollinearity is a statistical event where independent variables in a regression model are highly correlated and cause an unrealistic model to be developed with incorrect results. The Bartlett sphericity test [34] was applied to survey the correlation between the parameters and eliminate highly correlated variables. In the first step, the p-value resulting from the test statistic was much less than the significant level (α), equal to a value of 0.05, which indicated a dependency on the data. By using the correlation matrix, it was found that the maximum correlation factor was between Re_* and $\theta_{shields}$, with an r value of 79%. Therefore, Re_* was removed from the predictor variables.

Later, by reusing the test for the remaining variables, the p-value came to be 0.0003, which was also less than α. In the correlation matrix, it was determined that the parameters T/D_h and $\theta_{shields}$ had the highest correlation coefficient (r = 41%). Consequently, the variable T/D_h was eliminated from the other parameters. By readjusting the test, the p-value was equal to 0.1256, more than α. Thus, the assumption of data independence was established, and the final correlation coefficients between the variables were low enough, as shown in Table 2.

Table 2. Final correlation coefficients between the independent variables (%).

Variables	$SG-1$	$\theta_{shields}$	R_c/T	Fr
$SG-1$	100	26	−30	−18
$\theta_{shields}$	26	100	−23	−22
R_c/T	−30	−23	100	29
Fr	−18	−22	29	100

Accordingly, among the dimensionless groups, $SG-1$, $\theta_{shields}$, R_c/T, and Fr were considered independent variables to develop the sediment particle size relation. Before using the regression models, the data were split into training and testing sets. There is no particular criterion for splitting, and researchers have used various segmentations between the data [35]. This study used a repeated random sub-sampling technique based on the splitting data with 50 iterations. The data are split into 75% for training and 25% for testing.

3. Results

3.1. Power Regression Model

Multivariate power regression is one of the most common methods applied in dimensional analysis [36]. It is expressed as:

$$y = c_0 \prod_{i=1}^{N} x_i^{K_i} \tag{3}$$

where y is a dependent output variable $(\frac{T}{d_{50}})$, c_0 displays a constant term, x_i presents independent input variables (Table 2), k_i clarifies the power of the ith term, and N reflects a number of variables.

Fifteen combinations of the independent variables were considered using the power regression method, and they were evaluated by the Root-Mean-Square Error (RMSE) and the Mean Absolute Error (MAE) indices using the following expressions [37].

$$RMSE = \sqrt{\frac{\sum_{i=1}^{n}(O_i - P_i)^2}{n}} \tag{4}$$

$$MAE = \frac{\sum_{i=1}^{n}|O_i - P_i|}{n} \tag{5}$$

in which O_i is the real value for ith observation, P_i is the predicted value for ith observation, and n is the number of observations.

The best combination, based on testing data, was related to $SG-1$, $\theta_{shields}$, Fr, and R_c/T (Table 3).

Table 3. The values of RMSE and MAE indices for different combinations of independent variables.

Combination of Independent Variables	RMSE		MAE	
	Training	Testing	Training	Testing
$SG-1$	525.97	608.34	419.99	379.70
$\theta_{shields}$	416.95	632.43	270.43	473.38
Fr	551.72	689.03	389.27	523.99
R_c/T	462.39	512.70	300.02	315.26
$SG-1, \theta_{shields}$	582.82	407.16	252.70	441.27
$SG-1, Fr$	600.90	529.84	379.52	416.06
$SG-1, R_c/T$	436.20	556.10	284.49	364.76
$\theta_{shields}, Fr$	409.28	580.50	262.30	398.90
$\theta_{shields}, R_c/T$	325.61	440.31	224.85	329.45
$Fr, R_c/T$	456.88	515.11	291.77	310.20
$SG-1, \theta_{shields}, Fr$	369.94	541.79	254.82	392.73
$SG-1, \theta_{shields}, R_c/T$	311.40	365.47	214.74	281.36
$\theta_{shields}, Fr, R_c/T$	301.26	386.83	200.28	271.23
$SG-1, Fr, R_c/T$	437.61	549.43	282.11	367.02
$SG-1, \theta_{shields}, Fr, R_c/T$	287.65	320.99	197.53	247.61

For the prominent combination, the RMSE and MAE values of the training data were 287.65 and 197.53, respectively, and the values for the testing data in rank were 320.99

and 247.61. The model showed sufficient accuracy in its predictions and can be described as follows:

$$\frac{T}{d_{50}} = 0.96(SG-1)^{10.73}(\theta_{shields})^{-0.56}(Fr)^{-0.49}\left(\frac{R_C}{T}\right)^{-0.96} \quad (6)$$

In addition, sensitivity analysis was performed for the input variables using an analysis of variance (ANOVA) test to discover the significance of each parameter in the model [38]. As shown in Table 4, based on the p-value, it can be concluded that the $(SG-1)$ variable had no significant level. Therefore, it produced a minimum effect on Equation (6), so it could be ignored. The remaining variables R_c/T and $\theta_{shields}$ had high and F_r medium significant levels.

Table 4. p-values based on ANOVA test for each variable in power regression.

Variables	p-value	Result
$SG-1$	$0.31230 > 0.05$	No significant level
Fr	$0.05 < 0.05448 < 0.1$	Medium significant level
$\theta_{shields}$	$0.00021 < 0.001$	High significant level
R_c/T	$0.00002 \ll 0.001$	High significant level

Accordingly, an optimum power relation for predicting sediment particle size could be re-expressed without the $(SG-1)$ variable with a coefficient of determination (R^2) of 0.81 (Equation (7)).

$$\frac{T}{d_{50}} = 314.1(\theta_{shields})^{-0.53}(Fr)^{-0.51}\left(\frac{R_C}{T}\right)^{-1.03} \quad (7)$$

3.2. GAM Model

The Generalized Additive Model (GAM) is an algorithm based on generalizing a linear model with a series of predictors containing common features for each auxiliary variable, first introduced by Hastie and Tibshirani in 1990 [39]. This model has neither the complexity of advanced models nor the incompatibility of classical ones. Proper configuration of different functions in GAM may produce an accurate model more consistent with the data than the linear model [40]. Using nonparametric smoothing functions makes it possible to describe complex environmental phenomena adequately [23]. The basic form of the GAM model is represented by:

$$y = c_0 + \sum_{i=1}^{N} S_i\,(x_i) \quad (8)$$

where c_0 reflects a constant term, $S_i\,(x_i)$ is different from the smooth functions, and N displays a number of variables.

Generally, the GAM optimum model has been developed by selecting the best normal function for each auxiliary variable versus the objective variable separately and adding them together [41]. Here, we used four primary function forms, namely polynomial, power, exponential, and logarithmic, to obtain the most appropriate model for each auxiliary variable. Then, the best-fitted function form was selected based on the R^2 value (Figures 2 and 3).

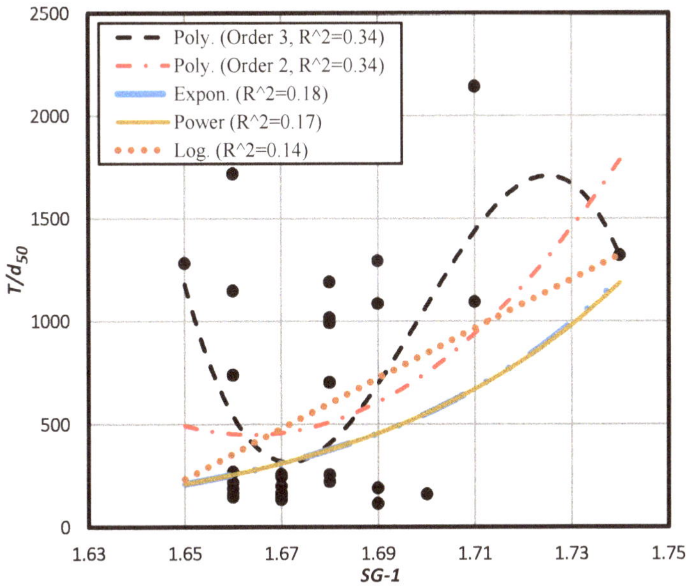

(a)

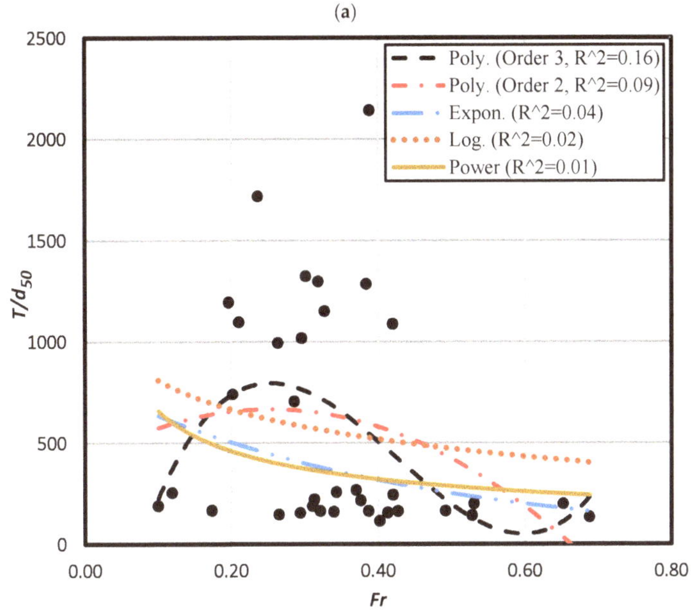

(b)

Figure 2. Primary function forms to obtain the most appropriate model for each auxiliary variable in the GAM model: (**a**) T/d_{50} versus $SG-1$; (**b**) T/d_{50} versus Fr.

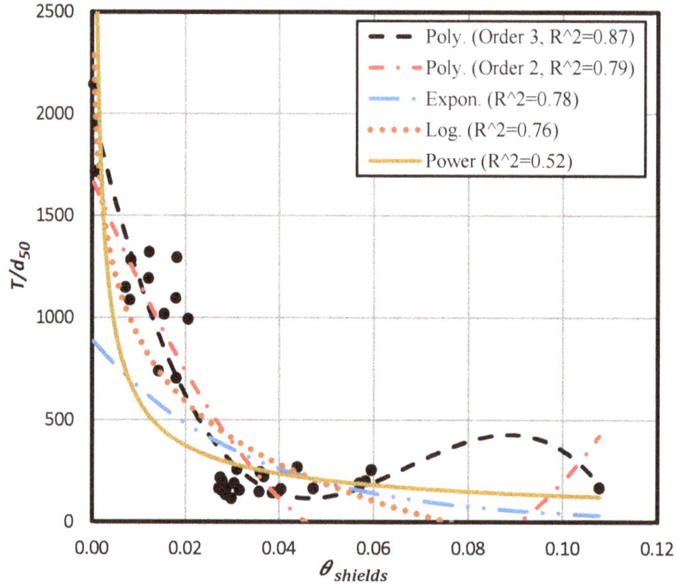

(**a**)

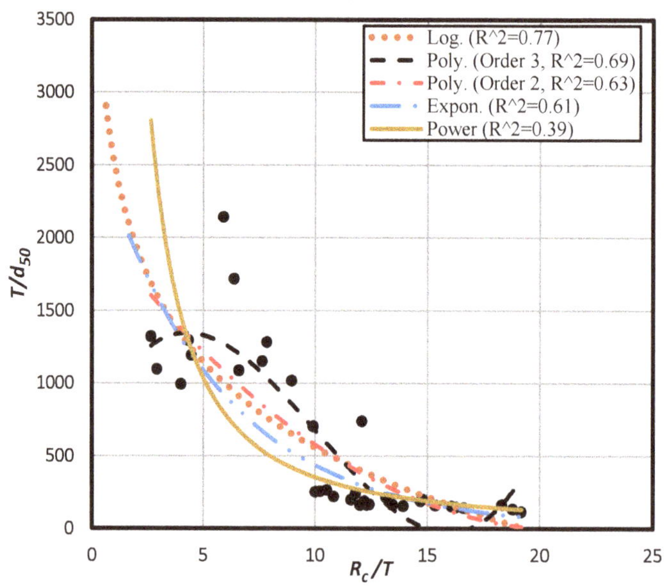

(**b**)

Figure 3. Primary function forms to obtain the most appropriate model for each auxiliary variable in the GAM model: (**a**) T/d_{50} versus $\theta_{shields}$; (**b**) T/d_{50} versus R_c/T.

Among the fitted functions, the polynomial form with degrees 2 and 3 seemed the best for T/d_{50} versus $SG - 1$. The value of R^2 for them was the same and almost equal to 0.34. Hence, the low-order polynomial function with degree 2 was picked as the best form.

A polynomial function with order 3 created the best match of T/d_{50} versus $\theta_{shields}$ and F_r with R^2 values of 0.87 and 0.16, respectively. Also, the best-fitted function for T/d_{50}

versus R_c/T appeared to be the logarithmic form with the highest R^2 equal to 0.77. Table 5 shows the best-matched functions with the corresponding R^2 for each auxiliary variable.

Table 5. Best-matched functions for auxiliary variables in GAM regression model.

Variables	$SG-1$	Fr	$\theta_{shields}$	R_c/T
Best form	Polynomial	Polynomial	Polynomial	Logarithmic
Order	2	3	3	-
R^2	0.34	0.16	0.87	0.77

Using the GAM model, the selected forms for each auxiliary variable were combined as an additive equation to create a single equation and predict the T/d_{50} with (Equation (9)). The RMSE and MAE equal 390.79 and 375.32, respectively, for the test data set.

$$\begin{gathered}\frac{T}{d_{50}} \approx 979.763 - 220.184 \times log\left(\frac{R_C}{T}\right) + f(F_r) + g(\theta_{Shields}) + k(SG-1)\\ f(F_r) = F_r \times (-351.188 + 70.58 \times F_r + 411.418 \times {F_r}^2)\\ g(\theta_{Shields}) = \theta_{Shields} \times (-771.595 + 827.056 \times \theta_{Shields} - 305.378 \times \theta^2_{Shields})\\ k\,(SG-1) = (SG-1) \times (672.23 + 534.36 \times (SG-1))\end{gathered} \tag{9}$$

In order to find the optimal combination of the variables (Equation (9)), the recursive elimination algorithm [42] was used to select the most effective variables so that T/d_{50} could be predicted. This algorithm applied a backward selection process and removed the variable(s) of minor importance based on the model evaluation metric. Here, the ANOVA test was applied, and the $SG-1$ with a p-value equal to 0.52 (more than 0.05) was omitted. After removing this parameter and re-examining the test, the p-value was less than 0.05 for other variables. Thus, the final model extracted for T/d_{50} was determined considering the functions $\theta_{shields}$, F_r, and R_c/T, with RMSE and MAE equal to 382.49 and 367.57, respectively. This relation, with R^2 equal to 0.72, is expressed as follows:

$$\begin{gathered}\frac{T}{d_{50}} = 1108.80 - 278.18 \times log\left(\frac{R_C}{T}\right) + f(F_r) + g(\theta_{Shields})\\ f(F_r) = F_r \times \left(-341.91 - 18.89 \times F_r + 434.69 \times {F_r}^2\right)\\ g(\theta_{Shields}) = \theta_{Shields} \times \left(-817.36 + 1004.84 \times \theta_{Shields} - 465.78 \times \theta^2_{Shields}\right)\end{gathered} \tag{10}$$

3.3. MARS Model

The multivariate adaptive regression spline (MARS) method is a nonparametric regression initially introduced by Friedman in 1991 and applied by various researchers [43]. It separates the data into several intervals and fits a spline to an interval. Each spline divides the predictors into subgroups for linear relations, automatically combining the relations between parameters to predict outputs [44]. The MARS general formula will be:

$$y = c_0 + \sum_{i=1}^{N} c_i S_i(x) \tag{11}$$

In which y, c_0, and x were described earlier, c_i means constant, S_i implies basis functions that may be one of the three forms: constant, hinge functions, and products of two or more hinge functions, while N indicates the number of basis functions of the model. Basis functions can have two different subcategories, one $max(0, x - constant)$ and the other $max(0, constant - x)$.

MARS uses two stages, the forward and backward processes, to get the optimum model. In the forward stage, the model is built by generating the basis functions based on Equation (9). Then, backward elimination is employed to simplify the model by removing the least effective basis function terms according to performance-evaluating indices (where RMSE, R^2, and MAE were used). Then, backward elimination is employed to simplify the model by removing the least effective basis function terms according to performance-evaluating indices (here, RMSE, R^2, and MAE were used). Two tunning parameters are associated with the MARS to avoid overfitting in the model: the maximum degree of interaction and the nprune, which shows the maximum number of expressions after removing some predictors [45]. The degree value can be considered one, two, and more, and the nprune is calculated using $2 \leq nprune < [min(200, \max((20, 2n_x)) + 1]$, where n_x presents the number of predictor variables.

Since increasing the degree can lead to error enhancement and instability in the model predictions, we experienced 1 and 2 degree values for the existing model. Also, four predictors, $SG - 1$, $\theta_{shields}$, R_c/T, and Fr, existed to form the model ($n_x = 4$). So, 19 possible values for nprune (nprune = 2, 3, . . . 20) were taken into account. The MARS model was initially created based on four basis functions with defined degree and nprune values, whereas all the developed models were evaluated considering the RMSE and MAE performance indicators. The results showed an optimum model achieved with a degree value of 1 and a nprune value of 3, with the lowest RMSE and MAE values equal to 140.47 and 84.80 based on the testing data set, respectively. This model had a high value of R^2 equal to 0.95, and T/d_{50} was predicted as follows:

$$\begin{gathered}\frac{T}{d_{50}} = 107.86 + 39457.31\, h(\theta_{shields}) + 40.71\, h\left(\frac{R_c}{T}\right) \\ h(\theta_{shields}) = max(0,\ 0.0291183 - \theta_{shields}) \\ h\left(\frac{R_c}{T}\right) = max\left(0,\ 13.4466 - \frac{R_c}{T}\right)\end{gathered} \quad (12)$$

4. Discussion

4.1. Comparison of the Models

Power, GAM, and MARS models were used to develop the relation for predicting $\frac{T}{d_{50}}$ versus the effective variables. To determine which model has the maximum adaptation with the observed data, the scatter graph was plotted for the training and testing data separately (Figure 4). Consequently, the MARS method matched the observed data better than the others, with the highest correlation coefficient of R^2 equal to 0.95 compared to power (0.81) and GAM (0.72). Also, the performance of the three developed models based on RMSE and MAE was calculated for both the training and testing data sets. In addition, the Mean Absolute Percentage Error, MAPE, was used from the following expression [46], to ensure that more accurate results were gained (Table 6),

$$MAPE = \frac{1}{n}\sum_{i=1}^{n}\left|\frac{O_i - P_i}{O_i}\right| \times 100 \quad (13)$$

where O_i and P_i is related to both the real and predicted values for ith observation, respectively, and n is the number of observations.

Table 6. Error analysis of predicted $\frac{T}{d_{50}}$ by power model against GAM and MARS.

	Training			Testing		
Index	MARS	Power	GAM	MARS	Power	GAM
R^2	0.96	0.85	0.74	0.95	0.81	0.72
RMSE	140.64	287.65	523.06	140.47	320.99	382.49
MAE	79.12	197.53	472.59	84.80	247.61	367.57
MAPE (%)	14.39	31.13	188.93	13.75	44.10	143.79

According to Table 6, all the error indicator values for MARS are less than the power and GAM models in both data sets. The value of RMSE was equal to 140.47 for MARS versus 320.99 and 382.49 for the two others in the testing data. In addition, the MAPE value of 13.75% was gained for the MARS in the testing situation. So, this model can present an adequate relation for predicting sediment particle size in the river bends with better results than the others. After the MARS, the power model gives proper outputs, indicating the ability of this traditional method to develop an empirical relation for sediment grain size determination.

Also, the performance of the three developed models based on RMSE and MAE was calculated for both the training and testing data sets. In addition, the mean absolute percentage error, MAPE, ref. [46], was applied to ensure that more accurate results were gained (Table 6).

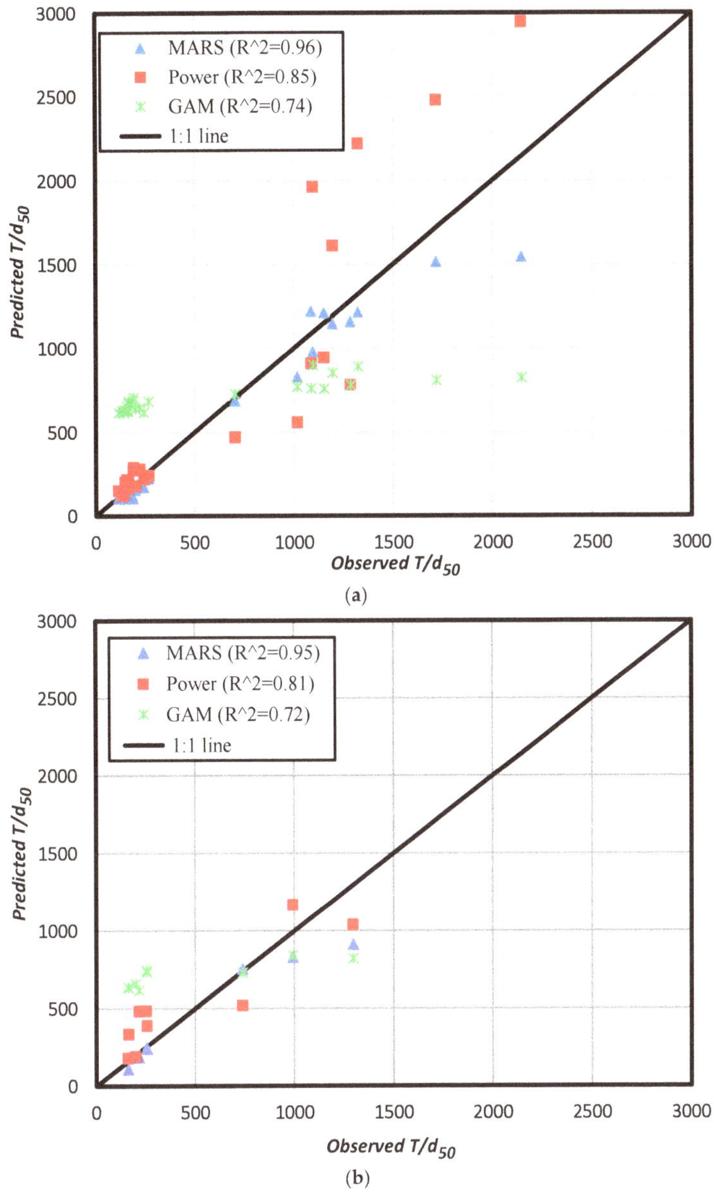

Figure 4. Predicted T/d_{50} versus observed with R^2 for training and testing data: (**a**) Training data, n = 26; (**b**) Testing data, n = 9.

4.2. Comparison of MARS with Analytical Method

To validate the superior model (MARS) with an analytical approach, Bridge's relation [5] was used. This relation calculates the median size of bed sediments in river bends for steady nonuniform flow and noncircular (sine-generated) bend conditions (Equation (14)).

$$D = \frac{3\tau_b}{2(\rho_s - \rho_w)g(\cos\alpha\tan\varphi\cos\gamma - \sin\gamma)} \tag{14}$$

where D is the median size of bed sediments, α and γ are the bed transverse and longitudinal slopes, φ shows the angle of internal friction, and other parameters have already been introduced.

The values of d_{50} calculated by Bridges' relation utilizing the field data were used to obtain T/d_{50} in the bends. Then, both analytical and MARS models' results were compared with the observed data, as shown in Figure 5. The MARS results presented a better fit than the analytical method with a higher R^2, which equals 0.96 against 0.89.

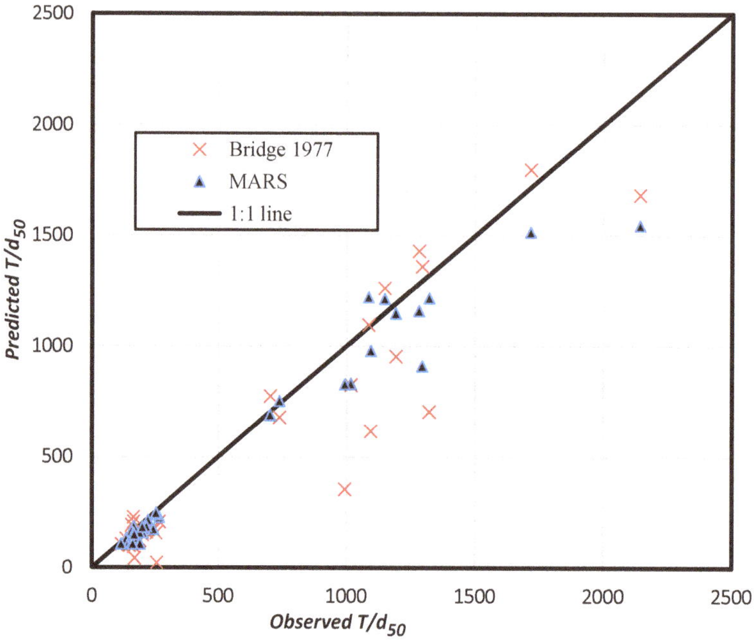

Figure 5. MARS and Bridge results against observed values of T/d_{50} for all the data, n = 35.

Furthermore, the outputs of the RMSE, MAE, and MAPE indices showed that the MARS performed better than the Bridge's relation (Table 7).

Table 7. Error analysis of predicted $\frac{T}{d_{50}}$ by MARS against the analytical model.

Index	MARS	Analytical Model
R^2	0.96	0.89
RMSE	140.64	200.21
MAE	78.78	116.44
MAPE (%)	14.22	23.46

The error percentage (MAPE) of the MARS model (14.22%) is about 9 percent less than the value of the analytical model (23.46%).

4.3. Sensitivity and Uncertainty Analysis for MARS Model

The sensitivity analysis for input parameters in the proposed MARS model was performed. Two parameters, $\theta_{shields}$ and $\frac{R_c}{T}$, affected the model's output. Figure 6 shows the changes in the percentage of $\frac{T}{d_{50}}$ against the variable's change percentage.

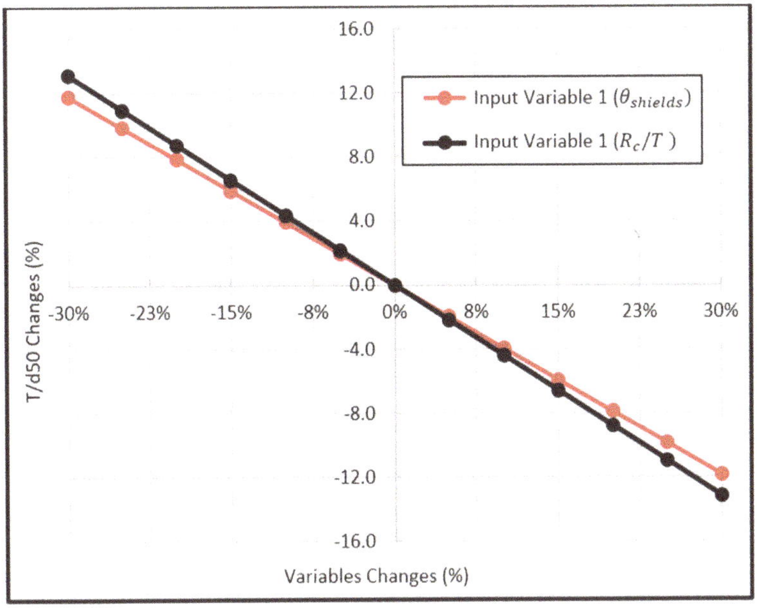

Figure 6. Changes percentages of T/d_{50} against $\theta_{shields}$ and R_c/T change percentage.

The figure shows that with a change of 5% to 30% of the values of $\theta_{shields}$ and R_c/T, T/d_{50} will change about 3.5% to 18% and 2% to 13%, respectively. This represents the sensitivity of the MARS model results to the input variables within the uncertainty in the model. Also, analysis of the uncertainty of the MARS along with the power, GAM, and analytical method for prediction of the bend of gravel-bed rivers is demonstrated in Table 8. This table reflects the mean prediction errors, the width of the confidence band, and the 95% confidence interval of the mean prediction errors. The mean prediction error was calculated based on the average error for each data set; the width of the confidence band was determined via the error and Standard Deviation (SD) according to the Wilson score method; and the 95% confidence interval of the mean prediction errors can be specified as ±1.96SD [28,47].

Table 8. Uncertainty analysis for the prediction of $\frac{T}{d_{50}}$ by MARS and other approaches.

Model	Mean Prediction Error	Width of Confidence Band	95% Confidence Interval of Mean Prediction Error
MARS	−63.27	±42.22	−105.49–21.05
Analytical model	−90.09	±62.43	−152.52–27.66
Power	91.71	±108.00	−16.29–199.71
GAM	145.76	±159.75	−13.99–305.51

The values of the mean prediction error and the minimum confidence band width for MARS are −63.27 and ±42.22, respectively. These demonstrate that the MARS is more suitable for the prediction of T/d_{50} than the other methods.

5. Conclusions

This study has developed an empirical relation capable of estimating the median sediment particle size (d_{50}) in gravel river bends. Field data were collected from different cross-sections placed at bend apex and crossovers in various rivers. By using these data and dimensional analysis procedures, $\frac{T}{d_{50}}$ was calculated at the bends' cross-sections. Applying correlation analysis between variables showed that the cross-sectional mean flow velocity, $\overline{u}$, bed shear stress, τ_b, radius curvature, R_c, as well as sediment density, ρ_s, were essential parameters to generate the nondimensional variables containing Froude number, Fr, Shields parameter, $\theta_{shields}$, the curvature ratio, $\frac{R_c}{T}$, and the submerged weight of sediment, $(SG-1)$, respectively. These parameters based on the collected field data were restricted to $Fr < 1$, $0.01 \leq \theta_{shields} \leq 0.11$, $2.62 \leq \frac{R_c}{T} \leq 19.18$, and $1.65 \leq SG-1 \leq 1.74$. In the following, we applied three regression models (power, GAM, and MARS) to develop an appropriate relation between the geometric, hydraulic, and sedimentary parameters. Then, statistical metrics were employed to compare and choose the best-fitted model. Finally, MARS, as the preferred one, was validated with the analytical model, and sensitivity and uncertainty analyses were performed for it. It seems that the relation developed here has enough robustness to predict the bed sediment size in river bends. Other conclusions have been established from this research, as considered below:

1. It was found that two parameters, R_c/T and $\theta_{shields}$, are the most important in affecting d_{50}. This means that τ_b and R_c from the flow hydraulic and channel geometry characteristics are the significant parameters to determine d_{50} in meandering river bends.
2. The MARS formula showed that it was a better match with the observed data than power and GAM and had less error compared with the analytical model of Bridge. Although this needs to be assessed in more rivers, it can be an appropriate relation to calculate d_{50} in gravel channel bends in engineering applications within parameter ranges.
3. There have been rare studies to determine the sediment particle sizes in river bends, and the existing relations, such as Bridge's model, do not provide physical insight on how bend parameters affect sediment size. The proposed relation in this current article provides a reliable evaluation of sediment sizes based on bend characteristics.
4. After MARS, the power model created better outputs. Even if this is a traditional approach, it presents a simpler relation with fairly good results for determining the size of sediment particles in bends.

Author Contributions: This article is the result of the shared contributions of all authors. A.N.D. has carried out the data analysis and established the manuscript. A.S. has supervised all the activities and not only reviewed the results but also edited the initial text of the paper. M.M. and S.A.H. have presented the research methodology and guided the field observations. All authors have read and agreed to the published version of the manuscript.

Funding: This research received no external funding.

Data Availability Statement: Data are available in the article.

Conflicts of Interest: The authors declare no conflicts of interest.

References

1. Rovira, A.; Núñez-González, F.; Ibañez, C. Dependence of Sediment Sorting on Bedload Transport Phase in a River Meander. *Earth Surf. Process. Landf.* **2018**, *43*, 2077–2088. [CrossRef]
2. Cordier, F.; Tassi, P.; Claude, N.; Crosato, A.; Rodrigues, S.; Pham Van Bang, D. Bar Pattern and Sediment Sorting in a Channel Contraction/Expansion Area: Application to the Loire River at Bréhémont (France). *Adv. Water Resour.* **2020**, *140*, 103580. [CrossRef]
3. Fernández, R.; Vitale, A.J.; Parker, G.; García, M.H. Hydraulic Resistance in Mixed Bedrock-Alluvial Meandering Channels. *J. Hydraul. Res.* **2021**, *59*, 298–313. [CrossRef]
4. Li, J.; He, X.; Wei, J.; Bao, Y.; Tang, Q.; de Nambajimana, J.D.; Nsabimana, G.; Khurram, D. Multifractal Features of the Particle-Size Distribution of Suspended Sediment in the Three Gorges Reservoir, China. *Int. J. Sediment Res.* **2021**, *36*, 489–500. [CrossRef]
5. Bridge, J.S. Flow, Bed Topography, Grain Size and Sedimentary Structure in Open Channel Bends: A Three-Dimensional Model. *Earth Surf Process.* **1977**, *2*, 401–416. [CrossRef]
6. Odgaard, A.J. Grain Size Distribution of River Bed Armor Layers. *J. Hydraul. Eng.* **1984**, *110*, 1479–1484. [CrossRef]
7. Milhous, R.T. *Effect of Sediment Transport and Flow Regulation on the Ecology of Gravel-Bed Rivers*; John Wiley & Sons: Chichester, UK, 1982; pp. 819–842.
8. Julien, P.Y.; Anthony, D.J. Bed Load Motion and Grain Sorting in a Meandering Stream. *J. Hydraul. Res.* **2002**, *40*, 125–133. [CrossRef]
9. Wright, S.; Parker, G. Modeling Downstream Fining in Sand-Bed Rivers. I: Formulation. *J. Hydraul. Res.* **2010**, *43*, 613–620. [CrossRef]
10. Jang, J.-H.; Ho, H.; Yen, C. Effects of Lifting Force on Bed Topography and Bed-Surface Sediment Size in Channel Bend. *J. Hydraul. Eng.* **2011**, *137*, 911–920. [CrossRef]
11. Kuhnle, R.A.; Wren, D.G.; Langendoen, E.J. Structural Changes of Mobile Gravel Bed Surface for Increasing Flow Intensity. *J. Hydraul. Eng.* **2019**, *146*, 04019065. [CrossRef]
12. McKie, C.W.; Juez, C.; Plumb, B.D.; Annable, W.K.; Franca, M.J. How Large Immobile Sediments in Gravel Bed Rivers Impact Sediment Transport and Bed Morphology. *J. Hydraul. Eng.* **2021**, *147*, 04020096. [CrossRef]
13. White, D.C.; Nelson, P.A. Flume Investigation Into Mechanisms Responsible for Particle Sorting in Gravel-Bed Meandering Channels. *J. Geophys. Res. Earth Surf.* **2023**, *128*, e2022JF006821. [CrossRef]
14. Yen, C.; Lee, K.T. Bed Topography and Sediment Sorting in Channel Bend with Unsteady Flow. *J. Hydraul. Eng.* **1995**, *121*, 591–599. [CrossRef]
15. Pitlick, J.; Mueller, E.R.; Segura, C.; Cress, R.; Torizzo, M. Relation between Flow, Surface-Layer Armoring and Sediment Transport in Gravel-Bed Rivers. *Earth Surf. Process. Landf.* **2008**, *33*, 1192–1209. [CrossRef]
16. Naito, K.; Ma, H.; Nittrouer, J.A.; Zhang, Y.; Wu, B.; Wang, Y.; Fu, X.; Parker, G. Extended Engelund–Hansen Type Sediment Transport Relation for Mixtures Based on the Sand-Silt-Bed Lower Yellow River, China. *J. Hydraul. Res.* **2019**, *57*, 770–785. [CrossRef]
17. Bateni, S.M.; Vosoughifar, H.R.; Truce, B.; Jeng, D.S. Estimation of Clear-Water Local Scour at Pile Groups Using Genetic Expression Programming and Multivariate Adaptive Regression Splines. *J. Waterw. Port Coast. Ocean. Eng.* **2019**, *145*, 04018029. [CrossRef]
18. Bazrkar, M.H.; Chu, X. Development of Category-Based Scoring Support Vector Regression (CBS-SVR) for Drought Prediction. *J. Hydroinform.* **2022**, *24*, 202–222. [CrossRef]
19. Rajesh, M.; Rehana, S. Prediction of River Water Temperature Using Machine Learning Algorithms: A Tropical River System of India. *J. Hydroinform.* **2021**, *23*, 605–626. [CrossRef]
20. Abbasi, M.; Farokhnia, A.; Bahreinimotlagh, M.; Roozbahani, R. A Hybrid of Random Forest and Deep Auto-Encoder with Support Vector Regression Methods for Accuracy Improvement and Uncertainty Reduction of Long-Term Streamflow Prediction. *J. Hydrol.* **2021**, *597*, 125717. [CrossRef]
21. Park, S.; Hamm, S.Y.; Jeon, H.T.; Kim, J. Evaluation of Logistic Regression and Multivariate Adaptive Regression Spline Models for Groundwater Potential Mapping Using R and GIS. *Sustainability* **2017**, *9*, 1157. [CrossRef]
22. Asquith, W.H. Regression Models of Discharge and Mean Velocity Associated with Near-Median Streamflow Conditions in Texas: Utility of the U.S. Geological Survey Discharge Measurement Database. *J. Hydrol. Eng.* **2013**, *19*, 108–122. [CrossRef]
23. Asquith, W.H.; Herrmann, G.R.; Cleveland, T.G. Generalized Additive Regression Models of Discharge and Mean Velocity Associated with Direct-Runoff Conditions in Texas: Utility of the U.S. Geological Survey Discharge Measurement Database. *J. Hydrol. Eng.* **2013**, *18*, 1331–1348. [CrossRef]
24. Adnan, R.M.; Parmar, K.S.; Heddam, S.; Shahid, S.; Kisi, O. Suspended Sediment Modeling Using a Heuristic Regression Method Hybridized with Kmeans Clustering. *Sustainability* **2021**, *13*, 4648. [CrossRef]
25. Faraway, J.J. *Extending the Linear Model with R: Generalized Linear, Mixed Effects and Nonparametric Regression Models*, 2nd ed.; Chapman and Hall/CRC: Boca Raton, FL, USA, 2016; pp. 1–395. [CrossRef]
26. Hagemann, M.; Kim, D.; Park, M.H. Estimating Nutrient and Organic Carbon Loads to Water-Supply Reservoir Using Semiparametric Models. *J. Environ. Eng.* **2016**, *142*, 04016036. [CrossRef]
27. Mehdizadeh, S.; Behmanesh, J.; Khalili, K. Using MARS, SVM, GEP and Empirical Equations for Estimation of Monthly Mean Reference Evapotranspiration. *Comput. Electron. Agric.* **2017**, *139*, 103–114. [CrossRef]

28. Mohanta, A.; Patra, K.C. MARS for Prediction of Shear Force and Discharge in Two-Stage Meandering Channel. *J. Irrig. Drain. Eng.* **2019**, *145*, 04019016. [CrossRef]
29. Ihaka, R.; Gentleman, R. R: A Language for Data Analysis and Graphics. *J. Comput. Graph. Stat.* **1996**, *5*, 299. [CrossRef]
30. Germaine, J.T.; Germaine, A.V. *Geotechnical Laboratory Measurements for Engineers*; John Wiley & Sons: Hoboken, NJ, USA, 2009; pp. 1–351. [CrossRef]
31. Buckingham, E. On Physically Similar Systems; Illustrations of the Use of Dimensional Equations. *Phys. Rev.* **1914**, *4*, 345. [CrossRef]
32. Ferreira da Silva, A.M.; Ebrahimi, M. Meandering Morphodynamics: Insights from Laboratory and Numerical Experiments and Beyond. *J. Hydraul. Eng.* **2017**, *143*, 03117005. [CrossRef]
33. Ferreira da Silva, A.M.; Ana, M. *Fluvial Processes (IAHR Monograph)*, 2nd ed.; CRC Press: Leiden, The Netherlands, 2017; pp. 1–289. [CrossRef]
34. Bartlett, M.S. Properties of Sufficiency and Statistical Tests. *Proc. R. Soc. London. Ser. A Math. Phys. Sci.* **1937**, *160*, 268–282. [CrossRef]
35. Froehlich, D.C. Neural Network Prediction of Maximum Scour in Bends of Sand-Bed Rivers. *J. Hydraul. Eng.* **2020**, *146*, 04020065. [CrossRef]
36. McCuen, R.H.; Leahy, R.B.; Johnson, P.A. Problems with Logarithmic Transformations in Regression. *J. Hydraul. Eng.* **1990**, *116*, 414–428. [CrossRef]
37. Najafzadeh, M.; Oliveto, G. Riprap Incipient Motion for Overtopping Flows with Machine Learning Models. *J. Hydroinform.* **2020**, *22*, 749–767. [CrossRef]
38. Finch, W.H. Multivariate Analysis of Variance for Multilevel Data: A Simulation Study Comparing Methods. *J. Exp. Educ.* **2022**, *90*, 173–190. [CrossRef]
39. Hastie, T.; Tibshirani, R. *Generalized Additive Models*, 1st ed.; ChapMan & Hall: Boca Raton, FL, USA, 1990; pp. 1–335.
40. Nejat Dehkordi, A.; Sharafati, A.; Mehraein, M.; Hoseini, S.A. Modelling of Sediment Grains Size Distribution in River Bend Using Generalized Additive Model. *Water Irrig. Manag.* **2022**, *11*, 713–724. [CrossRef]
41. Leathwick, J.R.; Elith, J.; Hastie, T. Comparative Performance of Generalized Additive Models and Multivariate Adaptive Regression Splines for Statistical Modelling of Species Distributions. *Ecol. Modell.* **2006**, *199*, 188–196. [CrossRef]
42. Witten, I.H.; Frank, E.; Hall, M.A. Data Transformations. In *Data Mining: Practical Machine Learning Tools and Techniques*; Morgan Kaufmann: Burlington, MA, USA, 2011; pp. 305–349. [CrossRef]
43. Moisen, G.G.; Frescino, T.S. Comparing Five Modelling Techniques for Predicting Forest Characteristics. *Ecol. Model.* **2002**, *157*, 209–225. [CrossRef]
44. Catalano, G.A.; D'Urso, P.R.; Maci, F.; Arcidiacono, C. Influence of Parameters in SDM Application on Citrus Presence in Mediterranean Area. *Sustainability* **2023**, *15*, 7656. [CrossRef]
45. Kuhn, M.; Johnson, K. *Applied Predictive Modeling*, 1st ed.; Springer Science & Business Media: New York, NY, USA, 2013; pp. 1–595. [CrossRef]
46. Amininia, K.; Saghebian, S.M. Uncertainty Analysis of Monthly River Flow Modeling in Consecutive Hydrometric Stations Using Integrated Data-Driven Models. *J. Hydroinform.* **2021**, *23*, 897–913. [CrossRef]
47. Sattar, A.M.A. Gene Expression Models for the Prediction of Longitudinal Dispersion Coefficients in Transitional and Turbulent Pipe Flow. *J. Pipeline Syst. Eng. Pract.* **2013**, *5*, 04013011. [CrossRef]

water

MDPI

Article

The Role of Stream Restoration in Mitigating Sediment and Phosphorous Loads in Urbanizing Watersheds

Kevin MacKenzie [1], Steve Auger [2], Sara Beitollahpour [1] and Bahram Gharabaghi [1,*]

[1] School of Engineering, University of Guelph, Guelph, ON N1G 2W1, Canada; kmacke06@uoguelph.ca (K.M.); sbeitoll@uoguelph.ca (S.B.)
[2] Lake Simcoe Region Conservation Authority, Newmarket, ON L3Y 3W3, Canada; s.auger@lsrca.on.ca
* Correspondence: bgharaba@uoguelph.ca; Tel.: +1-(519)-824-4120 (ext. 58451)

Abstract: Stream corridor erosion can majorly contribute to the overall sediment and phosphorus load in urbanizing watersheds. However, the relative contribution of stream bed and bank erosion, compared with upland watershed sources and the potential for stream restoration to mitigate total contaminant loads, is poorly understood. In this study, a new method was developed, using the percent impervious cover (PIC) indicator of urbanization to evaluate the relative contribution of the stream corridor versus upland watershed contributions to observed total sediment and phosphorus loads in the receiving watercourse. This method was used to develop a cost-optimized mitigation plan, including implementing low-impact development (LID) stormwater infrastructure for urban areas and agricultural best management practices (BMPs) for rural areas in the watershed and stream restoration for the degraded stream reaches. A new cost–benefit analysis methodology is developed and used to assess the relative benefits of the mitigation measures for the case study of the Tannery Creek sub-watershed of the East Holland River in Ontario, Canada. The novel contributions of this study include the development of three relationships to estimate sediment and associated phosphorus loading based on contributing catchment area and land use, as well as a method to optimize the costs and benefits of planned mitigation measures. The results support stream restoration as an essential and cost-effective part of a comprehensive water quality watershed management plan to help maintain healthy streams in urbanized watersheds.

Keywords: stream restoration; regime channel; suspended sediment; phosphorous load; percent impervious cover; cost–benefit analysis

1. Introduction

Altered hydrological flows from urbanizing watersheds can cause a range of harmful effects on aquatic life, which may manifest into what is commonly referred to as the urban stream syndrome [1–3]. Urban stream syndrome describes the effects of significant and unmitigated urbanization on stream systems [3]. The common symptoms include change in flow regime, erosion and sedimentation, and associated impacts on pollutant loading and habitat health [4,5]. In particular, bank erosion may contribute a significant fraction of total phosphorous (P) loads [6,7].

Regime conditions refer to a state of dynamic stability where energy expenditure by flowing water is at a minimum, erosive forces of the flowing water are in balance with substrate resistance to erosion, and the net transport of sediment into and out of the reach approaches zero [8]. In watercourses impacted by urban development, the distribution of flow increases and may cause the erosive forces of the flowing water to become out of balance with the channel geometry and substrate resistance to erosion. In a previous study, we assembled a database of channel geometry and sediment size for in-regime rivers and used machine learning to develop a model of regime conditions based on stream power [8]. The specific stream power (stream power per unit channel width) model was found to be a reliable indicator of stream reaches both in and out-of-regime conditions [8]. In a subsequent study, flow distribution and contaminant loading in the Southern Ontario, Canada, study area were investigated and found to be well correlated with the level of urbanization in the contributing watershed [9].

Urban stormwater management systems can partially mitigate the adverse effects of urbanization on streamflow and sediment regimes [3]. Systems prioritizing storage, infiltration, and evapotranspiration should be included to mitigate the effects of urbanization on the water budget and form determining peak flows. Efforts to reduce the component of the annual sediment load mobilized by larger flood flows, streambed and bank stabilization, and stormwater management systems that significantly control peak flows during large storm events will be most beneficial.

Low-impact development (LID) green infrastructure is increasingly utilized in urban stormwater management and has proven effective in both water quality and peak flow control [10–16]. LIDs are designed to mimic natural hydrologic processes and address hydrologic symptoms of the urban stream syndrome [17]. In rural areas in watersheds such as agricultural farmlands, several studies report the effectiveness of agricultural best management practices (BMPs) such as no-tillage, buffer strips, and cover crops in stream water quality improvements [18–22].

In recent stormwater management systems, LID and green infrastructure practices are often combined with agricultural BMPs and stream restoration projects [23–25]. However, in highly degraded streams, bank stabilization has the largest potentially beneficial effect in improving stream water quality [26]. Customized solutions show better results when selecting an approach for urban stream management [27]. Further research is required to identify cost-effective BMPs to restore the ecosystem's health in urbanizing streams.

Aquafor Beech Ltd. [28] completed a review of published total phosphorus (TP) concentrations in streambank materials. After screening data and limiting it to North American studies, they found that "the arithmetic average of total phosphorus concentration was approximately 400 mg of Total Phosphorus (TP) per kg of sediment [26]". This value can be used as a representative estimate of TP content in bank materials to estimate potential TP load that may be prevented with stream restoration work. However, if a stream restoration project is deemed worthy of phosphorus (P) offset credits, site-specific sampling of bank materials should be completed to improve the estimate of TP content before finalizing the dollar amount of credit to be awarded.

Recent studies have emphasized combining interrelated river functions when calculating stream restoration credits [29,30]. The Chesapeake Bay Stormwater Network recently reviewed and provided consensus recommendations for assigning stream restoration credits through an existing Prevented Sediment Protocol [31]. Credits given under the Prevented Sediment Protocol recognize that stream restoration can secure sediment and phosphorus at the source and prevent it from contributing to phosphorus loading downstream. Recent studies [29–34] have recommended that restoration credits seek a balance between sediment entrance and exit by acknowledging the combination of hydrologic contribution and bed and bank erosion to nutrient load. Conversely, stream restoration projects only built for infrastructure protection should not be eligible [33]. In addition, past research indicates that stream restoration projects are more effective at the watershed scale rather than the reach scale [35,36].

Several studies have identified stream corridor erosion as a significant source of sediment and P loads; however, a method to assess the magnitude of P loads using readily available data remains a knowledge gap [29–36]. Identifying a general relationship between stream channel stability or regime conditions and sediment and phosphorous load, based on readily available data, represents an additional knowledge gap. A readily applicable method to assess the relative role of stream corridor sources of sediment and P loads, compared with upland sources, including wash-off from impervious urban areas and sediment delivery from agricultural land, also represents a knowledge gap. In addition, a method to assess the relative benefits of common mitigation measures, including stream restoration, LID, and BMPs, on reducing total P load represents a knowledge gap.

Therefore, the main goal of this study is to develop a more accurate methodology to quantify the relative contribution of stream bed and bank erosion compared with the upland watershed sources of sediments and phosphorous loads. The sediment and phosphorous loads source contribution allocation results are then used in a catchment scale cost–benefit analysis of available mitigation measures, including BMPs for the rural areas of the watershed, green stormwater management infrastructures (LIDs) for the urban areas, and stream restoration projects for the degraded portions of the stream network in urbanizing watersheds.

2. Materials and Methods

This section provides an overview of the data sources and methods used to assess sediment and P loading, mitigation measures, including stream restoration, and a cost–benefit analysis. The results of this analysis are discussed in Section 3.

2.1. Data Sources

Tannery Creek is a tributary of the East Holland River, which flows to Lake Simcoe near Holland Landing, Ontario. Tannery Creek drains an approximately 40 km^2 watershed in an urbanizing area with a history of agricultural land use. In this study, we optimized mitigation alternatives to reduce sediment and phosphorous loads from Tannery Creek and its watershed. The Tannery Creek Watershed is shown in Figure 1.

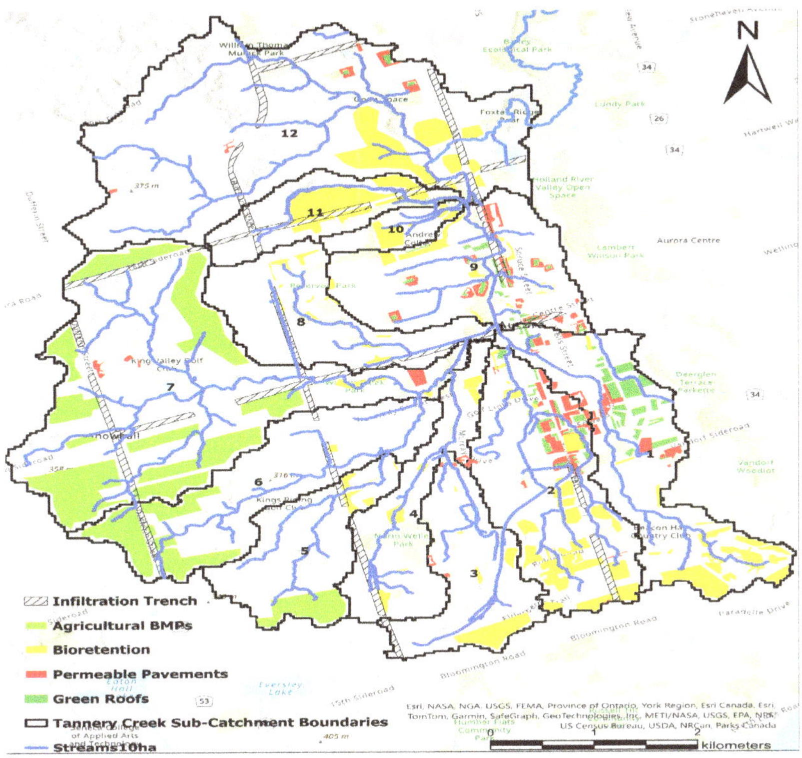

Figure 1. Tannery Creek Watershed.

The Tannery Creek watershed has a history of agricultural land use, with significant urban development over the last few decades, particularly in Aurora. The Ontario Watershed

Information Tool (OWIT) was used to delineate catchment areas contributing to Tannery Creek tributaries, find incremental and total upstream catchment areas contributing to each numbered reach, and assess the percentage of the incremental catchment area with urban or rural land use from the Ontario Land Cover layer. The incremental catchment area is the local area contributing to the study reach, not including areas contributing to upstream tributaries, while the total upstream catchment area includes both the local catchment and the catchment contributing to all upstream tributaries of the local reach. The results are illustrated in Figure 1 and summarized in Table 1.

Percentage impervious cover (PIC) for every reach was estimated using OWIT urban land use areas for the Tannery Creek watershed. Using PIC and the incremental catchment area, the area of impervious cover in each incremental catchment and the total contributing upstream catchment were calculated. Typical performance data for various common Low-impact development (LID) features were sourced from the Sustainable Technologies Evaluation Program (STEP) Low-Impact Development Stormwater Management Planning and Design Guide [37]. Three common LID features were considered in this study, including infiltration trenches, bioretention cells, and permeable pavement. Infiltration trenches are designed to capture stormwater runoff and allow it to infiltrate into the ground over an extended period of time [37].

Bioretention cells are vegetated stormwater storage facilities that are excavated into the ground and filled with filter media capable of supporting vegetation [37]. Bioretention cells may serve to detain and filter stormwater prior to discharge or retain and infiltrate stormwater, depending on the conductivity of the surrounding soil. Permeable pavement is porous to allow precipitation to infiltrate through the pavement and into an underlying storage area, where it may be detained and discharged or retained and infiltrated into the surrounding soil over an extended period of time [37]. These features are commonly used to intercept and treat stormwater from impervious areas.

The cost of implementing LID measures varies significantly depending on the type of LID, whether the LIDs are implemented as part of a new development or retrofit, the planned lifetime of the LID and the development it services, the existing level of urbanization of the development area, real estate costs, topography, surficial soils and depth to the groundwater table. However, representative lifecycle costs for various LID measures were required to compare the costs and benefits of common LIDs with stream restoration costs. A range of cost data was reviewed and used to identify representative LID life cycle unit costs, as summarized in Table 1 [38–41].

Table 1. Summary table for cost and performance of various LID features [38–41].

LID [1] SWM [2] Practice		Infiltration Trench	Bioretention Cells	Permeable Pavement
Ratio of DA [3] to Treatment Area (range)		5:1 to 20:1	5:1 to 15:1	1:1
Ratio of DA [3] to Treatment Area (typical)		20:1	15:1	1:1
Underdrain		Yes	No	No
STEP [5]	Life Expectancy (years)	25	25	25
	Lifecycle Cost (Cdn $/m^2)	$670	$433	$201
Chesapeake Bay Area	Life Expectancy (years)	23	23	23
	Life Cycle Cost (Cdn $/m^2) [4]	$529	$440	$274
Adopted Life Cycle Cost (Cdn$/m^2) [6]		$529	$440	$274

Notes: [1] LID—Low-Impact Development. [2] SWM—Storm Water Management. [3] DA—Drainage Area. [4] Based on the currency exchange rate of 1.00 Canadian $ = 0.76 US $. [5] STEP—Sustainable Technologies Evaluation Program. [6] One Canadian $ = 0.68 European EUR (average 2019 to 2023).

Stream restoration cost may vary significantly based on several factors: the length of the channel restored, the ease of access to the channel, the level of urban development and encroachment on the channel, the cost of negotiating and restoring access easement, topography, type of native and introduced substrate materials, and the design concept employed. For this study, representative stream restoration costs were obtained from Lake Simcoe Region Conservation Authority (LSRCA) records of stream restoration projects in the East Holland River watershed (Figure 2), which Tannery Creek is a part of, showing a linear increase in project cost stream restoration length.

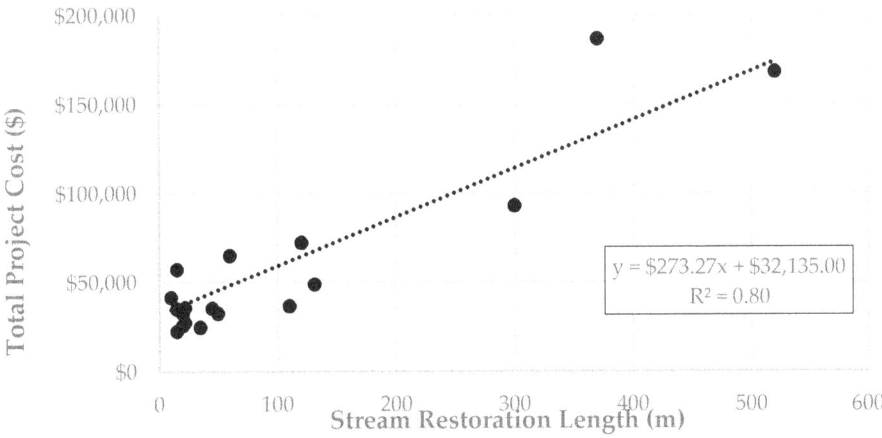

Figure 2. Lake Simcoe Region Conservation Authority records of stream restoration project costs.

Figure 2 shows a linear trend between the stream restoration length and total project cost, with an initial cost of about $32,000 (to cover the equipment, design, permitting, and mobilization) and about $274 per meter length of the stream restoration project. Data for Figure 2 were collected by the Lake Simcoe Region Conservation Authority (LSRCA) and provided to the authors for use in this study. These data include records of total costs to complete stream restoration projects in the Holland River watershed (which is within the Lake Simcoe Region Conservation Authority Jurisdiction).

Variability in these data is attributed to a range of project-specific factors, including ownership of the land, site access and mobilization costs, the required level of engineering design, and whether other infrastructure was affected by the restoration works. These costs are for past projects, and the inflation needs to be considered to estimate the cost for future projects. However, in this study, we are interested in the relative cost of stream restoration, compared with the cost of LIDs for the upland watershed and the cost of agricultural best management practices (BMPs) for a given snapshot in time.

2.2. Total Sediment and Phosphorus Loads

Total load estimates were calculated using the results of our recent studies [8,9], which were developed from fieldwork in the watershed as well as historic hydrometric data from the Water Survey of Canada and LSRCA, and water quality data from the Ontario Provincial Water Quality Monitoring Network. Observed total loads were found for six monitored watersheds within the study area, including both early and later stages of urbanization, resulting in 12 observed sets of observed loads and corresponding watershed land use distributions [9]. Previous studies have established the correlation between loading and contributing areas of specific land uses [42–52]. MacKenzie et al. also showed that total phosphorous load strongly correlates with suspended sediment load in the study area [8,9]. These observed data and established correlations with land use in the contributing watershed were used to develop linear relationships between land use and sediment loading at the outlet of the catchment areas included in the current study, as discussed in Section 3.4.

The estimated annual average P loads for Lake Simcoe is approximately 72 t/y, of which about 41 t/y is delivered to Lake Simcoe from tributaries [8,9]. The East Holland River drains approximately 15% of the total tributary load, indicating an average annual sediment load of approximately 2000 t/y (or 32 kg/d/km^2) and P load of approximately 4.5 t/yr (or 0.07 kg P/d/km^2). Total sediment and phosphorus load estimates for Tannery Creek tributaries were allocated based on the catchment area and PIC [8,9].

2.3. Upland Sediment Sources

Comparison between rural and urban area sediment loads reveals rural areas, with limited over-land-flow connectivity, have lower sediment loads than developed areas with higher levels of imperviousness. As imperviousness limits the availability of exposed soil, sources of sediment load move from upland areas to the stream corridor, destabilizing stream channels and causing erosion of the stream banks and bed [39,40].

2.4. Estimating Riverbank Sources

The specific stream power is the work of flowing water per unit area of the channel per unit time and is calculated using the following equation,

$$\omega = \frac{\gamma\, S\, Q_2}{T} \tag{1}$$

where ω (W/m^2) is the specific stream power, γ is the unit weight of water (kN/m^3), S is the slope of the channel, Q_2 (m^3/s) is the 2-yr return period flood flow, typically consistent with the bankfull discharge, and T (m) is the top width of the main channel. In previous studies, we developed a model for regime conditions using specific stream power [40]. The regime-specific stream power model was developed based on a large database of previously published regime channel geometry, flow, and substrate data using machine learning and considered channel slope, width, depth, substrate median particle size, and the 2-yr return period peak flow (Q_2) [40]. The model was tested against a range of observed stream reaches in the study area and found to identify reaches that were out-of-regime reliably [40]. Thayer et al. [41] discussed the effects of setting on the relative sensitivity of stream geometry parameters and noted that slope can only be considered a dependent variable on a geological time scale and is often not sensitive to change in shorter time scales typically associated with urbanization in stream power limited watercourses [41].

MacKenzie et al. [8,9] provided a discussion of the relative sensitivity of change in the key regime channel parameters and, based on typical conditions in the study area, identified flow as the most sensitive variable to urbanization-induced changes, followed by width and substrate size, then depth and lastly slope. The regime-specific stream power model provides a way to evaluate the stability of urbanizing watersheds and was used to evaluate regime-specific stream power for the Tannery Creek watershed.

Total stream corridor sediment sources were informed by data from the International Reference Group On Great Lakes Pollution From Land Use Activities (PLUARG) for stream bank erosion [53]. MacKenzie et al. [8,9] demonstrated the effects of percent impervious cover (PIC) on net sediment and phosphorous loads in urbanizing watersheds. Increasing PIC was shown to cause increases in both the flow duration curve and sediment rating curve, which can result in increases in sediment and phosphorous loading.

We defined regime conditions based on the ratio of the specific stream power (ω) during the Q_2 event to the regime-specific stream power ($\acute{\omega}$) (Equation (2)), following the methodology we developed in [8,9]. Channel reaches with $\omega/\acute{\omega} \approx 1.0$ were deemed to be within the regime, while reaches with $\omega/\acute{\omega} \geq 1.5$ were out of the regime, and reaches that were in between values with $1.0 \leq \omega/\acute{\omega} \leq 1.5$ were in transition. Out-of-regime channel density (ORD) is defined as the fraction of total stream length in a catchment that is out-of-regime as defined by Equation (2).

$$\frac{\omega}{\acute{\omega}} \geq 1.5 \tag{2}$$

Identifying out-of-regime channel reaches using Equation (2) requires knowledge of the existing channel geometry, slope and substrate size, as well as the 2-yr return period flow under existing conditions and estimated pre-development and future conditions. Hydraulic modeling was used to evaluate specific stream power under these conditions for comparison of the regime-specific stream power for each reach.

2.5. Hydraulic Modelling

The US Army Corps of Engineers, Hydrologic Engineering Center's River Analysis System (HEC-RAS) model was constructed for Tannery Creek and its tributaries to simulate streamflow and water level and validated against observed water level data. A range of estimated 2-yr return period flows spanning historic, current, and potential future levels of impervious cover were used to model hydraulic conditions in Tannery Creek.

Staff gauges equipped with logging water level transducers were installed at seven (7) locations in the Tannery Creek watershed to collect water level data from 2 September to 21 November 2021. In addition, a logging barometric pressure transducer (barologger) was installed at the same location as Staff Gauge No. 7 (SG7) to allow the collected water level data to be corrected for atmospheric pressure variations.

During the September to November 2022 monitoring period, one significant peak flow event on 13 October 2022 was used to validate the hydraulic model. The peak flows observed at LSRCA Tannery Creek at St. Andrew's College and Water Survey of Canada's Holland River East Branch at Holland Landing were prorated, based on drainage area, and used to populate flow change locations in the hydraulic model to run steady state simulations.

The modeled water level results from the steady state simulation were compared with the observed peak water levels at the staff gauge, and Manning's coefficients were adjusted in the model until the modeled flow depths reasonably approximated observed peak water levels during the 13 October 2022 event.

In addition to the model of current conditions in Tannery Creek, simulations of historical pre-development conditions and future fully built-out conditions were also completed. The 2-yr return period flows (Q_2) for the pre-development and future conditions scenarios were estimated using a developed relationship between Q_2 and PIC in the upstream catchment.

2.6. Mitigation Measures

The mitigation measures considered for this study included LID, BMPs, and stream restoration. LID measures considered in the study included infiltration trenches and bioretention facilities. Other LIDs, such as green roofs and permeable pavement, were initially considered but ultimately not included because of the relatively high cost and small treated area, especially when considered as retrofit opportunities. These measures should still be considered for future greenfield development, in addition to infiltration trenches and bioretention facilities. Further discussion of potential mitigation measures in the Tannery Creek setting is included in Sections 3.5–3.7.

2.7. Cost–Benefit Analysis

A cost–benefit analysis focused on P load reduction was completed for Tannery Creek considering the above mitigation measures. The objective of the cost–benefit analysis was to explore each mitigation measure's relative costs and benefits. The watershed-scale approach of this study will help to quantify the required funding to obtain the desired level of P load reduction. The results are specific to Tannery Creek and typical support funding mechanisms in place in 2023 but demonstrate a method that could be applied to mitigation planning in other study areas. The approach to mitigation and cost optimization is illustrated in Figure 3 and discussed further in Sections 3.4–3.8.

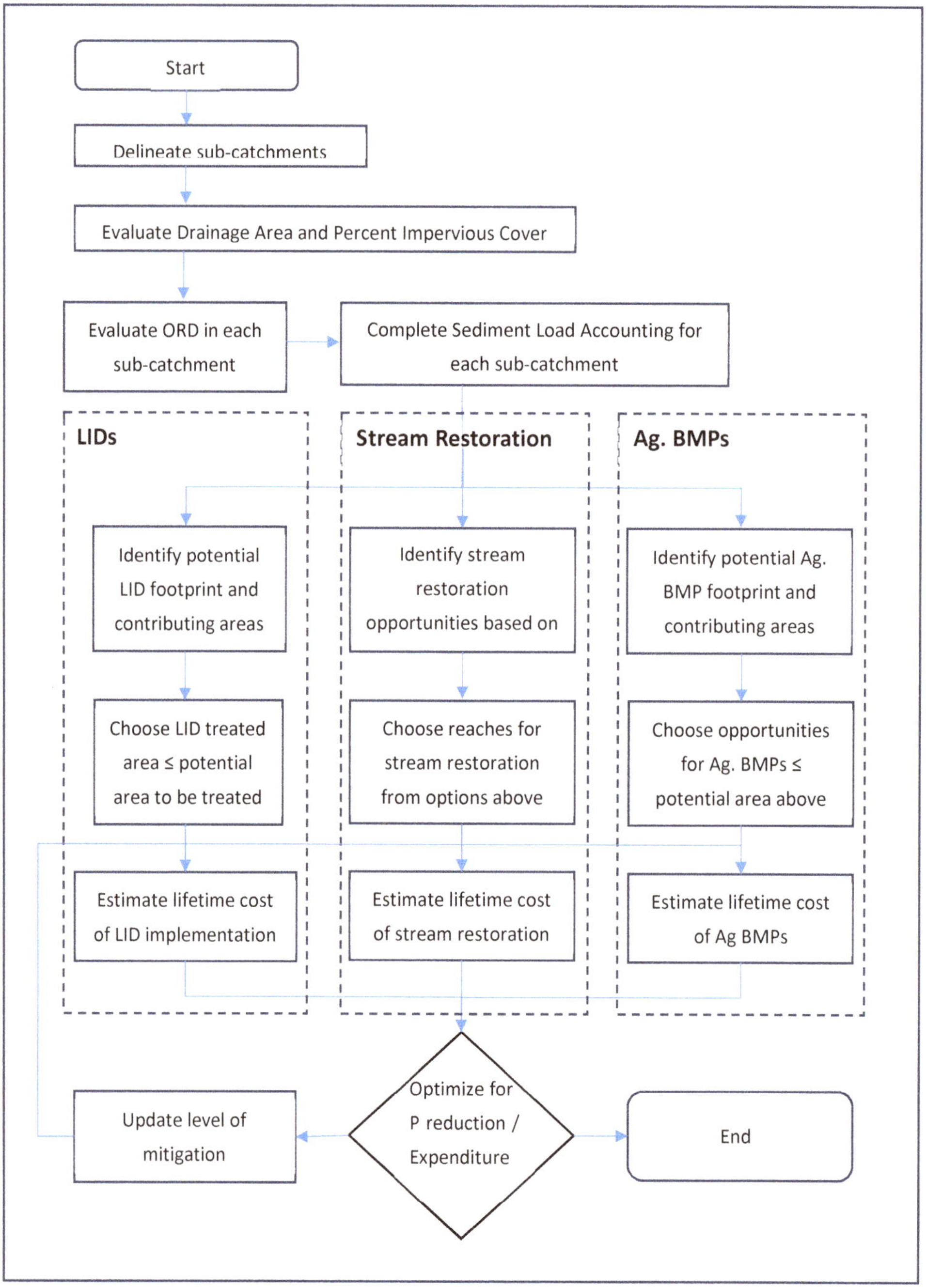

Figure 3. Mitigation implementation optimization algorithm.

3. Results and Discussions

The following sections provide a discussion of the results of this study, including the catchment characteristics, streambank erosion sources, hydraulic modeling, and sediment load accounting for Tannery Creek. The results of the cost-optimized plan for mitigating sediment and phosphorous loads using low-impact development (LID), stream restoration, and agricultural best management practices (BMPs) in the Tannery Creek watershed are also presented.

3.1. Catchment Characteristics

Tannery Creek catchment and sub-catchment characteristics were identified using a combination of site knowledge gained in the field, GIS analysis, Google Earth review, and spatial data available from the Lake Simcoe Region Conservation Authority (LSRCA) spatial data portal. Catchment areas were delineated using the Ontario Watershed Information Tool (OWIT) and calculated from urban area land use data. Tannery Creek sub-catchments are shown in Figure 1 and summarized in Table 2.

Table 2. Tannery Creek sub-catchment characteristics.

Reach	Area (km^2)	Incr. Area (ha)	Cumulative U/S Area (ha)	% Urban	PIC	Inc. Impervious Area (ha)	U/S Impervious Area (ha)	U/S PIC
TC 1	3.51	351	351	75%	25%	87.7	87.7	25%
TC 2	3.54	354	354	70%	23%	82.6	82.6	23%
TC 3	2.11	211	211	63%	21%	44.4	44.4	21%
TC 4	2.11	211	422	65%	22%	45.7	90.1	21%
TC 5	2.05	205	205	21%	7%	14.4	14.4	7%
TC 6	3.19	319	524	33%	11%	35.1	49.4	9%
TC 7	7.25	725	1249	26%	9%	62.8	112.2	9%
TC 8	2.29	229	1900	52%	17%	39.7	242.0	13%
TC 9	3.09	309	2913	50%	17%	51.4	463.7	16%
TC 10	0.51	51	51	78%	26%	13.2	13.2	26%
TC 11	1.47	147	147	32%	11%	15.7	15.7	11%
TC 12	8.10	810	3921	25%	8%	67.5	560.2	14%

Notes: PIC—Percent Impervious Cover. Inc.—Incremental. U/S—Upstream.

For some variables in Table 2, both an incremental (for the sub-catchment) and total upstream value are included as upland processes typically scale with incremental catchment area and land use characteristics. In contrast, stream corridor processes usually scale with the total upstream contributing area and land use characteristics.

3.2. Stream Bank Erosion Sources

Urbanization-induced changes to the hydrological flow regime include a significant increase in peak flow and total runoff volume during storm events. When peak flows increase, the channel typically responds by increasing erosion rates. Identifying the existing channel geometry and the projected regime channel geometry suitable for the changed hydrological flows allows the quantification of erodible sediment contributing to sediment and phosphorous loading without intervention.

In regime channels, bank erosion typically represents a small fraction (less than 5%) of the annual sediment load in undeveloped or rural/agricultural watersheds; however, the streamflow regime is altered in urbanized watersheds, resulting in elevated stream erosive power. For Southern Ontario streams, the International Reference Group On Great Lakes Pollution From Land Use Activities study found that the sediment yield rate from bank erosion sources was 5 to 223 $kg{\cdot}ha^{-1}.yr^{-1}$ (1.37 to 61.1 $kg{\cdot}km^{-2}{\cdot}d^{-1}$) [39].

In addition, the Ontario Ministry of Natural Resources (MNR, now MNRF) Technical Guide for River and Stream Systems: Erosion Hazard Limit notes that stream bank recession rates in Ontario typically range from 1 to 11.5 m/100-years (average of 0.01 to 0.12 m/yr). Stream bank erosion contributes approximately 5% of the annual sediment load in undisturbed watersheds and up to 30% in urbanized watersheds with significant lengths of channel out-of-regime, candidates for stream restoration projects [8,9]. The International

Reference Group On Great Lakes Pollution From Land Use Activities study results in constrained total stream corridor sediment sources for stream bank erosion [39].

3.3. *Hydraulic Modelling*

Hydraulic modeling of the river network is needed to assess the urbanizing stream's stability and estimate the sediment and phosphorus loads contributed by the out-of-regime stream reaches. For this study, a channel reach was deemed out-of-regime if the ratio of specific stream power to regime-specific stream power was more significant than or equal to a threshold of 1.5. Hydraulic modeling (HEC-RAS) was used to evaluate specific stream power (during the Q_2 event) and regime conditions in Tannery Creek. Each tributary's out-of-regime channel length was totaled and compared with the percent impervious cover (PIC) in the upstream catchment. The out-of-regime channel length, as a percentage of the total channel length in each tributary (ORD), was found to vary approximately linearly, with PIC for low to moderate values of PIC and, hence, was assumed to be roughly equal to PIC to identify the potential length of channel eligible for stream restoration.

A comparison between the length of the out-of-regime channel, identified in the current study using stream power (ω/ω'), and the rapid geomorphic assessment (RGA score) study conducted by the Aquafor Beech Ltd. [32] classifications is shown in Figure 4, that shows a good agreement between the two methods. Based on this finding, the ω/ω' criteria (<1 in regime; between 1 and 1.5 in transition; and >1.5 out-of-regime) may be used to assess regime conditions from topographic surveys and discharge records for short reaches of stream channels.

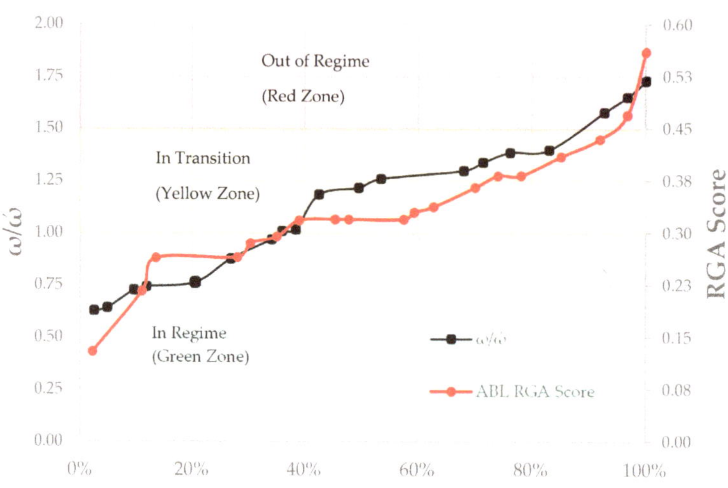

Figure 4. Comparison of Rapid Geomorphic Assessment (RGA) score vs. $\omega/\acute{\omega}$ channel out-of-regime channel length.

In addition to evaluating existing regime conditions, historical and future Q_2 estimates may be used to assess out-of-regime channel length under pre-urbanization and future full urban development scenarios. The ability to evaluate regime conditions under differing hydrological flows provides a supplemental tool to RGA, which is limited to evaluating the current state of a stream reach based on current and recent flow regimes. Figure 5 shows a comparison of modeled out-of-regime channel lengths in the Tannery Creek catchment approximately 20 years ago, at present conditions, and approximately 20 years in the future. These different scenarios are based on observed rates of change in Q_2 and PIC between the 1990s and the present day.

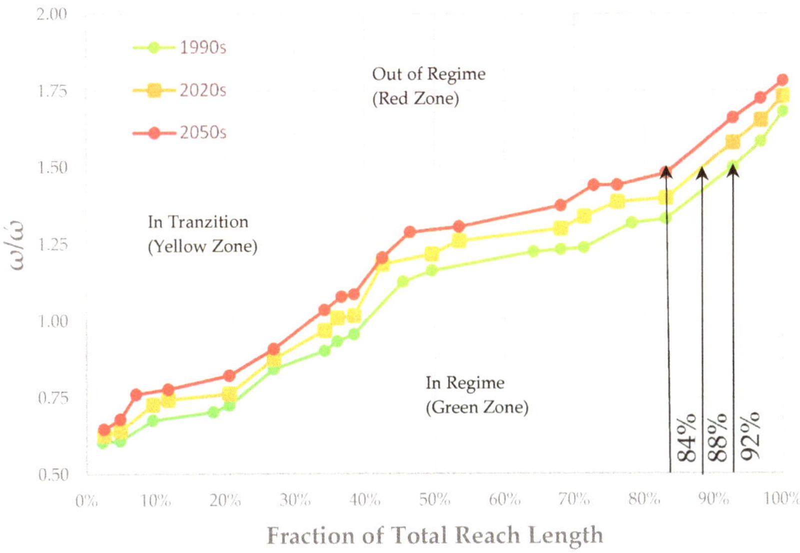

Figure 5. Tannery Creek fraction of total stream reach lengths, In Regime, In Transition, and Out-of-regime for the past (1990s), present (2020s), and future (2050s) scenarios.

3.4. Sediment and Phosphorus Load Accounting for Tannery Creek

The total suspended sediments (TSS) and total phosphorus (TP) loads in Tannery Creek, the East Holland River, and a dozen other nearby watersheds, summarized in Table 3 [8,9], were correlated with PIC and used to develop three linear equations (Equations (3)–(5)) for the area based estimates of the sediment or phosphorus loading in the current study. The linear relationships between sediment load and land use in the contributing catchment area were found using the Solver function in Microsoft Excel, and the observed TSS load data are shown in Table 3.

Table 3. Contribution of pervious, impervious, and stream TSS and TP loads.

		DA (km^2)	PIC (%)	TP Load ($kg\ d^{-1}\ km^{-2}$)	TSS Load ($kg\ d^{-1}\ km^{-2}$)	IMP (%)	PER (%)	STR (%)
02HC010	1966–1986	51	2.0	0.027	13.2	23%	74%	3%
02HC011	1966–1986	291	2.0	0.017	13.5	22%	73%	5%
02HC011	1998–2018	291	2.0	0.022	13.5	22%	73%	5%
02HC010	1998–2018	51	3.0	0.030	14.9	30%	65%	5%
02EC018	2003–2019	347	3.0	0.024	15.2	30%	64%	7%
02HB007	2003–2019	158	4.3	0.030	17.3	37%	55%	7%
02EC009	1984–2000	176	6.0	0.047	21.6	42%	44%	15%
02HC022	1968–1988	181	9.0	0.058	26.8	50%	34%	16%
02EC009	2003–2019	176	10.0	0.070	31.9	47%	28%	25%
Tannery	2003–2019	39	11.0	0.077	35.2	47%	25%	28%
02HC013	1956–1976	89	13.0	0.065	41.0	48%	21%	31%
02HC005	1965–1975	88	15.0	0.154	44.6	50%	19%	30%
02HC022	1998–2018	181	31.0	0.097	84.5	55%	8%	37%
02HC030	1966–1976	205	32.0	0.095	84.4	57%	8%	35%
02HC030	2008–2018	205	36.0	0.113	105.6	51%	6%	43%
02HC005	1985–1995	88	41.0	0.112	125.9	49%	5%	46%

Notes: DA—Drainage Area. PIC—Percent Impervious Cover. TP—Total Phosphorous. TSS—Total Suspended Solids. IMP—load contribution from impervious areas of the upstream catchment. PER—load contribution from permeable areas of the upstream catchment. STR—load contribution from stream corridors in upstream catchment.

Observed TSS loads were assumed to be sourced from three component land uses in the contributing catchment, impervious area, permeable area, and stream corridor, based on the findings of MacKenzie et al. (2022) [9]. The component TSS load contributions for all catchments in Table 3 were solved simultaneously by assuming the form of the linear relationship shown in Equations (3)–(5) and solving for the empirical coefficients that resulted in the best match between total predicted and total observed load.

To implement the solver function, the predicted component loads were summed for each catchment area, and the square error (difference squared) between the total predicted and observed loads was found. The coefficients in Equations (3)–(5) were changed iteratively until the sum of squared errors between observed and predicted total loads was minimized. The solution for the coefficient in Equation (4) for stream corridor contributions was constrained to the range of stream bank-sourced loads discussed in MacKenzie et al. [8,9] and the PLUARG study [9,53].

$$\text{IMP}\ (\text{kg}\cdot\text{d}^{-1}\cdot\text{km}^{-2}) = 150 \times \text{PIC} \tag{3}$$

$$\text{STR}\ (\text{kg}\cdot\text{d}^{-1}\cdot\text{km}^{-2}) = 50 \times \text{PIC} \tag{4}$$

$$\text{PER}\ (\text{kg}\cdot\text{d}^{-1}\cdot\text{km}^{-2}) = 10 \times (1 - \text{PIC}) \tag{5}$$

Equations (3)–(5) can be used to quantify the suspended sediment load with knowledge of the contributing catchment area and land use. The contribution from permeable sources (PER) included both urban and rural permeable areas. The resulting percentage contributions of each of the component TSS load contributions are summarised in Table 3.

Equations (3)–(5) relate the relative contributions of sediment in Tannery Creek to the fraction of impervious area in the sub-catchment. The method used in this study is based on the observed sediment load in the downstream receiving watercourse (i.e., measured at the outlet of the watershed) and estimates the net sediment and phosphorus yield considering the well-established concept of "sediment delivery ratio." The sediment delivery ratio is the ratio of sediment yield at the watershed outlet to gross erosion, affected by changes in storage in the drainage network and interception by infiltration, deposition, and plant uptake. While the sediment delivery ratio was not independently estimated using the current study methodology, the net effect of source load and sediment delivery ratio are implicitly included in the component load estimates provided by Equations (3)–(5). This approach differs from other methods that evaluate loads considering the export coefficient at or near the source [42,45].

Table 3 shows that as urbanization advances and pervious areas are converted into impervious areas, the TSS and TP loads per unit watershed area increase by an order of magnitude. In addition, the contribution of the stream corridor increases as rapidly urbanizing streams tend to go out-of-regime, and the bed and bank erosion loads become significant.

To estimate the sediment yield in a receiving watercourse, the loading per unit area estimated using Equations (3)–(5) is multiplied by the whole upstream contributing area. This approach differs from other methods that use the export coefficient, which is applied to the specific source area of interest [42,45]. Equations (3)–(5) may be used to explore the benefits of mitigation that adjust the effective catchment impervious cover or minimize bank erosion by reducing stream channel susceptibly to erosion.

3.5. Mitigating Sediment Loads Using Low-Impact Development

While evaluating the role of stream corridor sources of sediment and P loading is the focus of this study, the relative benefits of other mitigation measures, including catchment level low-impact development (LID) implementation, were included to allow a side-by-side comparison of costs and benefits. Typical LID features function by intercepting runoff from the footprint area of the LID and an upgraded catchment area. Intercepted water is retained within a typically underground storage unit and infiltrated for runoff events up to the design event. The typical ratio of the catchment area treated by an LID to the LID footprint area is 20:1 and 15:1 for the infiltration trench and the bioretention cell, as presented in Table 1. For permeable pavements, this ratio is 1:1. That is, for every m^2 of bioretention cell installed, 15 m^2 of impervious area is treated, reducing the PIC accordingly.

Upon intercepting and infiltrating runoff, LIDs moderate peak flows in receiving watercourses. In addition, contaminants in intercepted runoff are prevented from entering receiving watercourses and downstream waterbodies. For this study, a simplifying assumption was made that impervious areas contributing to infiltration LIDs effectively become permeable areas. This assumption was used to modify sub-catchment PIC and quantify Q_2 in Tannery Creek sub-catchments; however, relatively limited opportunities for LID implementation in the significantly urbanized sub-catchments limit the potential to mitigate Q_2 in those sub-catchments significantly. Similarly, contaminants in intercepted runoff were assumed to be contained in the LID and prevented from contributing to downstream receiving watercourses. This assumption implies that LIDs will be properly maintained to avoid fouling and contaminant breakthroughs.

- Available opportunities for four types of LID were identified in the Tannery Creek watershed based on previous studies implemented by the Lake Simcoe Region Conservation Authority: green roof, permeable pavement, bioretention, and infiltration trenches [42]. The costing example below disregarded the green roof and permeable pavement developments due to the high cost-to-treatment benefit ratio and predominantly private land opportunities. Google Earth Imagery was used to determine potential LID siting for infiltration trenches & bioretention cells, as shown in Figure 1.
- An analysis of LID implementation and benefits on phosphorus loading is presented in Table 4. For example, with reference to Table 4, the PIC for TC 3 before mitigation is 21%. If drainage from 5% of the available 15 ha of the catchment were diverted to bioretention LIDs, 5% of the 1.4 ha to infiltration trenches, and 5% of the available 1.8 ha to permeable pavement, the effective PIC after mitigation would drop to 15%, thereby reducing the TP load to Tannery Creek from that catchment by approximately 34.5 kg/y at a life cycle cost of $3,590,000.
- Based on incentive and offset funding available in 2023, P credits of approximately $3,080,000 would be available to fund the mitigation, resulting in a cost recovery ratio (P credit/life cycle cost of LIDs) of 0.9, meaning the proposed LID implementation in TC 3 is one of the more affordable and beneficial options. The focus of past stormwater management (SWM) and phosphorous load reduction has been on upland areas with less attention paid to riparian corridors; however, changed receiving watercourse hydrology increases erosive power of the flowing water and may destabilize regime channels and erode sediment and P from the riparian corridor as discussed in Section 3.7.
- Moderating (reducing) Q_2 by implementing LID measures in upstream catchments can partially reduce erosive forces in stream corridors, with corresponding reductions in sediment and P loads. For this study, the implementation of LIDs for this costing example is 5% of the estimated total available area for LID. The total treated area was defined as the footprint area of constructed LIDs plus the catchment area draining to the LID. LID implementation reduces the impervious cover of the catchments. In the costing example, the reduced PIC of the cumulative upstream catchment after the LIDs was calculated.

3.6. *Mitigating Sediment Loads Using Stream Restoration*

Stream restoration to stabilize eroding reaches and "immobilize sediments in place" represents an additional mitigation opportunity against sediment and phosphorous loading. As discussed in [8,9], urbanization increases PIC, resulting in increased peak flows and, if left unmitigated, development of the urban stream syndrome in receiving watercourses.

Table 4. Low-impact development implementation and benefits on sediment and phosphorous loading.

Reach	Permeable Pavement (ha)	Infiltration Trench (ha)	Bioretention (ha)	U/S PIC before	U/S PIC after	TSS Load Reduction (kg/d/km)	Whole Life LID_1 Cost ($)	Inc. TSS Load Reduction (kg/y)	Inc. P Load Reduction (kg/y)	Phosphorus Credit ($)	Cost Recovery Ratio
TC 1	12.4	0.6	33.0	25%	18%	10.8	7,380,000	13,900	69.3	6,190,000	0.8
TC 2	17.4	3.7	53.0	23%	11%	18.4	12,400,000	23,800	118.9	10,600,000	0.9
TC 3	1.8	1.4	15.0	21%	15%	8.9	3,590,000	6900	34.5	3,080,000	0.9
TC 4	–	5.0	20.0	21%	17%	7.1	5,460,000	5470	27.3	2,440,000	0.4
TC 5	–	2.0	5.5	7%	4%	4.5	1,630,000	3350	16.8	1,500,000	0.9
TC 6	–	3.9	1.0	9%	9%	1.3	1,040,000	1550	7.7	693,000	0.7
TC 7	5.7	5.0	4.0	9%	8%	1.0	1,940,000	2540	12.7	1,140,000	0.6
TC 8	–	5.0	7.0	13%	12%	0.8	2,600,000	676	3.4	302,000	0.1
TC 9	14.5	5.0	13.0	16%	15%	0.8	3,920,000	856	4.3	383,000	0.1
TC 10	–	2.0	12.0	26%	4%	32.4	3,060,000	6020	30.1	2,690,000	0.9
TC 11	–	2.0	14.0	11%	2%	12.8	3,500,000	6840	34.2	3,060,000	0.9
TC 12	3.5	7.5	46.0	14%	13%	1.6	11,700,000	4750	23.8	2,120,000	0.2

Notes: LID—Low-Impact Development. U/S—upstream. PIC—Percent Impervious Cover. Inc.—Incremental. TSS—Total Suspended Solids. P—Phosphorus.

As watercourses become unstable and the out-of-regime channel length increases, sediment and associated P loading from stream corridor sources also increase. If allowed to continue, this channel adjustment process would typically continue for decades until the channel cross section expands sufficiently to bring the erosive forces of increased hydrologic flows back into balance with substrate resistance to erosion. In addition to increased erosion and contaminant loading, changes to the substrate composition, low flow depths, and ecological function of impacted watercourses may never recover to pre-development conditions.

Stream restoration may reduce sediment and P loading. Still, it should be viewed as a second line of defense after catchment-level hydrological flow regime mitigation, which moderates both upland catchment and stream corridor sources by mitigating hydrological flows. Every stream reach that is restored would not contribute to the sediment and phosphorus load of the watershed. An analysis of stream restoration measures and benefits on sediment and P loading using Equations (3)–(5) is presented in Table 5.

Table 5. Summary of TSS load, TP load, and cost of stream restoration for the Tannery Creek catchment.

Reach	U/S PIC	Inc. Bank TSS Load (kg/d)	Inc. Whole Life Stream Restoration Cost ($)	Inc. Bank TSS Load (kg/y)	Inc. TP Load Reduction (kg/y)	Phosphorus Credit ($)	Cost Recovery Ratio
TC 1	25%	42.1	991,000	15,400	15.4	1,380,000	1.4
TC 2	23%	39.6	568,000	14,500	14.5	1,290,000	2.3
TC 3	21%	21.3	372,000	7780	7.8	696,000	1.9
TC 4	21%	21.9	588,000	8010	8.0	716,000	1.2
TC 5	7%	6.9	329,000	2520	2.5	225,000	0.7
TC 6	9%	16.8	859,000	6150	6.1	550,000	0.6
TC 7	9%	23.3	1,560,000	8490	8.5	760,000	0.5
TC 8	13%	25.9	644,000	9480	9.5	847,000	1.3
TC 9	16%	24.7	634,000	9020	9.0	807,000	1.3
TC 10	26%	6.3	275,000	2320	2.3	207,000	0.8
TC 11	11%	7.5	597,000	2750	2.7	246,000	0.4
TC 12	14%	32.4	914,000	11,800	11.8	1,060,000	1.2

Notes: Inc.—Incremental. TSS—Total Suspended Solids. TP—Total Phosphorus. U/S—upstream. PIC—Percent Impervious Cover.

For example, with reference to Table 5, the PIC for TC 1 is 25%, which is a reasonable proxy measure of the fraction of total channel length in the catchment that is out-of-regime. If all of the out-of-regime channels within the TC 2 catchment were restored to match the hydrologic flow regime, we can assume that the P load from that catchment would approach zero during typical flow conditions. As a result, the P load from that catchment would be reduced by 15.4 kg/y at an estimated life cycle cost of $991,000. Based on P load reduction incentives and offset funding available in 2023, P credits of approximately $1,380,000 would be available to fund the mitigation, resulting in a cost recovery ratio (P credit/life cycle cost of LIDs) of 1.4, meaning the proposed LID implementation in TC 1 is a highly worthwhile investment and an effective mitigation measure.

While close-to-source mitigation measures, such as LIDs and green and blue infrastructure, should remain the first line of defense, stream restoration should be considered as an additional tool to mitigate against the effects of urbanization on stream stability, contaminant loading, and ecological function. Stream restoration can be used to stabilize out-of-regime channel reaches by adjusting the channel cross section and profile to match the increased urban flow regime better or by armoring bed and bank materials with less erodible materials to resist erosion by flowing water in the creek. Where successfully implemented, stream restoration measures may reduce stream corridor sources of sediment and associated contaminants to a minimum, similar to regime conditions.

3.7. Mitigating Sediment Loads Using Agricultural Best Management Practices

While evaluating the role of stream corridor sources on sediment, P loading is the focus of this study. The relative benefits of other mitigation measures, including agricultural BMPs, were included to allow a side-by-side comparison of costs and benefits. As discussed in Section 3.4, loading from these areas decreases with urbanization as impervious land uses gradually replace them. The relative contribution of contaminant loading from permeable sources may become less significant with increasing urbanization; however, disturbed permeable areas such as construction sites and active agricultural areas represent readily attainable opportunities to mitigate sediment and P loads. This study considered BMPs to compare sources and mitigation options to address sediment and P loading to receiving watercourses.

Agricultural BMPs designed to address ongoing land disturbance include a range of land use practices intended to reduce soil loss, increase agricultural productivity, and reduce sediment yield to receiving watercourses. This study considered cover crops, vegetative buffer strips (VBSs), and conservation tillage practices (including no-till). VBSs are typically planted along the downgradient edge of agricultural fields and are not disturbed or harvested. By allowing dense ground covering vegetation to propagate in VBSs, eroded soil in runoff is trapped, and the vegetation consumes nutrients from upgradient areas. Typical sediment removal efficiencies for BMPs were sourced from [43], while typical annual costs per hectare were taken from [44]. Opportunities for BMPs were identified using OWIT and by reviewing historical Google Earth imagery.

A summary of the proposed BMPs and P removal for catchments with significant amounts of agricultural land use, is included in Table 6. For example, with reference to Table 6, the TSS load from TC 7 is 9.3 kg/d/km^2 (24,200 kg/y) before implementing BMPs. If 60% of the available 170 ha of active agricultural land starts using cover crops, 60% of the available 3.62 ha area for VBSs is established along the downgradient edge of fields, and 60% of the available 73 ha is moved to conservation tillage practices, the TSS load contributed to Tannery Creek from that catchment would drop by 12,600 kg/y with a corresponding reduction in P load of approximately 12.8 kg/y at a life cycle cost of $1,480,000. Based on incentives and offset funding available in 2023, P credits of approximately $1,130,000 would be available to fund the mitigation, resulting in a cost recovery ratio (P credit/life cycle cost of LIDs) of 0.76 meaning that the proposed LID implementation in TC 7 is one of the more affordable and beneficial options.

Table 6. Summary of TP load and cost of Agricultural Best Management Practices for Tannery Creek.

Reach	Inc. TSS Load (kg/d/km^2)	Cover Crops (ha)	Vegetative Buffer (ha)	No Tillage/ Cons. Tillage (ha)	Inc. TSS Removed (kg/y)	Phosphorus Credit ($/y)	Whole Life Cost ($)	Cost Recovery Ratio
TC 5	9.3	24	1.36	10	1920	172,100	294,000	0.59
TC 6	8.6	58	2.11	25	4190	375,000	592,000	0.63
TC 7	9.0	170	3.62	73	12,600	1,130,000	1,480,000	0.76
TC 8	8.0	30	1.52	13	962	86,100	351,000	0.24
TC 12	9.0	85	1.81	36.4	6310	565,000	742,000	0.76

Notes: TP—Total Phosphorus. Inc.—Incremental. TSS—Total Suspended Solids.

3.8. Cost–Benefit Analysis

A cost–benefit analysis was completed for Tannery Creek to provide a case study example of how P load mitigation funding may be applied to LID, Ag. BMP and stream restoration initiatives to maximize the benefits per dollar spent. The Tannery Creek sub-catchments were organized based on PIC and ORD (% of channel length with $\omega/\hat{\omega} > 1.5$). Sub-catchments with PIC > 10% and significant ORD were given priority for LID retrofit and stream restoration opportunities, while those with low values of PIC were considered for Ag. BMPs. Stream restoration opportunities were given priority ranking based on potential P-load reduction and cost.

Areas potentially available for LID retrofits were identified from existing mapping sources as discussed in Section 3.6 and prioritized for LID implementation based on potential P load reduction and cost. The cost of stream restoration versus LID and Ag. BMP implementation was compared and optimized based on P load reduction per dollar spent. Table 6 illustrates the relative costs and benefits of stream restoration, LID implementation, and Ag. BMPs in the Tannery Creek catchment. In addition to optimizing P load reduction, the cost recovery ratio was estimated for each of the three mitigation alternatives. Cost recovery is the ratio of available financial incentives for mitigation projects divided by the estimated lifetime cost of the projects.

To allow for side-by-side comparison, all mitigation measures were scaled to a twenty-five-year lifetime. However, some are funded and implemented more frequently and have shorter effective lifetimes (e.g., cover crops).

Table 7 shows that the urban LID retrofits have the highest potential benefit on annual P load reduction in Tannery Creek. However, they are also the most expensive option. Stream restoration opportunities in Tannery Creek result in the second highest potential benefit on annual P load reduction. They would cost approximately an order of magnitude less than LID, resulting in the highest percent reduction in annual P load and the highest cost recovery ratio of the three mitigation options. Opportunity, based on ORD, limits opportunities to further capitalize on the benefits of stream restoration to reduce P load in the Tannery Creek catchment.

Table 7. Cost–Benefit Analysis for Tannery Creek.

Method	Stream Restoration	Urban LID_1 Retrofits	Agricultural BMPs
Annual P Load Reduction (kg P/y)	76.0	304	26.5
Total Expenditure ($Cdn)	$4,710,000	$31,600,000	$3,460,000
Cost Recovery Ratio	1.44	0.86	0.68

Notes: LID—Low-Impact Development. BMP—Best Management Practice.

Agricultural BMP opportunities in the Tannery Creek catchment are limited because of the relatively low fraction of the total area in active agricultural land use; however, the annual P load reduction from agricultural BMPs is still significant; the lifetime cost is comparable to stream restoration, and significantly less than LID implementation. The percent reduction in annual P load is smaller for agricultural BMPs, primarily due to ongoing maintenance costs, and the cost recovery ratio is lower than the other two mitigation options considered.

However, BMPs have other significant benefits (e.g., increased crop yield, reduced irrigation requirements, reduced soil loss) that were not considered in this cost–benefit analysis of P load mitigation. Stream restoration and LID retrofits also have other benefits not considered in this cost–benefit analysis. LIDs are necessary for stormwater management and have other water quality treatment benefits and flood mitigation functions. The sediment and P load reduction benefits of stream restoration have been recognized in the literature in recent years. The reasons for conducting stream restoration (e.g., ecological function, erosion control, infrastructure protection) remain important.

Of the three groups of mitigation considered in this study, LID and stream restoration provided the most significant opportunities to mitigate the effects of urban development partially; however, implementation of these measures was limited by the amount of out-of-regime stream reaches and access to suitable areas for LID implementation in already significantly developed sub-catchments. Agricultural BMPs remain a vital part of the overall approach to minimizing sediment and phosphorus loads but represent a small opportunity in the Tannery Creek Catchment because of the relatively small fraction of the catchment in active agricultural land use.

Mitigation using LIDs has two modes of effect on sediment and phosphorus loads. First, LIDs may be used to intercept runoff from upland urban areas and retain and infiltrate it, resulting in the removal of suspended contaminants, including phosphorus, from surface runoff to receiving water courses. The secondary mode of the effect of LIDs is to moderate peak flows in the receiving watercourse by effectively causing impervious surfaces to respond to precipitation more similar to impermeable surfaces. Modifying peak flows in the receiving watercourse can mitigate stream corridor erosion and associated suspended sediment and phosphorus loads.

4. Conclusions

This study presents a novel methodology to evaluate the relative contribution of the stream corridor versus upland watershed contributions to total sediment and phosphorus loads in receiving watercourses. The new method can be used to develop a cost-optimized mitigation plan for the impacted watersheds, including the implementation of low-impact development (LID) for urban areas and best management practices (BMPs) for rural areas of the watershed as well as stream restoration projects for degraded stream reaches. We showcased the new method for the Tannery Creek watershed, which has gone through significant urbanization in recent decades.

The case study analysis of sediment and phosphorus loading in Tannery Creek illustrated the effects of urbanization and the potential benefits of three groups of mitigation measures (LIDs, stream restoration, and BMPs). The results showed that hydromodification caused by urbanization has resulted in increases in sediment and P loads to downstream receivers. Sediment and associated phosphorus loading were attributed to urban wash off from introduced impervious surfaces, stream corridor erosion resulting from increased hydrological flows, and some residual load from diminishing agricultural areas with frequent disturbance. Stream corridor erosion was primarily attributed to out-of-regime stream reaches in urbanizing sub-catchments.

The novel methodology developed in this study can assess the costs and benefits of mitigation measures, including low-impact development (LID), stream restoration, and agricultural best management practices (BMPs), and may be applied in other watersheds to facilitate mitigation planning. The fraction of total channel length that is out-of-regime (ORD) in an unmitigated urbanizing catchment is approximately equal to the percent impervious cover (PIC) of the contributing catchment area for low to moderate levels of PIC.

The cost–benefit analysis for the Tannery Creek Watershed showed that stream restoration should be included in water quality management programs as a cost-effective mitigation against increased contaminant loads associated with urbanization. Implementing various LIDs, BMPs, and stream restoration projects should be complemented with ongoing water quality monitoring near the watershed outlet and a re-analysis about five years after each significant implementation of mitigation measures to assess the water quality improvements achieved.

Author Contributions: Conceptualization, K.M. and B.G.; data preparation, K.M., S.B. and B.G.; methodology, K.M., S.A. and B.G.; modeling, K.M. and B.G.; analysis, K.M., S.A. and B.G.; writing—original draft preparation, K.M.; writing—review and editing, K.M., S.B., S.A. and B.G.; visualization, K.M. and B.G.; supervision, B.G.; funding acquisition B.G. All authors have read and agreed to the published version of the manuscript.

Funding: This work was supported by the Ontario Ministry of Environment, Conservation and Parks under Grant No. 053571 and the Lake Simcoe Region Conservation Authority under contract No. LS1718010.

Data Availability Statement: The datasets used in this study are in the public domain and are available for download from the Water Survey of Canada (for the Discharge and Water level data), the Ontario Ministry of the Environment Provincial Water Quality Monitoring Program (for the sediment and phosphorous datasets) and the Lake Simcoe Region Conservation Authority GIS Open Data (for the land use and GIS data).

Acknowledgments: The authors wish to thank our research partners Tim Krsul, Senior Program Advisor at the Ontario Ministry of the Environment, Conservation, and Parks, as well as Lake Simcoe Region Conservation Authority representatives, including Christa Sharp, Kenneth Cheney, David Lembcke, and Glenn MacMillan for providing insightful comments, feedback, shared data and support for this study. In addition, we would like to thank research assistants Corinne Ruthenberg, Mohammadreza Mahdavi, and Samaneh Nezhadhaghighi for their assistance in analyzing spatial and numerical data to support this study.

Conflicts of Interest: The authors have no conflict of interest.

References

1. Paul, M.J.; Meyer, J.L. Streams in the Urban Landscape. *Annu. Rev. Ecol. Syst.* **2001**, *32*, 333–365. [CrossRef]
2. Russell, K.L.; Vietz, G.J.; Fletcher, T.D. How Urban Stormwater Regimes Drive Geomorphic Degradation of Receiving Streams. *Prog. Phys. Geogr. Earth Environ.* **2020**, *44*, 746–778. [CrossRef]
3. Walsh, C.J.; Roy, A.H.; Feminella, J.W.; Cottingham, P.D.; Groffman, P.M.; Morgan, R.P. The Urban Stream Syndrome: Current Knowledge and the Search for a Cure. *J. N. Am. Benthol. Soc.* **2005**, *24*, 706–723. [CrossRef]
4. Bonakdari, H.; Ebtehaj, I.; Samui, P.; Gharabaghi, B. Lake Water-Level Fluctuations Forecasting Using Minimax Probability Machine Regression, Relevance Vector Machine, Gaussian Process Regression, and Extreme Learning Machine. *Water Resour. Manag.* **2019**, *33*, 3965–3984. [CrossRef]
5. Halstead, J.A.; Kliman, S.; Berheide, C.W.; Chaucer, A.; Cock-Esteb, A. Urban Stream Syndrome in a Small, Lightly Developed Watershed: A Statistical Analysis of Water Chemistry Parameters, Land Use Patterns, and Natural Sources. *Environ. Monit. Assess.* **2014**, *186*, 3391–3414. [CrossRef] [PubMed]
6. Margenot, A.J.; Zhou, S.; McDowell, R.; Hebert, T.; Fox, G.; Schilling, K.; Richmond, S.; Kovar, J.L.; Wickramarathne, N.; Lemke, D.; et al. Streambank Erosion and Phosphorus Loading to Surface Waters: Knowns, Unknowns, and Implications for Nutrient Loss Reduction Research and Policy. *J. Environ. Qual.* **2023**, *52*, 1063–1079. [CrossRef] [PubMed]
7. Bertora, A.; Grosman, F.; Sanzano, P.; Rosso, J.J. Combined Effects of Urbanization and Longitudinal Disruptions in Riparian and In-Stream Habitat on Water Quality of a Prairie Stream. *Knowl. Manag. Aquat. Ecosyst.* **2022**, *423*, 15. [CrossRef]
8. MacKenzie, K.M.; Gharabaghi, B.; Binns, A.D.; Whiteley, H.R. Early Detection Model for the Urban Stream Syndrome Using Specific Stream Power and Regime Theory. *J. Hydrol.* **2022**, *604*, 127167. [CrossRef]
9. MacKenzie, K.M.; Singh, K.; Binns, A.D.; Whiteley, H.R.; Gharabaghi, B. Effects of Urbanization on Stream Flow, Sediment, and Phosphorous Regime. *J. Hydrol.* **2022**, *612*, 128283. [CrossRef]
10. Smith, J.S.; Winston, R.J.; Wituszynski, D.M.; Tirpak, R.A.; Boening-Ulman, K.M.; Martin, J.F. Effects of Watershed-Scale Green Infrastructure Retrofits on Urban Stormwater Quality: A Paired Watershed Study to Quantify Nutrient and Sediment Removal. *Ecol. Eng.* **2023**, *186*, 106835. [CrossRef]
11. Jarden, K.M.; Jefferson, A.J.; Grieser, J.M. Assessing the Effects of Catchment-scale Urban Green Infrastructure Retrofits on Hydrograph Characteristics. *Hydrol. Process.* **2016**, *30*, 1536–1550. [CrossRef]
12. Spraakman, S.; Rodgers, T.F.M.; Monri-Fung, H.; Nowicki, A.; Diamond, M.L.; Passeport, E.; Thuna, M.; Drake, J. A Need for Standardized Reporting: A Scoping Review of Bioretention Research 2000–2019. *Water* **2020**, *12*, 3122. [CrossRef]
13. Søberg, L.C.; Al-Rubaei, A.M.; Viklander, M.; Blecken, G.-T. Phosphorus and TSS Removal by Stormwater Bioretention: Effects of Temperature, Salt, and a Submerged Zone and Their Interactions. *Water. Air. Soil Pollut.* **2020**, *231*, 270. [CrossRef]
14. Singer, M.N.; Hamouda, M.A.; El-Hassan, H.; Hinge, G. Permeable Pavement Systems for Effective Management of Stormwater Quantity and Quality: A Bibliometric Analysis and Highlights of Recent Advancements. *Sustainability* **2022**, *14*, 13061. [CrossRef]
15. Hua, P.; Yang, W.; Qi, X.; Jiang, S.; Xie, J.; Gu, X.; Li, H.; Zhang, J.; Krebs, P. Evaluating the Effect of Urban Flooding Reduction Strategies in Response to Design Rainfall and Low Impact Development. *J. Clean. Prod.* **2020**, *242*, 118515. [CrossRef]
16. Garbanzos, S.; Maniquiz-Redillas, M. Evaluating the Peak Flow and Runoff Coefficient Reductions of Bioretention, Infiltration Trench, and Permeable Pavement LID Using Stormwater Management Model. *Chem. Eng. Trans.* **2022**, *97*, 433–438. [CrossRef]
17. Askarizadeh, A.; Rippy, M.A.; Fletcher, T.D.; Feldman, D.L.; Peng, J.; Bowler, P.; Mehring, A.S.; Winfrey, B.K.; Vrugt, J.A.; AghaKouchak, A.; et al. From Rain Tanks to Catchments: Use of Low-Impact Development to Address Hydrologic Symptoms of the Urban Stream Syndrome. *Environ. Sci. Technol.* **2015**, *49*, 11264–11280. [CrossRef]
18. Ricci, G.F.; D'Ambrosio, E.; De Girolamo, A.M.; Gentile, F. Efficiency and Feasibility of Best Management Practices to Reduce Nutrient Loads in an Agricultural River Basin. *Agric. Water Manag.* **2022**, *259*, 107241. [CrossRef]
19. Lee, T.; Rister, M.E.; Narashimhan, B.; Srinivasan, R.; Andrew, D.; Ernst, M.R. Evaluation and spatially distributed analyses of proposed cost-effective BMPs for reducing phosphorous level in Cedar Creek Reservoir, Texas. *Trans. ASABE* **2010**, *53*, 1619–1627. [CrossRef]
20. Rodriguez, H.G.; Popp, J.; Maringanti, C.; Chaubey, I. Selection and Placement of Best Management Practices Used to Reduce Water Quality Degradation in Lincoln Lake Watershed. *Water Resour. Res.* **2011**, *47*, 2009WR008549. [CrossRef]
21. Jamshidi, S.; Imani, S.; Delavar, M. Impact Assessment of Best Management Practices (BMPs) on the Water Footprint of Agricultural Productions. *Int. J. Environ. Res.* **2020**, *14*, 641–652. [CrossRef]

22. Ha, M.; Wu, M. Land Management Strategies for Improving Water Quality in Biomass Production under Changing Climate. *Environ. Res. Lett.* **2017**, *12*, 034015. [CrossRef]
23. Napora, K.; Noe, G.; Ahn, C.; Fellows, M.Q.N. Urban Stream Restorations Increase Floodplain Soil Carbon and Nutrient Retention along a Chronosequence. *Ecol. Eng.* **2023**, *195*, 107063. [CrossRef]
24. Mrozińska, N.; Glińska-Lewczuk, K.; Burandt, P.; Kobus, S.; Gotkiewicz, W.; Szymańska, M.; Bąkowska, M.; Obolewski, K. Water Quality as an Indicator of Stream Restoration Effects—A Case Study of the Kwacza River Restoration Project. *Water* **2018**, *10*, 1249. [CrossRef]
25. Gordon, B.A.; Dorothy, O.; Lenhart, C.F. Nutrient Retention in Ecologically Functional Floodplains: A Review. *Water* **2020**, *12*, 2762. [CrossRef]
26. Lammers, R.W.; Bledsoe, B.P. What Role Does Stream Restoration Play in Nutrient Management? *Crit. Rev. Environ. Sci. Technol.* **2017**, *47*, 335–371. [CrossRef]
27. Parr, T.B.; Smucker, N.J.; Bentsen, C.N.; Neale, M.W. Potential Roles of Past, Present, and Future Urbanization Characteristics in Producing Varied Stream Responses. *Freshw. Sci.* **2016**, *35*, 436–443. [CrossRef]
28. Amos, R.; (Aquafor Beech Limited, Newmarket, ON, Canada); Ursulak, J.; (Aquafor Beech Limited, Newmarket, ON, Canada); MacMillan, G.; (Lake Simcoe Region Conservation Authority, Newmarket, ON, Canada). Personal communication, 2012.
29. Rice, J.L.; Woodhouse, C.A.; Lukas, J.J. Science and Decision Making: Water Management and Tree Ring Data in the Western United States. *JAWRA J. Am. Water Resour. Assoc.* **2009**, *45*, 1248–1259. [CrossRef]
30. Sudduth, E.B.; Meyer, J.L. Effects of Bioengineered Streambank Stabilization on Bank Habitat and Macroinvertebrates in Urban Streams. *Environ. Manag.* **2006**, *38*, 218–226. [CrossRef]
31. Zhang, Q.; Blomquist, J.D.; Fanelli, R.M.; Keisman, J.L.D.; Moyer, D.L.; Langland, M.J. Progress in Reducing Nutrient and Sediment Loads to Chesapeake Bay: Three Decades of Monitoring Data and Implications for Restoring Complex Ecosystems. *WIREs Water* **2023**, *10*, e1671. [CrossRef]
32. Berg, J.; Burch, J.; Cappuccitti, D.; Filoso, S.; Fraley-McNeal, L.; Goerman, D.; Hardman, N.; Kaushal, S.; Medina, D.; Meyers, M. *Recommendations of the Expert Panel to Define Removal Rates for Individual Stream Restoration Projects*; Chesapeake Stormwater Network and Center for Watershed Protection: Ellicott City, MD, USA, 2014. Available online: https://repository.library.noaa.gov/view/noaa/37925 (accessed on 28 November 2023).
33. Altland, D.; Berg, J.; Brown, B.; Burch, J.; Cook, R.; Fraley-McNeal, L.; Meyers, M.; Running, J.; Starr, R.; Sweeney, J.; et al. Consensus Recommendations for Improving the Application of the Prevented Sediment Protocol for Urban Stream Restoration Projects Built for Pollutant Removal Credit. 2019. Available online: https://static1.squarespace.com/static/603fd5769f97e558ed58da64/t/60637af70321681b9743e200/1617132282380/201912+-+Chesapeake+Bay+Expert_s+Panel+-+Protocol+1+change+requiring+measurement+of+nutrient+concentrations.pdf (accessed on 28 November 2023).
34. Boleman, P.; Jacobson, M. Nitrogen Credit Trading as an Incentive for Riparian Buffer Establishment on Pennsylvania Farmland. *Agrofor. Syst.* **2021**, *95*, 1033–1045. [CrossRef]
35. MacBroom, J.G.; Schiff, R. Classifying Stream Restoration Projects and Understanding Created Channel Types—Two New River Management Design Tools. In Proceedings of the World Environmental and Water Resources Congress 2007, Tampa, FL, USA, 15–19 May 2007; American Society of Civil Engineers: Tampa, FL, USA, 2007; pp. 1–10. [CrossRef]
36. Kaushal, S.S.; Fork, M.L.; Hawley, R.J.; Hopkins, K.G.; Ríos-Touma, B.; Roy, A.H. Stream Restoration Milestones: Monitoring Scales Determine Successes and Failures. *Urban Ecosyst.* **2023**, *26*, 1131–1142. [CrossRef]
37. Credit Valley Conservation Authority and Toronto and Region Conservation Authority. Low Impact Development Stormwater Management Planning and Design Guide. 2010. Available online: https://files.cvc.ca/cvc/uploads/2014/04/LID-SWM-Guide-v1.0_2010_1_no-appendices.pdf (accessed on 28 November 2023).
38. Ontario Ministry of Natural Resources and United States Department of Agriculture; Knap, K.M.; Mildner, W.F. *Streambank Erosion in the Great Lakes Basin: Joint Summary Technical Report*; International Joint Commission, Great Lakes Regional Office: Windsor, ON, Canada, 1978. Available online: https://ijc.org/sites/default/files/ID515.pdf (accessed on 28 November 2023).
39. Center for Watershed Protection and Chesapeake Bay Program, 2016. Recommendations of the Expert Panel to Define Removal Rates for Disconnecting Existing Impervious Area Runoff from Stormwater Drainage Systems. Available online: http://www.chesapeakebay.net/publications/title/impervious_disconnection_to_amended_soils_bmp_expert_panel_recommendations (accessed on 1 January 2024).
40. Sustainable Technologies Evaluation Program (TRCA, CVC, LSRCA). LID Life Cycle Costing Tool v.3.0. Available online: https://sustainabletechnologies.ca/lid-lcct/ (accessed on 2 August 2023).
41. Uda, M.; van Seters, T.; Graham, C.; Rocha, L. *Evaluation of Life Cycle Costs for Low Impact Development Stormwater Management Practices*; Sustainable Technologies Evaluation Program, Toronto and Region Conservation Authority: Vaughan, ON, Canada, 2013. Available online: https://sustainabletechnologies.ca/app/uploads/2013/06/LID-LCC-final-2013.pdf (accessed on 28 November 2023).
42. LSRCA [Lake Simcoe Region Conservation Authority]. *Equitable Responsibility for Transformative Design: A Systems-Based Approach to Stormwater Management—Technical Report*; Lake Simcoe Region Conservation Authority: Newmarket, ON, Canada, 2021; p. 43. Available online: https://www.lsrca.on.ca/Shared%20Documents/reports/Stormwater-Management-Summary.pdf (accessed on 28 November 2023).

43. Stang, C.; Gharabaghi, B.; Rudra, R.; Golmohammadi, G.; Mahboubi, A.A.; Ahmed, S.I. Conservation Management Practices: Success Story of the Hog Creek and Sturgeon River Watersheds, Ontario, Canada. *J. Soil Water Conserv.* **2016**, *71*, 237–248. [CrossRef]
44. Hoesen, J. *Retrospective Analysis of Stewardship Projects in the Lake Simcoe Watershed*; Ontario Ministry of Agriculture and Rural Affairs (OMAFRA): London, UK, 2018; pp. 31–49. Available online: https://www.ontariosoilcrop.org/wp-content/uploads/2022/08/Lake-Simcoe-FINAL-REPORT.pdf (accessed on 28 November 2023).
45. Karst-Riddoch, T.; Sinclair, D.; Nesbitt, R.; Hutchinson, N. *Managing New Urban Development in Phosphorus-Sensitive Watersheds*; Hutchinson Environmental Sciences Ltd.: Kitchener, ON, Canada, 2014. Available online: https://www.nvca.on.ca/wp-content/uploads/2023/08/NVCA-Phosphorus-Loading-Tool-Final-Report-2014.pdf (accessed on 2 December 2023).
46. Thayer, J.B.; Phillips, R.T.J.; Desloges, J.R. Downstream channel adjustment in a low-relief, glacially conditioned watershed. *Geomorphology* **2016**, *262*, 101–111. [CrossRef]
47. Winter, J.G.; Duthie, H.C. Export coefficient modelling to assess phosphorous loading in an urban watershed. *JAWRA J. Am. Water Resour. Assoc.* **2000**, *36*, 1053–1061. [CrossRef]
48. Corsi, S.R. *Unit-Area Loads of Suspended Sediment, Suspended Solids, and Total Phosphorus from Small Watersheds in Wisconsin*; US Department of the Interior: Washington, DC, USA; US Geological Survey: Reston, VA, USA, 1997.
49. Li, Q.; Huang, J.; Zhang, J.; Gao, J. A raster-based estimation of watershed phosphorus load and its impacts on surrounding rivers based on process-based modelling. *J. Environ. Manag.* **2023**, *339*, 117846. [CrossRef]
50. Soranno, P.A.; Hubler, S.L.; Carpenter, S.R.; Lathrop, R.C. Phosphorus loads to surface waters: A simple model to account for spatial pattern of land use. *Ecol. Appl.* **1996**, *6*, 865–878. [CrossRef]
51. USEPA (U.S. Environmental Protection Agency). BASINS Technical Note 8: Sediment Parameter and Calibration Guidance for HSPF; Office of Water 4305: 2006. 2006. Available online: https://www.epa.gov/sites/default/files/2015-08/documents/2006_02_02_basins_tecnote8.pdf (accessed on 1 January 2024).
52. Lang, M.; Li, P.; Yan, X. Runoff concentration and load of nitrogen and phosphorus from a residential area in an intensive agricultural watershed. *Sci. Total Environ.* **2013**, *458*, 238–245. [CrossRef]
53. International Joint Commission. *International Reference Group on Great Lakes Pollution from Land-Use Activities (PLUARG) Detailed Study Plan*; International Joint Commission: Washington, DC, USA, 1974. Available online: https://legacyfiles.ijc.org/publications/ID1061.pdf (accessed on 1 January 2024).

water

MDPI

Article

Back-Calculation of Manning's Roughness Coefficient by 2D Flow Simulation and Influence of In-Channel Physical Parameters in a Mountain River, Japan

Hiroshi Takata [1], Shogo Obata [2], Tatsuro Sato [3,*] and Yukihiro Shimatani [4]

[1] Organization for Regional Co-Creation of Sustainable Communities and Watershed Disaster Resilience, Prefectural University of Kumamoto, 3-1-100 Kumamoto, Higashi-ku, Kumamoto 862-8920, Japan; takata-h@pu-kumamoto.ac.jp
[2] River Planning Department, Nippon Koei Co., Ltd., 5-4 Kojimachi, Chiyoda-ku, Tokyo 102-8539, Japan
[3] Faculty of Architecture and Civil Engineering, Kyushu Sangyo University, 2-3-1 Matsukadai, Higashi-ku, Fukuoka 813-8503, Japan
[4] Laboratory for Watershed Disaster Resilience, Prefectural University of Kumamoto, 3-1-100 Kumamoto, Higashi-ku, Kumamoto 862-8920, Japan; yukihiro.shimatani@gmail.com
* Correspondence: sato@landform.jp

Abstract: This study attempts to back-calculate Manning's roughness coefficients by repeating a two-dimensional flow simulation to fit the spatially and temporally dense river water-level data observed in Japan's Yamatsuki River, a typical mountainous river with an average riverbed gradient of 1/50 and an average river width of 17.9 m. Furthermore, we aim to clarify the influence of the in-channel physical parameters on the coefficient of roughness obtained through the above method. In the Yamatsuki River, 16 water-level gauges were installed at intervals of about 40~80 m in the longitudinal direction in the study reach. Manning's roughness coefficients were back-calculated by repeating two-dimensional flow simulations to match the observed water levels of a flood in 2021 (the estimated maximum flow rate is 11.5 m^3/s). The back-calculated roughness coefficients approached a constant value in the range of 0.05 to 0.1 $s/m^{1/3}$ as the relative water depth increased, indicating that the roughness coefficient can be considered a constant value when performing plane two-dimensional flow calculations for flooding. The roughness coefficient during flooding was found to be correlated with the slope and step height (H)-step length (L)- channel slope (S) ratios (H/L/S). An equation for predicting the roughness coefficient during flooding based on the physical parameters of the channel is also proposed.

Keywords: mountainous river; river restoration; step pool; unsteady two-dimensional flow calculation; Manning's roughness coefficient; resistance characteristics

1. Introduction

Mountainous rivers are sources of flooding, sediment, and land-use disturbances [1,2] and are essential segments providing critical aquatic habitats [3,4]. There is a need to establish an integrated river management approach for mountainous rivers that implements flood protection while considering water and sediment continuity, biological habitats, and landscape. However, mountainous rivers flow quickly during floods and undergo large bed fluctuations, making field observations difficult. Consequently, primary field data and information on the geometry of the channel are lacking, and resistance characteristics such as the roughness coefficients of mountain river channels, which are essential for runoff analyses and riverbed fluctuation calculations in mountainous watersheds, are poorly understood.

A standard method for calculating the channel resistance of mountainous rivers is to establish several cross sections in the longitudinal direction of the river; determine the cross-sectional average velocity, water level, and flow rate in the study section at a specific rainfall event timing; and back-calculate Manning's roughness coefficient n and Darcy-Weisbach's friction coefficient f [5–10]. Asano et al. [5] organized the relationship between relative water depth and channel resistance using Manning's roughness coefficient based on the previous literature. They reported that channel resistance, expressed as Manning's roughness coefficient, converges to a constant value when the relative water depth exceeds 1. To back-calculate channel resistance, Comiti et al. [6] measured the flow and channel profiles in 10 reaches (slope 0.08 to 0.21 and length 16.1 to 76.4 m) of the Rio Cordon River with a catchment area of about 5 km^2. The measurement section was set for each channel configuration to be obtained. Gabrielle et al. [7] calculated mean velocities using tracers for a total of 15 reaches (9 in the step-pool channel, 5 in the cascade channel, and 1 in the plane–bed channel) to determine channel resistance to outflow events from 0.28 m^3/s to 0.68 m^3/s. Reid et al. [8] established 13 basins in five tributaries with basin areas ranging from 12 km^2 to 77 km^2 and calculated the friction coefficient f by observing flow velocity and water depth during outflow events in 61 cross sections. Yochum et al. [10] estimated Manning's roughness coefficient n and Darcy–Weisbach roughness in mountain rivers and reported that n ranged from 0.048 to 0.30 $m^{1/3}/s$ at high water and 0.057 to 0.96 $m^{1/3}/s$ at low flow. Marcus et al. [11] conducted a field study at Ptarmigan Glacier in the Juneau Icefield of Alaska with a mean precipitation of 2300 mm and slopes ranging from 0.02 to 0.16. Although the transect spacing was unknown, the authors reported that the flow rate during the observations ranged from 0.16 to 1.39 m^3/s and the estimated roughness n ranged from 0.064 to 0.183 $m^{1/3}/s$. Previous studies have also attempted to identify correlations between the calculated roughness or friction coefficients and hydraulic parameters, such as channel morphology, slope, and physical parameters, such as the grain size of the bed material and flow rate. The channel resistance of mountainous rivers under natural conditions not only varies spatially with channel morphology and gravel size [7,12] but also varies significantly with flow changes [5,10]. Montgomery et al. [12] showed that, for similar gradients, step-pool channels have higher roughness coefficients than plane–bed channels [13]. Yarahmadi et al. [14] analyzed laboratory data using soft computing models and indicated that the most important parameters affecting n were the energy grade line, the flow Froude number, the bed form dimensionless parameter.

As described above, field studies in various mountainous rivers have reported complex channel resistance in response to the complexity of channel geometry and flow fluctuations. However, the following issues remain. Since it is difficult to maintain uniform roughness conditions and channel geometry in mountainous rivers with complex channel geometries [11], to obtain a more accurate roughness coefficient, it is necessary to calculate roughness coefficients using spatially and temporally dense observed water-level data. However, observations of longitudinally dense data have rarely been conducted on mountainous rivers. Roughness coefficients for mountain rivers have been measured from low to very high flows, but little spatial variability has been reported. Takemura et al. [15] conducted 2D and 3D hydraulic analyses of a mountain river and reported that the possible roughness coefficients for 2D and 3D hydraulic calculations are different. Here, the three-dimensional calculations explicitly reflect the three-dimensional flow and non-hydrostatic distribution around boulders; the roughness coefficients representing the hydrograph converge to a constant value, while in the two-dimensional calculations, the roughness coefficients vary significantly from place to place. The authors noted that planar two-dimensional calculations cannot consider the three-dimensional flow and non-hydrostatic distribution around boulders. Additionally, the wrinkles are included in the roughness coefficients. However, although the reproducibility of the phenomenon is an issue for the roughness coefficient in 2D calculations, there is significant social demand for the roughness coefficient in planar 2D calculations, as models based on 2D calculations are widely used in practical applications from the perspective of computational resources.

This study attempts to obtain a roughness coefficient that fits the spatially and temporally dense river water-level data observed in Japan's Yamatsuki River, a typical mountainous river, by repeating a two-dimensional flow simulation. Furthermore, we aim to clarify the influence of the physical parameters of the river channel on the coefficient of roughness obtained via the above exploratory method.

2. Materials and Methods

2.1. Summary of Methods

In the Yamatsuki River, a typical mountainous river in Miyazaki Prefecture, Japan, a 667 m section was established for analysis. In this section, 16 water-level gauges were installed at intervals of about 40~80 m in the longitudinal direction. The computed section was divided into sections, and Manning's roughness coefficients were back-calculated by repeating unsteady two-dimensional flow simulations to conform to the results of water-level observations during flooding. The temporal variation of the roughness coefficient was also examined alongside the relative water depth, which is an important parameter. Several physical parameters of the river were extracted for the spatial variation of the roughness coefficient, and the relationship between the roughness coefficient and each parameter under high flow conditions was confirmed via a single correlation analysis. Multiple regression analysis was then conducted using Manning's roughness coefficient as the objective variable and the physical parameters with high relationships as explanatory variables, and a prediction equation for the roughness coefficient was proposed.

Details of the survey methodology, survey items, and analysis methods are presented in the following subsections.

2.2. Study Field

The study site is a section of the Yamatsuki River, a tributary of the Gokase River that flows through Miyazaki Prefecture. Figure 1 shows the location of the Yamatsuki River. The Yamatsuki River is a typical mountainous river with a drainage area of 8.2 km^2, a channel length of 4 km, an average riverbed gradient of 1/50, and an average river width of 17.9 m flowing through a valley. A 2.1 km span of the Yamatsuki River was damaged by Typhoon No. 14 in 2005 (Figure 2a). This flood caused significant damage, including the collapse of the revetment and the discharge of soil and sand from the rice paddies (Figure 2b).

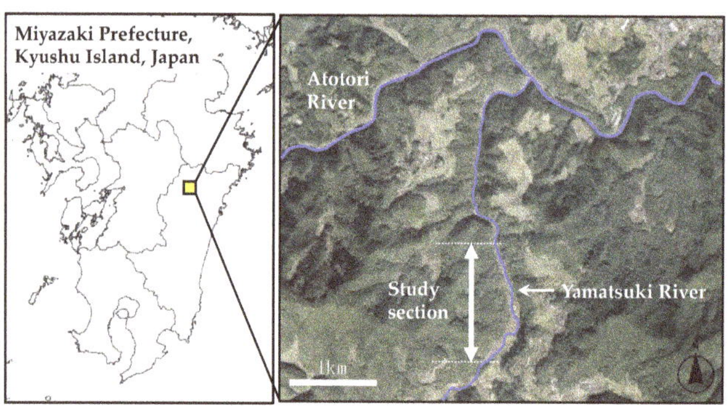

Figure 1. The location of the Yamatsuki River, Miyazaki Prefecture, Kyushu Island, Japan.

Figure 2. The river under study: (**a**) the river during the flood of 2005; (**b**) the river after the flood of 2005; (**c**) the river after restoration in 2009.

After the flood of Typhoon No.14, river restoration was undertaken for flood protection and restoration of the river environment in response to residents' opinions. At the time, there were many unknowns regarding mountainous rivers and no clear standards. Therefore, incorporating the opinions of river researchers, the river restoration was restored based on several innovations in addition to river flood calculations. The primary goal of the restoration was to restore the step-pool morphology (which is dominant in the riverbed morphology of mountainous rivers) to its natural state. To achieve this restoration, the river was not straightened but meandered, and the longitudinal gradient was maintained. Where the riverbanks were destroyed and the river's width eroded, the river was kept widened, with the revetment using stones washed away by the flood. In addition, boulders larger than 1 m in diameter were left on the riverbed in anticipation of the formation of the step pool (Figure 2c). The restoration was completed in 2008. A follow-up survey conducted three years after the restoration reported the development of the riverbed's topography and the recovery of plant and fish species [16].

For the study, we selected a 1190–1867 m section where the restorations were made, a 1 km section downstream of the river where the riverbed was not fixed by concrete, and a 677 m section where the influence of tributaries was minimal. In addition, the flow rate was expected to be somewhat constant.

2.3. Field Observations

The field survey consisted of continuously observing the water level using water-level gauges installed at 16 locations along the river in the study section and photogrammetric surveying using a UAV (Unmanned Aerial Vehicle).

Pressure water-level gauges (HOBO U20L, Onset, USA 470 MacArthur Blvd., Bourne, MA, USA) were installed for the water-level observations. The observation period was from 6 August to 22 August 2021. The water-level gauges were installed by fastening the gauges to pegs with a cable tie and driving the pegs into the roots of huge rocks and gaps in the seawall. The water-level gauges were labeled wl01–wl16 with sequential numbers starting with the gauges installed downstream. A pressure gauge measuring atmospheric pressure was also installed near the river for compensation. Figure 3 shows the locations of the pressure gauges in plane view.

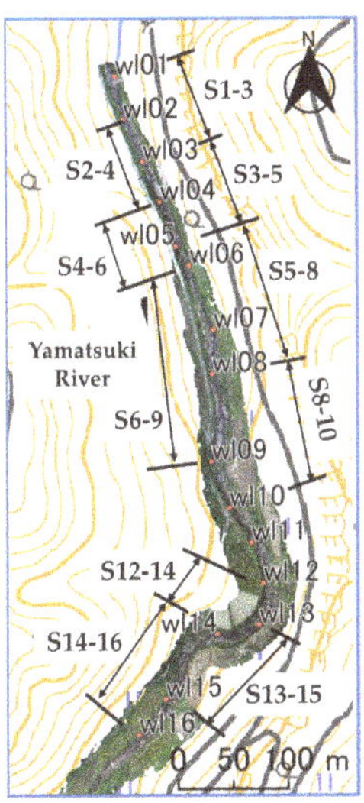

Figure 3. The location of water level gauges in the river and the individual sections in which the Manning's roughness coefficients were estimated with 2D hydraulic analyses.

Topographic surveying was conducted with a UAV (Phantom 4 Pro V2.0, DJI) in April 2021 when topographic surveying was optimal because the river flow was low during the year. Aerial photographs were taken with Ground Control Points (GCPs) covering the entire study section. The aerial images were then used to obtain a 0.5 m resolution digital elevation model and orthographic projection map of the river channel via structure from motion (SfM) analysis using Metashape Professional ver. 1.7.1 (Agisoft).

2.4. *Analysis Method*

Many unsteady 2D flow calculations were performed while gradually changing the Manning's roughness coefficient (given uniformly within a computed section), and the optimal Manning's roughness coefficient that fit the water-level data obtained from the field survey was back-calculated. Ten computed sections were established so that each computed section contained three longitudinally contiguous water-level gauges, for each of which an inverse roughness estimation was performed using 2D flow calculations. Figure 3 shows the calculation sections. Based on the preliminary analysis, the back-calculated roughness coefficients for sections 9–11, 10–12, and 11–13 were excluded from the back-calculation analysis because there was no water depth response to roughness changes. The sections were located in a meandering section, so water depth was primarily determined by the large-scale river channel morphology not affected by local riverbed roughness.

The following sections describe the data sets, computational methods, and methods used for calculating the roughness coefficients and physical parameters required for unsteady 2D flow calculations.

2.4.1. Preparation of Data Set for Unsteady Two-Dimensional Flow Calculations

For the unsteady two-dimensional flow calculations, we used free software called iRIC Nays 2DH [17], which can capture nonlinear, unsteady hydrodynamics and subsequent bed and bank morphodynamics. The data required for unsteady 2D flow calculations were the water level at the upstream and downstream ends, the flow discharge, and the ground elevation. The pressure difference between the installed water-level gauge and a pressure gauge measuring atmospheric pressure was determined, and the longitudinal water depth was calculated from the pressure difference. Based on the river channel cross section and observed water levels, the flow discharge was estimated using the trapezoidal weir overflow Formula (1) for the cross section (at wl04 in Figure 4), where the exposed bedrock forms a drop-off structure and the dominant cross section is assumed to exist. The following equation was used to estimate the flow discharge based on data from erosion control facilities provided by the Ministry of Land, Infrastructure, Transport, and Tourism [18]:

$$Q = \frac{2}{15} C \sqrt{2g} \left(3B^1 + 2B^2\right) h^{\frac{3}{2}}, \tag{1}$$

where Q is the estimated flow discharge (m^3/s), C is the flow coefficient (0.6–0.66), g is the acceleration of gravity (9.8 m/s^2), B^1 is the width of the through-flow bottom (m), B^2 is the overflow surface width (m), and h is the overflow depth (m). Figure 4 shows the estimated flow discharge. A maximum flow discharge of 11.5 m^3/s was observed during the observation period. Since the maximum flow discharge calculated by David et al. [7], who previously analyzed relatively large rivers, was 0.68 m^3/s, the roughness coefficients used in this study were considered sufficient to represent the flood. As the ground elevation data for the 2D flow calculations, we used a 0.5 m resolution digital elevation model obtained through photogrammetry.

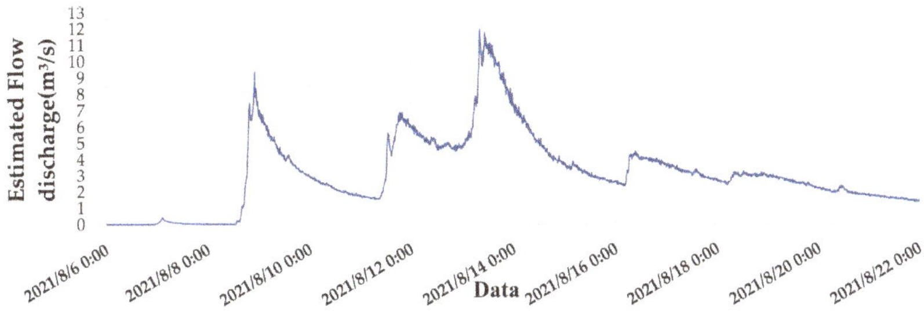

Figure 4. Hydrograph during the observation period.

2.4.2. Two-Dimensional Flow Calculation Method

(1) Overview of the software (Nays2DH)

Nays2DH is a planar two-dimensional solver developed for river flow calculations. The flow field calculation model uses a general curvilinear coordinate system, which allows complex boundaries and riverbed topography to be directly considered. The friction on the bottom is evaluated using Manning's roughness coefficient. The basic equations in Cartesian coordinate form (x, y) before mapping to the general curve coordinate system are shown below. The basic equations are taken from the solver manual [17]:

$$\frac{dh}{dt}+\frac{d(hu)}{dx}+\frac{d(hv)}{dy}=0 \tag{2}$$

$$\frac{d(uh)}{dt}+\frac{d\left(hu^2\right)}{dx}+\frac{d(huv)}{dy}=-gh\frac{dH}{dx}-\frac{\tau_x}{\rho}+D^x-\frac{F_x}{\rho} \tag{3}$$

$$\frac{d(vh)}{dt}+\frac{d(huv)}{dx}+\frac{d(hv)}{dy}=-gh\frac{dH}{dy}-\frac{\tau_y}{\rho}+D^y-\frac{F_y}{\rho} \tag{4}$$

$$\frac{\tau_x}{\rho}=C_f u\sqrt{u^2+v^2} \tag{5}$$

$$\frac{\tau_x}{\rho}=C_f u\sqrt{u^2+v^2} \tag{6}$$

$$Dx=\frac{d}{dx}\left(v_t h\frac{du}{dx}\right)+\frac{d}{dy}\left(v_t h\frac{du}{dy}\right) \tag{7}$$

$$Dy=\frac{d}{dx}\left(v_t h\frac{dv}{dx}\right)+\frac{d}{dy}\left(v_t h\frac{dv}{dy}\right) \tag{8}$$

$$\frac{F_x}{\rho}=\frac{1}{2}C_D a_s h_v u\sqrt{u^2+v^2} \tag{9}$$

$$\frac{F_y}{\rho}=\frac{1}{2}C_D a_s h_v v\sqrt{u^2+v^2}, \tag{10}$$

where h is water depth; t is time; u and v are depth-averaged velocities in the x and y directions, respectively; g is the acceleration of gravity; H is the water level; τx and τy are the riverbed shear forces in the x and y directions, respectively; ρ is the density of water; D^x and D^y are diffusion terms in the x and y directions, respectively; F_x and F_y are the resistance forces of the plant in the x and y directions, respectively; C_f is the riverbed shear coefficient; v_t is the eddy kinematic viscosity coefficient; C_D is the drag coefficient of the plant, a_s is the intercepted area of the plant in unit volume; and h_v is the smallest plant height and water depth. Using the above equations as a basis, the calculation mesh can be set to any shape by converting it into general coordinates.

(2) Calculation conditions

In the present study, the resistance forces of the plant were specified as zero, which is reflected in the roughness coefficient. "+Advanced" was selected as the solver type for the calculation conditions. The calculation time step was determined by considering the Courant number derived from the following equation:

$$C = v\frac{\Delta t}{\Delta l}, \tag{11}$$

where C is the Courant number, v is the flow velocity, Δt is the computation time step, and Δl is the grid size. The Courant condition requires that the Courant number be smaller than 1 for a stable analysis. The computation time step was set to 0.01 s to satisfy the Courant condition. Isocurrent calculations gave the initial water surface profile, and the number of CPUs used for parallel calculations was set to 8, with default values for other settings.

Ten calculation sections were established with three longitudinally contiguous water-level gauges. The water-level gauge wl07 was washed out through the flood. Therefore, the river was divided into calculation sections at the 15 water-level gauge locations from which the water-level data were obtained (Figure 3). The section names are given by the numbers of the water-level gauges located at the downstream and upstream ends of the section. For example, if wl01 is located at the downstream end and wl03 at the upstream end, the section is named section "S1–3". The 0.5 m ground elevation obtained from the field survey was used to create the computational grid, and the mesh size was also set to 0.5 m. The center line of the river channel and the right and left bank lines were created according to aerial photographs taken by the Geospatial Information Authority of Japan (GSI). The "Generic Lattice Generation Tool" was employed as the lattice generation algorithm, which solved the Laplace equation to generate a smooth lattice. The lattice was created so that the lattice points would overlap with the water-level gauge locations in the middle of the section. This arrangement allowed us to compare the observed and calculated water levels at the water-level gauge locations.

(3) Boundary conditions

Ten time series data with varying flow levels were extracted for 2D flow calculations and roughness estimation to obtain roughness coefficients for each runoff phase. For each time series data, 2D indeterminate flow calculations were performed using the upstream end discharge data and downstream end water-level data for 10 min as boundary conditions for the time series data. To stabilize the calculations, an additional 10 min steady flow calculation was performed at the beginning of the calculation using the initial flow and water level values, followed by the 10-minute unsteady flow calculation. Table 1 shows the hydraulic boundary conditions for the periods used in the calculations.

Table 1. Boundary conditions of the ten time periods.

Date and Time	Estimated Discharge at wl04 (m^3/s)
8 August 2021 13:00	0.062
8 August 2021 15:20	0.416
8 August 2021 17:00	1.261
8 August 2021 17:20	1.876
8 August 2021 18:40	3.611
8 August 2021 19:10	5.617
8 August 2021 21:00	6.754
8 August 2021 22:20	8.056
13 August 2021 7:50	9.894
13 August 2021 8:20	11.519

(4) Back-calculation of Manning's roughness coefficient

In this study, Manning's roughness coefficient was back-calculated using the local indeterminate flow back-calculation method. In this method, the river channel is divided into several sections, and the roughness coefficients of each section are changed by trial and error to calculate the indeterminate flow. The Nays2DH solver in iRIC was used for the unsteady two-dimensional flow calculations. The Manning's roughness coefficients ranged from 0.1 to 1 in increments of 0.1, and calculations were performed by further varying the coefficients at intervals of 0.01 between values that were close to each other. The calculated and observed water levels were compared; the result with the smallest difference between the calculated and observed water levels was used as the Manning's roughness coefficient for that section. In this study, we calculated 10 times × 10 sections x 20 times the number of trials = 2000 patterns. The iRIC project data were recorded in the CGNS file format, which is used in computational fluid dynamics, making it challenging to edit the data directly. The PyAutoGUI library in Python was used to create project files with different conditions in RPA. The roughness coefficient for the calculation pattern with the most minor difference between the calculated water depth obtained from the simulation in the section and the observed water depth was determined as the back-calculated roughness coefficient.

(5) Accuracy verification

To confirm the accuracy of the unsteady two-dimensional flow calculation simulation, a comparison was made between the water depth obtained from the flow observations and the calculated water depth obtained from the simulation. Table 2 shows the mean squared error between each section's calculated and observed water levels for the back-calculated roughness coefficient. Here, the mean squared error is small, indicating that the calculated water level accurately reproduced the observed water level when the back-calculated roughness coefficient was used (see also Table 2).

Table 2. The mean squared error between observed and calculated water levels in each calculation section.

Section Name	Mean Squared Error
S1–3	0.00017
S2–4	0.00031
S3–5	0.00003
S4–6	0.00008
S5–8	0.00023
S6–9	0.00048
S8–10	0.00183
S12–14	0.00012
S13–15	0.00021
S14–16	0.00003

2.4.3. In-Channel Physical Parameters Related to the Roughness Coefficient

Previous studies have reported that riverbed roughness varies with water depth [19]. Indeed, roughness is expected to vary with water depth, even in unsteady two-dimensional calculations. To calculate Manning's roughness coefficient during a flood, we compared the roughness coefficient and relative water depth at each of the times indicated in the boundary conditions, with reference to previous studies. The relative water depth was calculated by dividing the water depth by D_{84}, which was obtained from the grain size accumulation curve. From among the ten values set as boundary conditions, water depth under maximum flow was used. The D_{84} grain size of the riverbed material was calculated by creating a grain size distribution curve using the image-based line grid method.

In predicting the back-calculated roughness coefficient during flooding based on the abovementioned physical channel parameters, the following relevant channel physical parameters were selected: river ratio of width to depth, slope, D_{84} grain size of riverbed materials, relative water depth, ratio of plants, ratio of habitat per section, and H/L/S (H:

step height; L: step length; S: channel slope). H/L/S, as proposed by Abrahams et al. [20], is widely used to test the occurrence of the maximum flow resistance condition in mountainous streams. These parameters were calculated using QGIS 3.1.6 with a 0.5 m grid DEM and ortho imagery.

The width-to-depth ratio was selected as a dimensionless value divided by the water depth to assess the extent to which the river width, a parameter that can be controlled during design, affects the roughness coefficient. The width–depth ratio was calculated by dividing the river's average width by the river's depth. The average river width per section was divided into five equal sections, and the average value was used as the average river width. The water depth was selected as the water depth value when the relative water depth reached its maximum value. Gradient and relative water depth values were selected because they were related to the roughness coefficients of mountainous rivers in previous studies [5]. The slope was calculated by dividing the channel length at the upper and lower ends of each set section by the difference in elevation between the upper and lower ends, i.e., channel length/elevation difference = slope. The relative water depth was calculated for each of the ten time periods described above, and the value with the most significant back-calculated roughness coefficient was selected. The D_{84} grain size of the riverbed material, ratio of habitat per section, and maximum flow resistance index H/L/S were selected as parameters to investigate the effectiveness of the step-pool as the main roughness factor for mountainous rivers, which was already known from existing studies. The ratio of habitat, as an occupation of each habitat in each section, was calculated by classifying the river channel into four habitats: step, pool, rapid, and riffle. In the field, we classified step-pools, characteristic river bed forms in mountain river channels. During the step-pool sequence, we classified areas with little wave action as riffle and areas with fast flow and whitecaps as rapid. For each habitat roughly classified during the field survey, polygons were created using ortho-images in GIS 3.2 software. The ratio of each habitat was calculated by dividing the polygon area of each habitat by the area of each section.

H/L/S is calculated as H: step height, L: Step length from the step to the end of the pool, and S: section longitudinal gradient. A longitudinal gradient was created from the obtained ground elevation, and the average value of H/L/S was calculated for each set section. The ratio of plants, as an area occupation of plants in each section, was selected based on the expectation that the effect of the plant would be included in the roughness coefficient since the plant density term was omitted in the unsteady two-dimensional flow calculation in this study. The distribution of vegetation in the Yamatsuki River is closely related to the development of troughs (Sato, unpublished data) and should be interpreted as implicitly including the effects of flow inhibition by troughs. Polygons surrounding the plant areas were created using ortho-images in GIS software. The ratio of plants was calculated by dividing the polygon area of plants by the area of each section.

2.4.4. Regression Analysis to Examine In-Channel Physical Parameters Most Strongly Affecting the Roughness Coefficient

A single regression analysis between the physical parameters described above and Manning's roughness coefficient was applied to investigate the factors influencing the back-calculated inverse roughness coefficient. The corresponding data set is shown in Table 3. In addition, multiple regression analysis was conducted to develop an equation to predict Manning's roughness coefficient in mountainous rivers, selecting the physical parameters with significant correlation coefficients.

Table 3. Data set used in the single and multiple regression analysis.

Cal. Section Name	Manning's Roughness Coefficient ($s/m^{1/3}$)	Ratio of Width to Depth (m/m)	Slope (m/m)	D_{84} (m)	Relative Depth (m/m)	Ratio of Plant (m^2/m^2)	Step (%)	Riffle (%)	Rapid (%)	Pool (%)	H/L/S
S1–3	0.11	6.46	0.005	0.40	2.32	0.00	3.07	36.87	2.90	1.50	4.02
S2–4	0.06	4.94	0.025	1.25	0.87	0.00	3.38	22.70	3.83	3.04	1.06
S3–5	0.06	6.95	0.028	1.45	0.61	0.00	3.22	21.87	2.22	6.34	1.41
S4–6	0.09	6.51	0.018	0.55	1.88	0.01	3.79	31.39	0.82	5.80	3.11
S5–8	0.02	9.00	0.035	0.60	1.42	0.22	7.08	16.04	3.18	5.77	2.36
S6–9	0.05	10.29	0.035	0.75	1.31	0.39	5.84	20.04	2.35	4.71	2.06
S8–10	0.05	18.23	0.040	0.55	1.34	0.63	4.24	24.62	2.26	2.32	1.46
S12–14	0.10	9.44	0.023	0.70	1.84	0.30	6.96	17.45	2.02	8.06	3.13
S13–15	0.07	11.47	0.042	0.90	1.08	0.20	9.02	18.32	3.18	4.65	2.13
S14–16	0.07	12.39	0.041	1.05	0.76	0.14	9.03	16.88	3.95	7.27	2.48

3. Results and Discussion

3.1. The Back-Calculated Manning's Roughness Coefficients and Their Relationship with the Relative Water Depth

We were able to observe spatially and temporally dense water-level data for a mountainous river and back-calculated Manning's roughness coefficients at various discharge stages using several results from two-dimensional flow calculations. Since it is known from previous studies that the coefficient of roughness varies with changes in flow rate, we first estimated the final coefficient of roughness during a flood event and then assessed whether the coefficients differed from section to section.

The relationship between the obtained back-calculated roughness coefficients and relative water depths are shown in Figure 5. The Manning's roughness coefficients ranged from 0.11 to 1.28 $m^{1/3}/s$ in the early stages of computation in each computed section, i.e., when the riverbed gravels were not submerged in the early stages of the flood. In particular, sections 2–4 and 12–14 had coefficients of roughness greater than 0.96, with values exceeding the range of 0.057 to 0.96 $m^{1/3}/s$ reported by Yochum et al. [10] for roughness under a low flow. As the relative water depth increased, Manning's roughness coefficients generally reached their maximum when the relative water depth fell within the range of 0.5 to 0.1 $m^{1/3}/s$. Asano et al. [5] reported a minimum roughness coefficient of 0.16 $m^{1/3}/s$ for the step pool, Montgomery et al. [12] reported a roughness coefficient of 0.1~0.2 $m^{1/3}/s$ for the step pool, and Yochum et al. [10] reported a coefficient of 0.048~0.3 $m^{1/3}/s$ for high water. Therefore, we maintain that a Manning's roughness coefficient with a high relative water depth is generally reasonable.

Step height significantly correlates with the D_{50} and D_{84} values of the constituent gravel diameter [21]. Since the relative water depth in this study is assumed to be h/D_{84}, the step is expected to be submerged below the water surface where the relative water depth exceeds 1 [21,22]. Asano et al. [5] reported that Manning's roughness coefficient calculated from steady flow calculations converges to a constant value with an increase in relative water depth. This study calculated the roughness coefficient backward from unsteady two-dimensional flow calculations for each section. The results confirmed that Manning's roughness coefficient decreases with an increase in relative water depth and converges to a constant value. However, Manning's roughness coefficient was found to vary from section to section. The longitudinal distributions of the back-calculated roughness coefficient, calculated water level, and observed water level at peak flow (i.e., when the relative water depth is maximum) are shown in Figure 6. It is suggested that the convergence of Manning's roughness coefficient may change depending on the river channel's spatial factors and physical parameters.

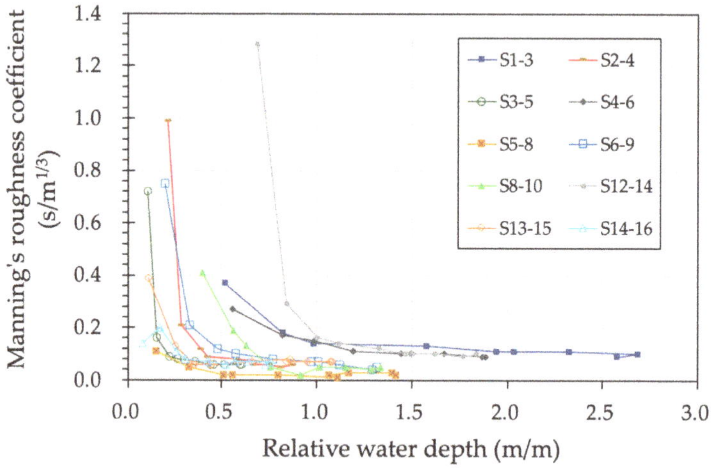

Figure 5. Relationship between Manning's roughness coefficient and relative water depth.

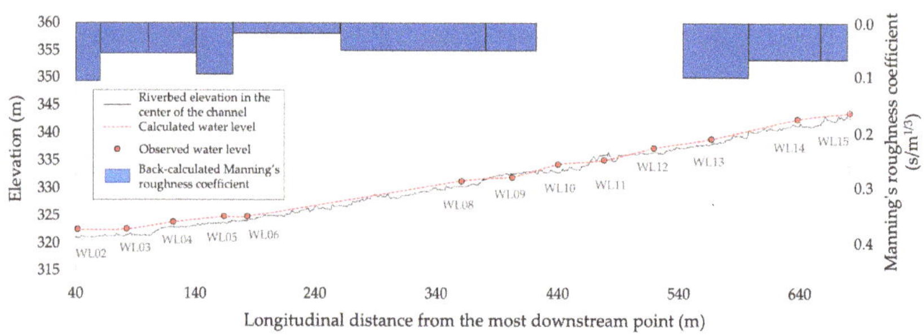

Figure 6. Longitudinal distribution of back-calculated Manning's roughness coefficients, calculated water level, and observed water level at peak flow. Calculated water levels in sections 9–11 (WL10), 10–12 (WL11), and 11–13 (WL12) are for reference.

3.2. Effects of In-Channel Physical Parameters on the Roughness Coefficient

Single regression analysis was performed to clarify the relationship between the physical parameters and roughness coefficient. In the single regression analysis, Manning's roughness coefficients at the maximum relative water depth were used to represent channel resistance during the flood. The results are shown in Figure 7. Furthermore, multiple regression analysis was performed between the parameters with the strongest correlation and Manning's roughness while considering multicollinearity.

The previous section confirmed that Manning's roughness coefficient during a flood with high water levels is spatially varied. Figure 7 shows that the slope was negatively correlated with Manning's roughness coefficient, and H/L/S was selected as the positively correlated parameter. On the other hand, the ratio of width to depth, D_{84} grain size of the streambed material, relative water depth, plant rate, and habitat ratio per section did not correlate well. The D_{84} grain size of the riverbed material and the ratio of habitat per section were selected to explain the effects of the step pool in dispersing the energy of the flow, and H/L/S, which was selected for the same reason, had a high correlation. Thus, H/L/S was selected as the number to express the step pool for the roughness coefficient. H/L/S is likely the most appropriate number to express the step pool for the roughness coefficient. Manning's roughness coefficient, calculated using a planar two-dimensional calculation, expresses the effects of friction roughness and shape roughness. The river roughness is considered to contain the roughness caused by friction due to plants and the grain of the riverbed material and the roughness due to shapes such as boulders, troughs, gradients, and H/L/S. In the sections where the gradient is low, the D_{84} grain size of the riverbed is small, and the water depth is significant. However, the H/L/S is considered to have increased due to the development of a step pool caused by flood flow, suggesting that the step pool developed regardless of the D_{84} grain size and that Manning's roughness coefficient is increasing. This result indicates that the roughness due to the shape is larger than the roughness due to the grain size. This result is also consistent with a previous study that found that roughness caused by shape contribute more to flow drag than roughness caused by friction, although such factors depend on the flow regime during flooding [6].

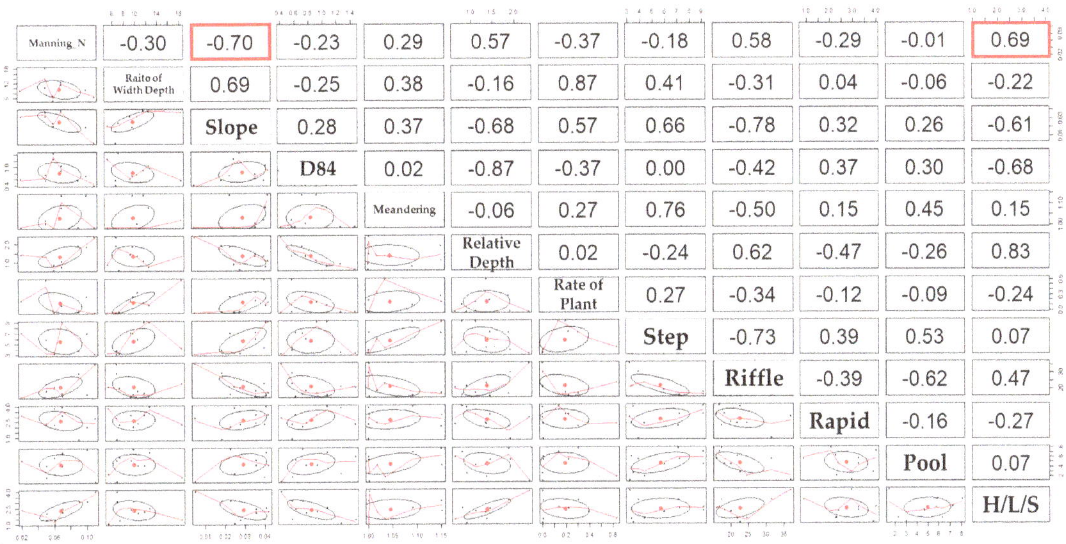

Figure 7. Correlation coefficients between measured parameters (red boxes: correlation coefficients exceeding 0.6 for physical parameters that showed a strong correlation with Manning's roughness coefficient).

The plants in the channel of the Yamatsuki River are mainly vine reeds covering the sandbars. Many of the crane reeds fell in the downstream direction during the field survey after the river's outflow. In this analysis, the simulated water depth in the section with the plants was confirmed to be around 1 m (Figure 8). Reeds collapsed at a rate of 0.4 m/s or greater, and the coarse roughness coefficient dropped to around 0.05 [23]. The vital force of the water during flooding likely caused the reeds to collapse, which reduced the roughness coefficient of the plant section and did not have a substantial effect on the Manning's roughness coefficient.

Because of the collinearity of the physical parameters, the selection of parameters was narrowed down to slope and H/L/S, which have robust correlation coefficients, and a single regression equation with the roughness coefficient was created for each parameter. Figure 9 shows that Manning's roughness coefficient decreased with an increase in slope, and the trend of an increasing roughness coefficient with increasing H/L/S was very clear. In a previous study, Spearman's rank correlation coefficient was used to evaluate the influence of slope, watershed area, and size of streambed material on the calculated roughness coefficient [5]. As a result, we confirmed that the slope, bed grain size, and coefficient of roughness were significantly correlated with the riverbed morphology found in gentle gradients, such as plane–bed and pool–riffle gradients, while slope, catchment area, and bed grain size had no effects on roughness in steep gradients, such as step-pool and cascade gradients. However, this study showed that the effects of gradients on roughness coefficients are strong even in channel sections where step pools are observed.

In addition, an increase in H/L/S means that the step pool is more developed, and since the loss of flow energy increases as the step pool develops [24,25], an increase in H/L/S is considered to have contributed to the increase in the roughness coefficient. An increase in H/L/S also increases the coefficient of roughness.

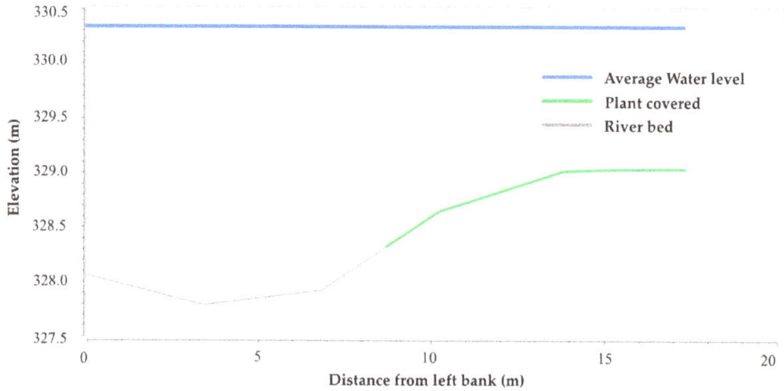

Figure 8. A cross section at the midpoint of wl08–wl09 with the highest water level.

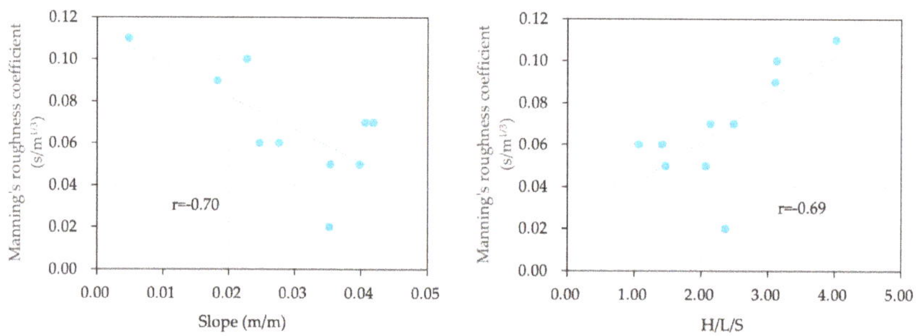

Figure 9. Relationships between Manning's roughness coefficients, slope, and H/L/S with high correlation coefficients exceeding 0.6.

To construct an equation to predict the roughness coefficient in mountainous rivers using physical parameters, a multiple regression analysis was conducted with Manning's roughness coefficient n $[m^{1/3}/s]$ as the objective variable and slope and H/L/S as the explanatory variables featuring high correlation coefficients with the roughness coefficient while taking multi-collinearity into account. As a result, the following regression equation was derived:

$$n = -0.984777 \times Slope + 0.012368 \times \mathrm{H/L/S} + 0.067915. \quad (12)$$

The coefficient of determination was 0.60, and the adjusted coefficient of determination was 0.48. Since the coefficient of determination is smaller than 0.5, the accuracy is considered poor. Nevertheless, we believe that these results demonstrate the possibility to create a more accurate prediction equation by increasing the number of samples and conducting similar analyses in the future.

4. Conclusions

This study aimed to collect the spatially and temporally dense water-level data in a mountainous river to back-calculate Manning's roughness coefficients via a two-dimensional unsteady flow calculation analysis and determine how these coefficients are affected by in-channel physical parameters. The two-dimensional calculation of unsteady flow based on longitudinally observed water levels showed that the roughness coefficient was within the range of the Manning's roughness coefficient estimated in existing studies. The roughness coefficient calculated by the plane two-dimensional calculations of unsteady flow approached a constant value in the range of 0.05 to 0.1 $s/m^{1/3}$ as the relative water depth increased. This result indicates that the roughness coefficient can be considered a constant value when performing plane two-dimensional flow calculations for flooding. The roughness coefficient during flooding was found to be correlated with the slope and H/L/S. An equation for predicting the roughness coefficient during flooding based on the physical parameters of the channel was also proposed.

Author Contributions: Conceptualization, T.S. and Y.S.; methodology, H.T., T.S. and S.O.; validation, H.T., S.O. and T.S.; investigation, H.T., S.O. and T.S.; flow calculation, S.O. and T.S.; writing—original draft preparation, H.T.; writing—review and editing, Y.S. and T.S.; supervision, T.S. and Y.S.; project administration, T.S. and Y.S. All authors have read and agreed to the published version of the manuscript.

Funding: This work was supported by JSPS KAKENHI Grant Numbers JP21K04316, JST Grant Number JPMJPF2109 from JAPAN SOCIETY FOR THE PROMOTION OF SCIENCE and the "River Ecology Research Group of Japan" on the Chikugo River from the Ministry of Land, Infrastructure, Transport, and Tourism, Japan.

Data Availability Statement: Date are contained within the article.

Acknowledgments: We wish to thank Taichi Sakabe, Ryota Nakamura, Masaki Oome, and Mizuki Takesue for their field survey assistance and Toshimitsu Kai for providing information on the field.

Conflicts of Interest: Author Shogo Obata was employed by the company Nippon Koei Co., Ltd. The remaining authors declare that the research was conducted in the absence of any commercial or financial relationships that could be construed as a potential conflict of interest.

References

1. Milliman, J.D.; Syvitski, J.P.M. Geomorphic/Tectonic Control of Sediment Discharge to the Ocean: The Importance of Small Mountainous Rivers. *J. Geol.* **1992**, *100*, 525–544. [CrossRef]
2. Reid, L.M. Research and Cumulative Watershed Effects. In *General Technical Report*; US Department of Agriculture, Forest Service, Pacific Southwest Research Station: Berkeley, CA, USA, 1993; Volume 141, 118p. [CrossRef]
3. Nehlsen, W.; Williams, J.E.; Lichatowich, J.A. Pacific Salmon at the Crossroads: Stocks at Risk from California, Oregon, Idaho, and Washington. *Fisheries* **1991**, *16*, 4–21. [CrossRef]
4. Frissell, C.A. Topology of Extinction and Endangerment of Native Fishes in the Pacific Northwest and California (U.S.A.). *Conserv. Biol.* **1993**, *7*, 342–354. [CrossRef]
5. Asano, Y.; Uchida, T.; Nishiguchi, Y. Measured Channel Resistance at Mountain Rivers during Flood. *Jpn. Soc. Eros. Control Eng.* **2018**, *70*, 12–23.
6. Comiti, F.; Mao, L.; Wilcox, A.; Wohl, E.E.; Lenzi, M.A. Field-Derived Relationships for Flow Velocity and Resistance in High-Gradient Streams. *J. Hydrol.* **2007**, *340*, 48–62. [CrossRef]
7. David, G.C.L.; Wohl, E.; Yochum, S.E.; Bledsoe, B.P. Controls on Spatial Variations in Flow Resistance along Steep Mountain Streams. *Water Resources Research* **2010**, *46*, W03513. [CrossRef]
8. Reid, D.E.; Hickin, E.J. Flow Resistance in Steep Mountain Streams. *Earth Surf. Process. Landf.* **2008**, *33*, 2211–2240. [CrossRef]
9. Rickenmann, D.; Recking, A. Evaluation of Flow Resistance in Gravel-Bed Rivers through a Large Field Data Set. *Water Resour. Res.* **2011**, *47*, W0758. [CrossRef]
10. Yochum, S.E.; Comiti, F.; Wohl, E.; David, G.C.L.; Mao, L. Photographic Guidance for Selecting Flow Resistance Coefficients in High-Gradient Channels. In *General Technical Report RMRS-GTR*; GTR-323; USDA Forest Service: Washington, DC, USA, 2014; pp. 1–91.
11. Marcus, W.A.; Roberts, K.; Harvey, L.; Tackman, G. An Evaluation of Methods for Estimating Manning's n in Small Mountain Streams. *Mt. Res. Dev.* **1992**, *12*, 227–239. [CrossRef]
12. Montgomery, D.R.; Buffington, J.M. Channel-Reach Morphology in Mountain Drainage Basins. *Geol. Soc. Am. Bull.* **1997**, *109*, 596–611. [CrossRef]
13. Barnes, H.H., Jr. Roughness Characteristics of Natural Channels. In *U.S. Geological Survey Water-Supply Paper*; US Government Printing Office: Washington, DC, USA, 1967.
14. Yarahmadi, M.B.; Parsaie, A.; Shafai-Bejestan, M.; Heydari, M.; Badzanchin, M. Estimation of Manning Roughness Coefficient in Alluvial Rivers with Bed Forms Using Soft Computing Models. *Water Resour. Manage.* **2023**, *37*, 3563–3584. [CrossRef]

15. Takemura, Y.; Nigo, S.; Ohno, J.; Fukuoka, S. Problem of Conventional 2D Analysis Method for Flood Flows in Mountain Streams with Boulders and Proposal of a New Analysis Method. *Adv. River Eng.* **2019**, *25*, 267–272.
16. Liu, Y.; Shimatani, Y.; Yamashita, T.; Sato, T.; Ikematsu, S. Stepped-Bed Morphology Changes in Restoredmountain Stream Inuced by Flood. *Adv. River Eng.* **2012**, *18*, 83–88.
17. Shimizu, Y.; Takebayashi, H.; Inoue, T.; Hamaki, M.; Iwasaki, T.; Nabi, M. Nays2DH_SolverManual_Japanese_v4.Pdf. Available online: https://i-ric.org/webadmin/wp-content/uploads/2023/06/Nays2DH_SolverManual_Japanese_v4.pdf (accessed on 1 December 2023).
18. Ministry of Land, Infrastructure, Transport and Tourism Civil Engineering Work Design Guidelines Part II River Edition Chapter 4 Sabo Facilities. Available online: https://www.qsr.mlit.go.jp/site_files/file/s_top/h22doboku/kasen/2-05.pdf (accessed on 15 December 2023).
19. Asano, Y.; Uchida, T. Detailed Documentation of Dynamic Changes in Flow Depth and Surface Velocity during a Large Flood in a Steep Mountain Stream. *J. Hydrol.* **2016**, *541*, 127–135. [CrossRef]
20. Abrahams, A.D.; Li, G.; Atkinson, J.F. Step-pool Streams: Adjustment to Maximum Flow Resistance. *Water Resour. Res.* **1995**, *31*, 2593–2602. [CrossRef]
21. Shawn, M.; Chartrand, P.J.W. Alluvial Architecture in Headwater Streams with Special Emphasis on Step–Pool Topography. *Earth Surf. Process. Landf.* **2000**, *25*, 583–600.
22. Hasegawa, K.; Kanbayashi, S. Formation Mechanism of Step-Pool Systems in Steep Rivers and Guide Lines for the Design of Construction. *Proc. Hydraulic. Eng.* **1996**, *40*, 893–900. [CrossRef]
23. Fukuoka, S.; Shimatani, Y.; Tamura, H.; Tomari, K.; Nakayama, M.; Takase, A.; Iuch, T. Field Experiments on Reed Deformation and Falling and Roughness Coefficient on Floodchannel by Flows. *Adv. River Eng.* **2003**, *9*, 49–54.
24. Ashida, K.; Egashira, S.; Takayuki, N. Structure and friction law of flow over a step-pool bed form. *Disaster Prev. Res. Inst. Annu.* **1986**, *29*, 391–403.
25. Wilcox, A.C.; Wohl, E.E.; Comiti, F.; Mao, L. Hydraulics, Morphology, and Energy Dissipation in an Alpine Step-Pool Channel. *Water Resour. Res.* **2011**, *47*. [CrossRef]

water

MDPI

Article

Comparative Experimental Study of Geotube Groins and Mixed Clay–Geotube Groins under Various Flow Conditions

Xiaobing Tu [1], Hongxiang Gao [2], Kejie Fan [2], Jiahao Xu [3], Jianxiong Li [1], Chunchen Xia [3,*] and Xinjie Pang [3]

1 POWER CHINA HUADONG Engineering Co., Ltd., Hangzhou 310000, China; tu_xb@hdec.com (X.T.); li_jx2@hdec.com (J.L.)
2 Zhejiang Jiangnanchun Construction Group Co., Ltd., Hangzhou 311241, China; 13758147277@163.com (H.G.); 15858267311@163.com (K.F.)
3 College of Civil Engineering, Zhejiang University of Technology, Hangzhou 310023, China; 2112106127@zjut.edu.cn (J.X.); pangxj795@163.com (X.P.)
* Correspondence: xiacc@zjut.edu.cn

Abstract: During the cofferdam construction of the toe reinforcement project at the Qiantang River Estuary, the scouring of the riverbed at the groin head often led to the collapse of geotube groins due to strong tidal currents. Based on field experience, employing a combination of clay and geotubes proved to be a more effective solution to this problem. This study adopted a flume model experiment to investigate the scouring and deposition around geotube groins and mixed clay–geotube groins. The results indicated that the influence of tidal surges on geomorphic changes surrounding the groins was more pronounced during spring tides than during neap tides. Under the same flow conditions, the scour depth at the head of the geotube groin was notably deeper than that of the mixed clay–geotube groin. Additionally, sediment silting behind the mixed clay–geotube groin was significantly greater than that behind the geotube groin. The clay component of the mixed clay–geotube groin served to mitigate the head scour, enhancing the overall structural stability to a certain extent. The geotube groin, with its surrounding scour pits expanding over time, experienced increasing tensile strain. This resulted in the rupture of the geotextile material, the loss of internal sand and, ultimately, groin collapse. It was found that mixed clay–geotube groins were better suited for cofferdam construction in strong tidal estuaries compared to geotube groin alternatives.

Keywords: geotube groin; flume model experiment; Qiantang River Estuary; scour pits; sedimentation

1. Introduction

The Qiantang River Estuary, spanning from the Fuchun River Power Station in Tonglu to the line connecting Luchao Port in Shanghai with Zhenhai in Ningbo, is a globally renowned strong tidal estuary, stretching approximately 290 km [1]. The north embankment of the Qiantang River is an essential barrier safeguarding the Hangjiahu Plain and the southern edges of Suzhou and Songjiang in Shanghai against typhoon storm surges. The Haining section of the Qiantang River's northern embankment, measuring about 55.7 km, is composed of segments like the Yancang Embankment, the historical Qing Dynasty Embankment, and the Jianshan Embankment. Recent monitoring has revealed that parts of the sea embankment between Yangtian Temple and Tashan do not meet the original design standards for overall stability and scour protection. Under a 300-year event standard, 16.475 km of the embankment toe fails to meet the specified requirements [2]. To ensure flood and tide defense, there is an urgent need for a provisional cofferdam, facilitating reinforcement work for the embankment toe between Yangtian Temple and Tashan.

Given the intense tidal surges of this section of the Qiantang River, coupled with rapid flow velocities and significant river bed changes, constructing a cofferdam here is exceptionally challenging. Ensuring the temporary cofferdam's stability during the 3–4-month construction period is paramount. Drawing from construction experience, the groins, on the one hand, act as flow deflectors, reducing the water current's impact on the main cofferdam body and ensuring safety. On the other hand, they promote sediment deposition behind the groins, raising the riverbed to save construction costs, and can function as transverse cofferdams, enhancing construction efficiency. Thus, this project adopted a strategy of building vertical groins along the embankment, followed by the construction of the longitudinal cofferdam (as shown in Figure 1). Initially, geotube groins were used as transverse cofferdams. However, approximately 50–60 m offshore, significant scouring of the riverbed sand at the groin's toe was observed. Progressive undercutting led to eccentric loading on the groin, causing local tensile deformation in the geotubes. When localized tensile forces surpassed the geotube's tensile strength, it would be pulled to failure, resulting in a loss of fill material and the eventual collapse (as depicted in Figure 2), which incurred significant losses. Subsequently, an approach to integrating clay with geotube groins (referred to as mixed clay–geotube groins) was employed. The region near the embankment (within 50 m offshore) used geotube groins, while areas with faster water currents on the river side employed natural clay (as illustrated in Figure 3). The mixed clay–geotube groins were able to maintain their overall structural integrity without significant deformation during strong tidal surge periods, ensuring the smooth running of the project. Moreover, the sediment deposition efficiency behind the mixed clay–geotube groins was notably higher than that behind the geotube groins, offering convenience for subsequent longitudinal cofferdam construction.

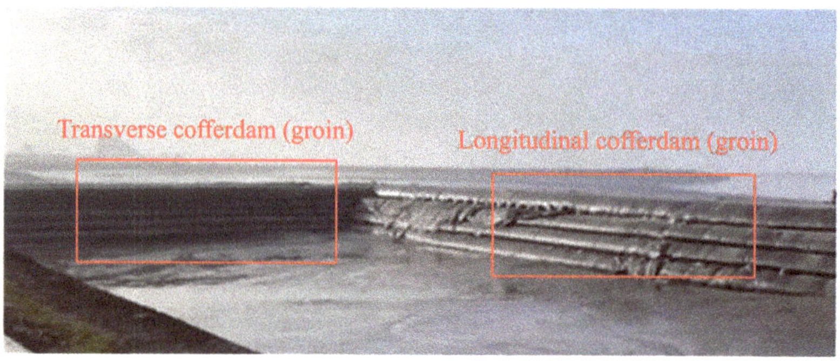

Figure 1. Transverse cofferdam and longitudinal cofferdam.

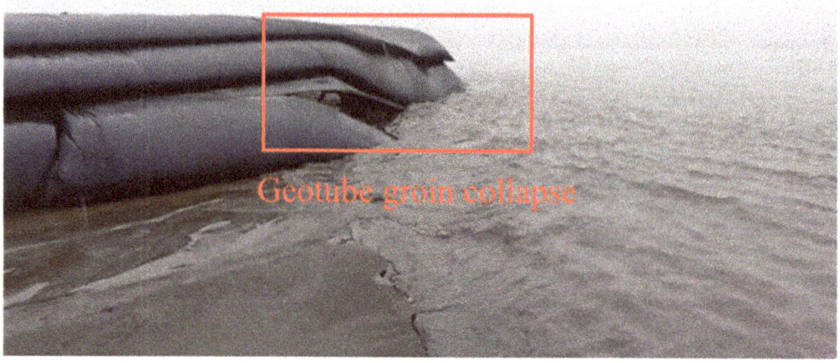

Figure 2. Geotube groin collapse.

Figure 3. Mixed clay–geotube groin.

The groin is one kind of spur dike that is often used in temporary water-related projects. Comprehensive studies on scouring and protection around groins are vital in actual engineering scenarios [3]. Considerable progress has been made in understanding the scouring and deposition around groins [4–10]. However, the fluid dynamics around these structures becomes intricate in estuaries with strong tidal actions, and studies in this area remain insufficient. In the context of field measurements, Zhao et al. [11–13] carried out observations and comparative analyses of pile-supported groins intended for dike protection. Field measurements of scour holes around the groins and observations of their influence on the elevation and tidal bores of nearby tidal flats revealed that, compared to traditional rock-filled groins, pile-supported groins demonstrated less local scouring around the structure, better stability, and notable effectiveness in moderating tidal bore velocities. Experimentally, Qian et al. [14] combined holistic physical modeling to simulate the behavior of groins in bends and flume modeling to study groins in straight sections, investigating differences in localized scouring under strong tidal action. Their results indicated that groins located on bends experienced deeper local scour holes than those on straight sections when subjected to strong tidal action. Zhang et al. [15] took the spur dike in the Qiantang River Estuary's Jianshan section as a prototype and conducted flume experiments on submerged groin scouring induced by tidal bores. The results showed that the scour depth was directly proportional to the relative current intensity and the dike height, but inversely proportional to the water depth before the bore. Pan and Li [16] used flume models to investigate the hydraulic structures, scour hole morphology, and depth around the widely used sheet-pile groins in the Qiantang River Estuary. They obtained the fitted maximum scour depth calculation formulae at the groin head and the upstream side of a groin using the least squares method. Numerically, conventional models struggled to accurately simulate the water process under the influence of tidal bores, due to the discontinuity of physical quantities before and after the tidal bores. Initial efforts focused on exploring tidal bores' propagation [17–20], followed by studies on sediment transport under the influence of groins [21–26]. However, most of these studies centered around permanent groins (e.g., rock-filled dikes, sheet-pile groins), with limited investigations of the often-used geotube groins in temporary water-related projects, and those available predominantly discussed their stability [27,28]. Research on scour and deposition around geotube groins, especially under tidal environments, is notably lacking. Additionally, studies on the innovative mixed clay–geotube groins, deduced from the latest engineering experiences, are virtually nonexistent.

Based on the embankment engineering project on the north bank of the Qiantang River, from Yangtian Temple to Tashan (specifically focusing on the toe of the dike), this study addresses the different performances of geotube groins and mixed clay–geotube groins (i.e., the instability and potential collapse of geotube groins and the relatively stable mixed clay–geotube groins under intense water flow). The remainder of this paper is structured as follows: Section 2 articulates the materials and methods adopted in this study, including the experimental setup and measurement equipment, the determination of the scale ratio, the selection of the model sediment, and the experimental design. Section 3 provides a detailed analysis of the experimental results. Section 4 discusses the potential reasons for the experimental results. Finally, conclusions and prospects for future works are presented in Section 5.

2. Materials and Methods

2.1. Experimental Setup and Measurement Equipment

The physical model experimentation was conducted in a multifunctional experimental flume with dimensions of 30 m in length, 10 m in width, and 1 m in height (as depicted in Figure 4). The flume featured a flat, ungraded bottom surface along its entire length. The experimental water supply was facilitated through a closed-loop water supply system, with a maximum flow rate of 400 L/s achievable through precision control by an intelligent flow control system. In the outflow section of the flume, adjustable sluice gates and baffles were installed to regulate both the outflow rate and the water depth. The primary flume was characterized by its flat-bottomed profile, rendering it unsuitable for dynamic bed experiments. To overcome this limitation, a custom-built, portable sediment flume was constructed within the experimental flume. This sediment flume measured 4 m in length, 3 m in width, and 0.2 m in depth. Transition sections measuring 4.8 m and 3.6 m in length, respectively, were positioned at the inlet and outlet ends of the sediment flume to ensure a relatively stable flow profile as the water entered the dynamic bed testing segment.

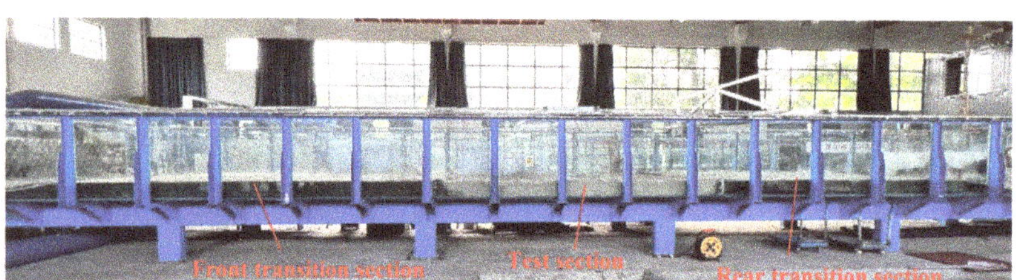

Figure 4. Experimental flume and sediment flume.

Utilizing our independently developed high-precision automated topographic instrument for sediment–water physical experiments (as depicted in Figure 5), we monitored changes in the terrain. This instrument is capable of stable operation, facilitating measurements of the experimental terrain in varying sediment–water environments. It is equipped with a high-precision probe that records the descent height as soon as the probe makes contact with the riverbed after being lowered from its initial height. For a given point, the difference between the descent heights recorded by the topographic instrument in its initial and final terrain states represents the terrain change at that point.

Figure 5. High-precision topographic instrument.

2.2. Similarity Conditions and Scale Ratio

The physical experiment primarily focused on the local scour and sedimentation variations around the groins, characterized by significant three-dimensionality. As the variations in scour and sedimentation at the Qiantang River Estuary are predominantly influenced by the suspended sediment transport dynamics [29,30], the experimental design adhered to the principles of the normal sediment transport bed model. Complying with the specifications outlined in hydraulic model testing standards [31], it needed to simultaneously satisfy conditions of hydraulic similarity and sediment transport similarity. Additionally, to ensure minimal perturbation of the flow patterns around the model groin by the opposite flume sidewall, the length of the model groin should not exceed approximately one-third of the flume's width according to numerical simulations. The selected experimental scale ratios are presented in Table 1.

Table 1. Scale ratio determination.

Nomenclature of Scale Ratios	Symbols	Number
Planar scale ratio	λ_l	80
Vertical scale ratio	λ_h	80
Velocity scale ratio	λ_v	8.94
Time scale ratio	λ_t	8.94
Incipient velocity scale ratio	λ_{vc}	8.94
Settling velocity scale ratio	λ_ω	8.94

2.3. Selection of Model Sediment

This model adhered to the principles of a normal sediment transport bed model for suspended sediment, with the selection of the model sediments primarily governed by considerations of sediment transport similarity and incipient motion similarity. In the construction zone of the groin, the average water depth typically fell within the range of 2 to 5 m, and the median particle size d_{50} of the riverbed sediment was within the range of 0.03 mm to 0.06 mm. Within this specified range of water depth and median particle size, the selection of the model sediments aligned with Zhang Ruijin's formulae for incipient flow velocity (Equation (1)) and settling velocity (Equation (2)).

$$U_e = \left(\frac{h}{d}\right)^{0.14}\left[17.6\frac{\rho_s-\rho}{\rho}d + 0.000000605\left(\frac{10+h}{d^{0.72}}\right)\right]^{1/2} \quad (1)$$

where U_e refers to the initiation velocity of the sediment, h refers to the depth of the water in the construction area, d refers to the median particle size of the sediment, and ρ_s and ρ refer to the density of the sediment and water, respectively.

$$w = \sqrt{\left(13.95\frac{v}{d}\right)^2 + 1.09\frac{\rho_s-\rho}{\rho}gd} - 13.95\frac{v}{d} \quad (2)$$

where w refers to the settling velocity of the sediment, v refers to the flow velocity of the water, and g stands for the gravitational acceleration.

It could be determined that the average incipient flow velocity for sediment transport in this region fell within the range of 0.47 m/s to 0.71 m/s, while the settling velocity ranged from 0.06 cm/s to 0.18 cm/s. Based on the calculations of prototype sediment settling, incipient flow velocity, and scale analysis, the model sediment should exhibit incipient flow velocities ranging from 5.2 cm/s to 8.3 cm/s and settling velocities between 0.008 cm/s and 0.03 cm/s. Following the comparative analysis of various model sediments' incipient flow velocities by experimentation, a model sediment consisting of wood powder (Ziyang Wood Powder Factory, Ziyang, Shandong, China) with a density of 1.2 t/m^3 and a median particle size of 0.1 mm was selected for the dynamic bed experiments. The model sediment exhibited an incipient flow velocity of 6.5 cm/s and a settling velocity of 0.028 cm/s.

2.4. Experimental Design

2.4.1. Groin Models

To investigate and compare the bed changes in the surrounding topography induced by both geotube groins and mixed clay–geotube groins under varying flow conditions, it was necessary to make the groin models prior to the flume experiments. These models were designed based on the dimensions of the field prototype groins (with a length of 48 m, a base width of 36 m, a top width of 12 m, and a height of 12 m) and the prescribed geometric scaling ratios. This ensured that the constructed groin models maintained the same overall dimensions, with respective lengths, widths, and heights of 60 cm, 45 cm, and 15 cm. For the model of the geotube groin, a prototype geotextile material from the construction site was used for stitching. The geotubes were internally filled with riverbed sediment. The construction comprised three segments with dimensions of 60 cm × 45 cm × 5 cm, 60 cm × 30 cm × 5 cm, and 60 cm × 15 cm × 5 cm, respectively. These were stacked in descending order by size from bottom to top, as illustrated in Figure 6. In the case of the mixed clay–geotube groin model, it included both a section with geotubes and a section with clay material. The three geotube models had dimensions of 47 cm in length, 45 cm in width, and 5 cm in height; 47 cm in length, 30 cm in width, and 5 cm in height; and 47 cm in length, 15 cm in width, and 5 cm in height, respectively. The stacking configuration of these models was consistent with that of the geotube groin. Additionally, a layer of clay was applied at the head of the geotube models, closely adhering to the groin head. This clay layer had a height of 15 cm and extended outward by 13 cm, as depicted in Figure 7. Subsequently, the fabricated models were individually placed in flumes filled with saturated model sediment for experimental testing.

2.4.2. Measurement Point Configuration

Topographical measurements were systematically arranged across 37 longitudinal monitoring sections oriented perpendicular to the direction of the water flow. These sections were designated as sections 1-1 to 37-37, as illustrated in Figure 8. Upstream of the model groin, 10 longitudinal sections (Sections 1 to 10) were established, with intersectional spacings of 15, 10, 5, 5, 5, 5, 5, 2, 2, and 2 cm, respectively. Each of these longitudinal sections consisted of 15 points spaced 10 cm apart. At the groin head, 10 additional longitudinal sections (Sections 11 to 20) were set up, with 5 cm of spacing between adjacent sections, and each section contained 18 points with 5 cm intervals between points. Downstream of the model groin, there were 17 longitudinal sections (Sections 21 to 37), with respective intersectional spacings of 2, 2, 2, 5, 5, 5, 5, 5, 5, 10, 10, 10, 10, 10, 15, 15, and 20 cm. Each of these sections incorporated 15 points, with a spacing of 10 cm between the individual points.

Figure 6. Geotube groin model (model dimensions: length 60 cm, width 45 cm, height 15 cm).

Figure 7. Mixed clay–geotube groin model (model dimensions: length 60 cm, width 45 cm, height 15 cm; the geotube part: 47 cm long, 45 cm wide, and 15 cm high; the clay part: 13 cm long, 45 cm wide, and 15 cm high).

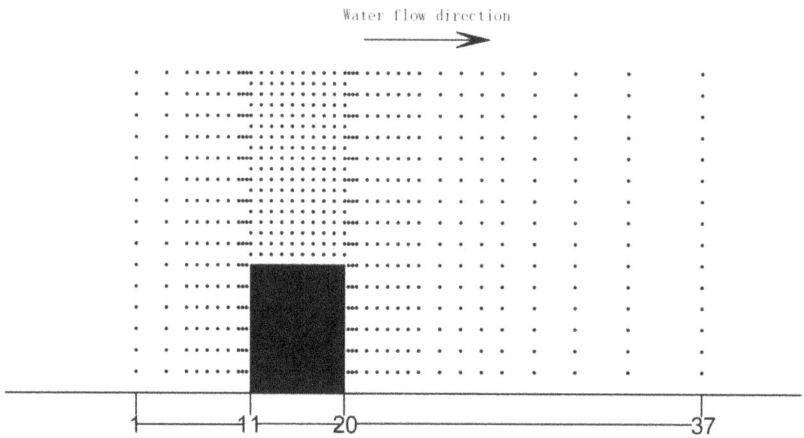

Figure 8. Measurement point distribution map (overall dimensions of the groin model shown as the black box: length 60 cm, width 45 cm, height 15 cm).

2.4.3. Experimental Conditions

The primary construction phase of the on-site engineering project spanned from October to May of the following year. This study focuses on the month of April as a representative case, examining the morphological changes due to sediment deposition and erosion around two types of groins under both spring and neap tidal flow conditions. A summary of the experimental conditions is presented in Table 2.

Table 2. Summary table of experimental conditions.

Experiment Conditions	Experimental Flow Conditions	Representative Tidal Elevation Process	Tiered Tidal Elevation Process	Measurement Variables	Measurement Methodology
One	Neap tide inflow	Flood tide on 7 April	Figure 9	Scouring at the head of the groin and sediment deposition downstream of the groin	High-precision topographic instrument; high-definition photography
Two	Spring tide inflow	Flood tide on 1 April	Figure 10		

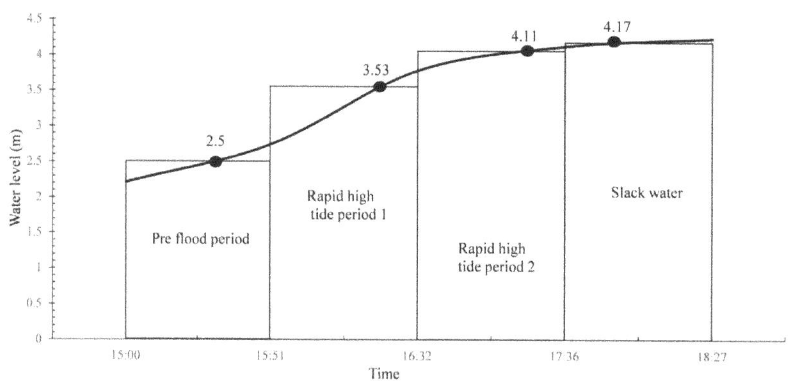

Figure 9. Elevation gradation during the neap tide inflow period.

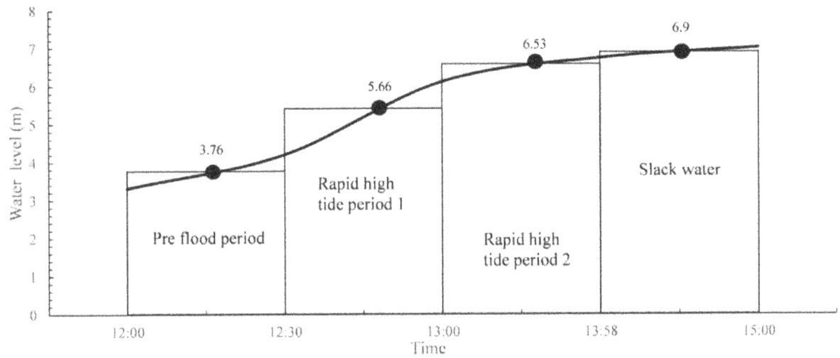

Figure 10. Elevation gradation during the spring tide inflow period.

Based on tidal data obtained from the Yanguan tide gauge station, representative hydrodynamic conditions were selected from two distinct days in April: one with the maximum tidal range (1 April, representing spring tide), and the other with the minimum tidal range (7 April, representing neap tide). Due to the inherently unsteady nature of the tidal cycle, the experimental approach utilized the concept of 'quasi-steady flows'. In this method, a complete unsteady inflow and sediment transport process was approximated by several stepped steady processes. Consequently, for a complete tidal cycle, three periods were identified, and the flow conditions were graded based on the extent of the variations in water level and flow velocity in each period. These periods were the pre-flood period (where water levels rose gradually, divided into one grade), the rapid high-tide period (featuring significant changes in water level and flow velocity, divided into two grades), and the slack-water period (with minimal changes in water level, divided into one grade).

According to the data for water level and flow velocity provided by the Yanguan tide gauge station, the averaged water level divisions for neap and spring tides are illustrated in Figures 9 and 10, respectively. Using the scale relationships established in Table 1, the variables are listed in Tables 3 and 4, where V is the averaged flow velocity, H is the averaged water level, and variables of the prototype and the model are denoted by the subscripts p and m, respectively. Q is the pump discharge, and T is the scaled time for each period in the flume.

Table 3. Parameters for each classification in the neap tide experiment.

Phase	Time Interval	V_p (m/s)	H_p (m)	V_m (cm/s)	H_m (cm)	T (min)	Q (L/s)
Pre-flood period	15:00–15:51	1.2	3.5	13.4	4.4	6	18
Rapid high-tide period	15:51–16:32	2.5	4.53	28	5.7	6	48
	16:32–17:36	2	5.11	22.4	6.4	7	43
Slack-water period	17:36–18:27	0.5	5.17	5.6	6.5	7	11

Table 4. Parameters for each classification in the spring tide experiment.

Phase	Time Interval	V_p (m/s)	H_p (m)	V_m (cm/s)	H_m (cm)	T (min)	Q (L/s)
Pre-flood period	12:00–12:30	1.5	4.76	16.8	5.9	4	30
Rapid high-tide period	12:30–13:00	4	6.66	45	8.4	4	113.4
	13:00–13:58	3	7.53	33.5	9.4	7	95
Slack-water period	13:58–15:00	0.8	7.9	8.9	10	7	26.7

3. Results

3.1. Experimental Condition 1: Neap Tide Inflow

Under the inflow conditions of the neap tide, the initial and final topographies of both types of groins can be seen in Figure 11, where GG signifies the geotube groin, while MGG denotes the mixed clay–geotube groin.

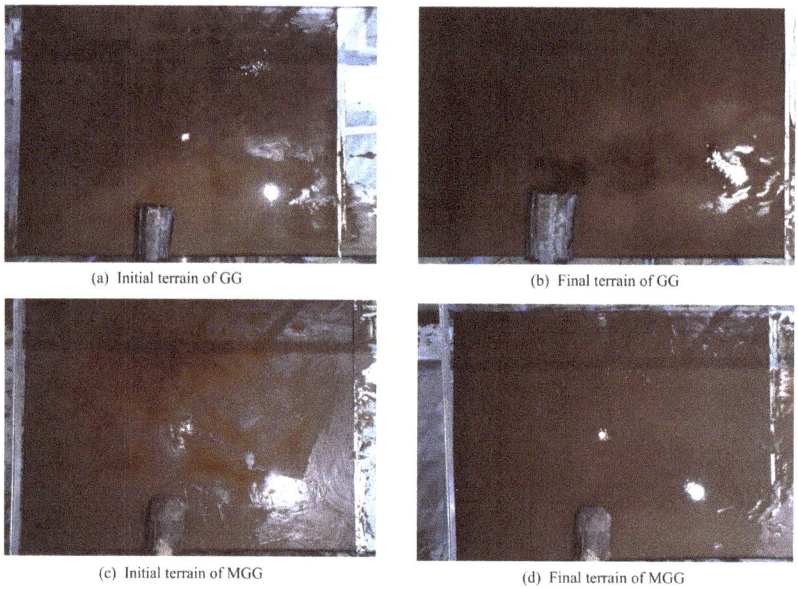

(a) Initial terrain of GG (b) Final terrain of GG (c) Initial terrain of MGG (d) Final terrain of MGG

Figure 11. Topographical changes around both types of groins under neap tide inflow conditions.

According to the topographic variations measured by the topographic instrument, the final contour lines of sedimentation and erosion for both types of groins were as illustrated in Figure 12, with both the horizontal and vertical axes scaled in mm. The model groins were positioned within the flume at coordinates ranging from 560 mm to 1010 mm on the x-axis and from 0 mm to 600 mm on the y-axis. The x-axis represents the longitudinal direction along the flume, while the y-axis represents the direction perpendicular to the flume.

By synthesizing the overall topographic contour maps with high-resolution experimental images, it was evident that under conditions of neap tide inflow, the downstream terrain of both types of groins exhibited an increase in elevation. This observation corroborated that both groins were effective in promoting sedimentation in their respective downstream zones.

To offer a more detailed assessment of the extent of sediment accumulation behind the groins, elevational profiles from three selected downstream longitudinal section measurement points were analyzed, revealing the longitudinal variations, as shown in Figure 13. The distances from CS 22, CS 28, and CS 34 to the backwater side of the groin (CS 20) were 4 cm, 31 cm, and 86 cm, respectively. An elevation of 0 refers to the relative initial topography. An elevation greater than 0 indicates sedimentation, while elevation below 0 means bed erosion. In the downstream zone of the model groins located within approximately 600 mm from the shore, the elevation increase in the topography was more pronounced for the three longitudinal sections downstream of the mixed clay–geotube groin as compared to their counterparts downstream of the geotube groin. The relative sediment accretion exceeded 10 mm for the mixed clay–geotube groin, while it remained below 5 mm for the geotube groin. It was noted that areas in closer proximity to the mixed clay–geotube groin exhibited a higher degree of sedimentation. This was attributed to the partial dispersal of clay from the head of the mixed clay–geotube groin during erosive processes, which was then transported downstream and accumulated.

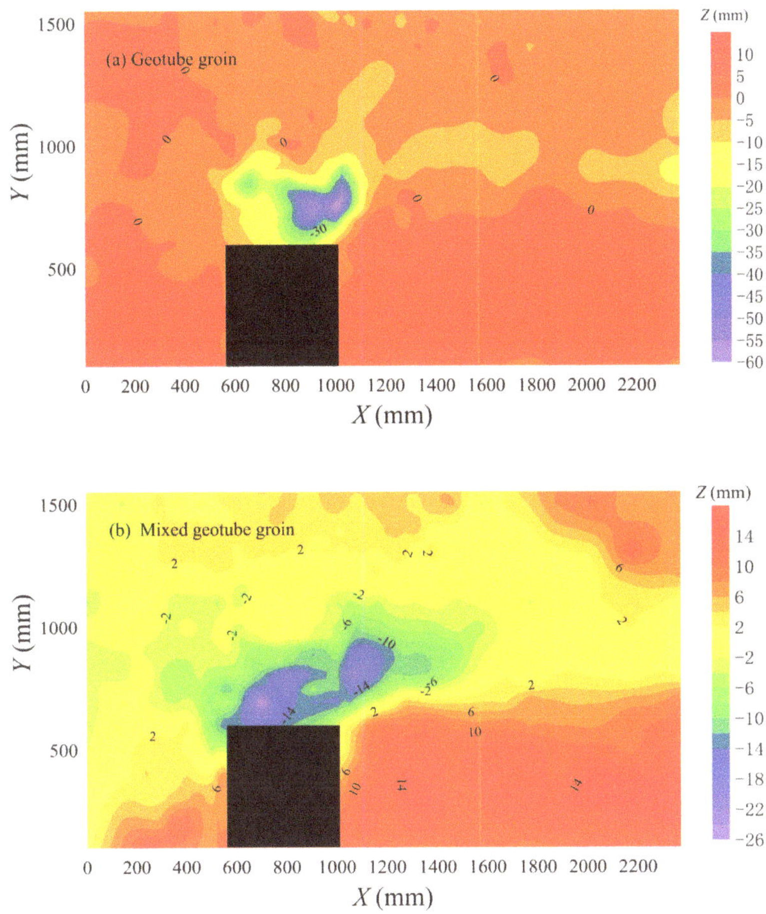

Figure 12. Contour maps of sedimentation and erosion for two different groin materials during the neap tide inflow.

In regions situated at a distance greater than 600 mm from the shore, the riverbeds surrounding both types of groins generally exhibited erosive tendencies, with the extent of erosion intensifying as the proximity to the groins increased. For any given longitudinal section, the erosion depth adjacent to the geotube groin was notably greater than that near the mixed clay–geotube groin. For instance, at section CS 22, the maximum erosion depths for both groins were located approximately 700 mm from the flume's sidewall. The geotube groin exhibited an erosion depth of 40 mm, whereas the mixed clay–geotube groin showed a comparatively shallower depth of 20 mm. When comparing the topographic contour maps (as shown in Figure 12), the scour holes around the geotube groin were, on the whole, deeper than those around the mixed clay–geotube groin. The deepest erosion near the head of the geotube groin reached up to 60 mm, while that for the mixed clay–geotube groin was limited to 25 mm.

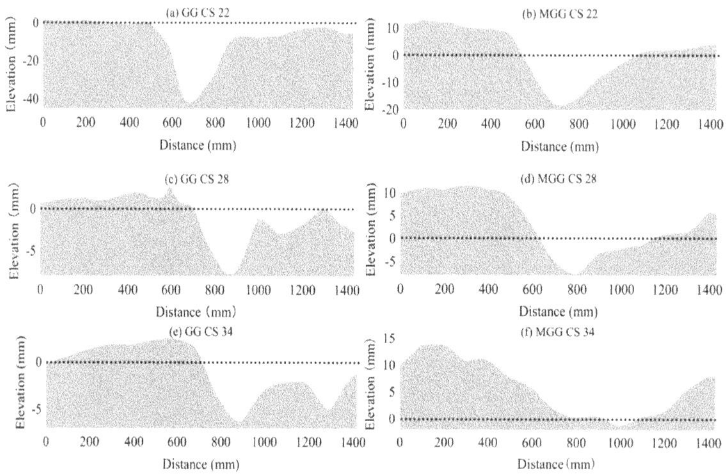

Figure 13. Elevation change profile of the longitudinal section behind the groins.

To further explore the erosional patterns at the heads of the two different groins, cross-sectional variations at distances of 5 cm ($Y = 650$ mm), 15 cm ($Y = 750$ mm), and 25 cm ($Y = 850$ mm) from the groin head were plotted. It can be seen in Figure 14 that the scour depth at the head of the geotube groin exceeded that of the mixed clay–geotube groin at the same cross-section. Specifically, the maximum scour depth observed for the geotube groin reached up to 60 mm, whereas it was limited to 20 mm for the mixed clay–geotube groin. However, the spatial extent of the scour pit associated with the mixed clay–geotube groin was broader than that associated with the geotube groin. The scour dimensions for the geotube groin spanned a range of 400 mm to 1200 mm, compared to 300 mm to 1400 mm for the mixed clay–geotube groin. In summary, during the period of neap tide inflow, the geotube groin exhibited greater scour depths, but within a narrower range, whereas the mixed clay–geotube groin manifested lower scour depths, but over a more expansive spatial domain.

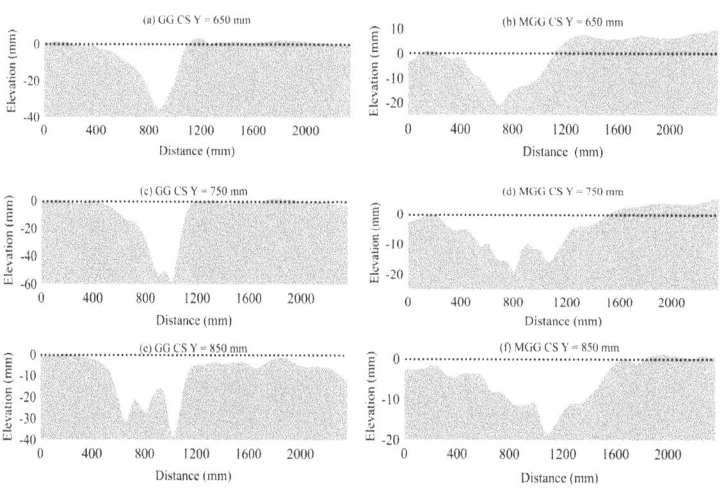

Figure 14. Topographic elevation variation diagrams of cross-sections at various distances from the groin.

3.2. Experimental Condition 2: Spring Tide Inflow

Under the inflow conditions of the spring tide, the initial and final topographies of both types of groins can be seen in Figure 15.

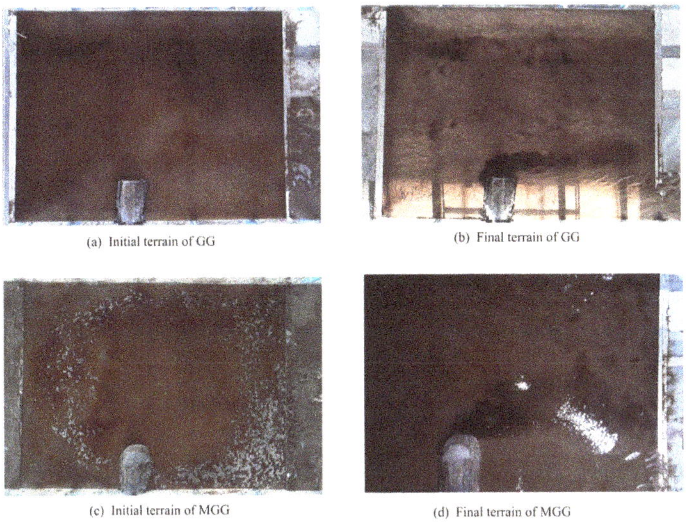

Figure 15. Topographical changes around both types of groins under spring tide inflow conditions.

According to the topographic variations measured by the topographic instrument, the final contour lines of sedimentation and erosion for both types of groin were as illustrated in Figure 16. Both the x- and y-axes are represented in mm, and the location of the model groins in the flume correspond to conditions during the neap tide phase.

By comparing the overall contour lines of both groins, it can be observed that after the passage of the spring tide inflow, the post-groin topography of the geotube groin showed relative sedimentation between 5 mm and 10 mm (as shown in Figure 16a), while the sedimentation behind the mixed clay–geotube groin was between 20 mm and 30 mm (Figure 16b). This indicates that both types of groins enhanced sediment deposition in the downstream region, with the mixed clay–geotube groin demonstrating superior siltation promotion compared to the geotube groin. When compared with the neap tide inflow conditions, where the geotube groin had a relative sedimentation thickness of less than 5 mm (Figure 12a) and the mixed clay–geotube groin was between 10 mm and 15 mm (Figure 12b), it was evident that the spring tide inflow had a more pronounced effect on the post-groin topographic changes than the neap tide for both groins.

To offer a more detailed assessment of the extent of sediment accumulation behind the groins, elevational profiles from three selected downstream longitudinal section measurement points were analyzed, revealing the following longitudinal variations:

From Figure 17, it was evident that during the high tide of the spring tide, within the area behind the model groins (within 600 mm from the shoreline), the change in elevation across the three cross-sections behind the mixed clay–geotube groin was more pronounced compared to the corresponding positions behind the geotube groin. Relative to the conditions during the neap tide inflow, there was greater sediment accretion behind the mixed clay–geotube groin during the spring flood tide, with siltation approximately 20 mm thick, while the increase in sediment accretion behind the geotube groin during the spring tide inflow was rather modest compared to the neap tide conditions.

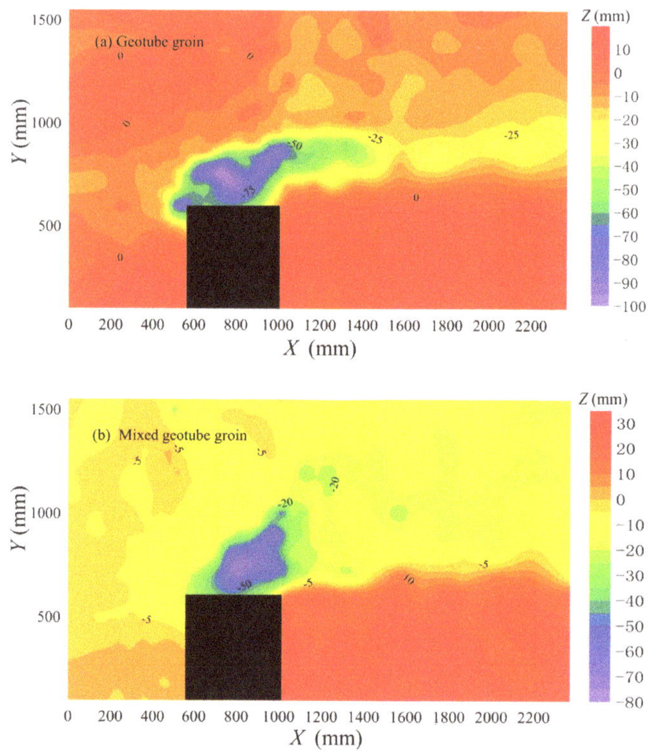

Figure 16. Contour maps of sedimentation and erosion for two different groin materials during the spring tide inflow.

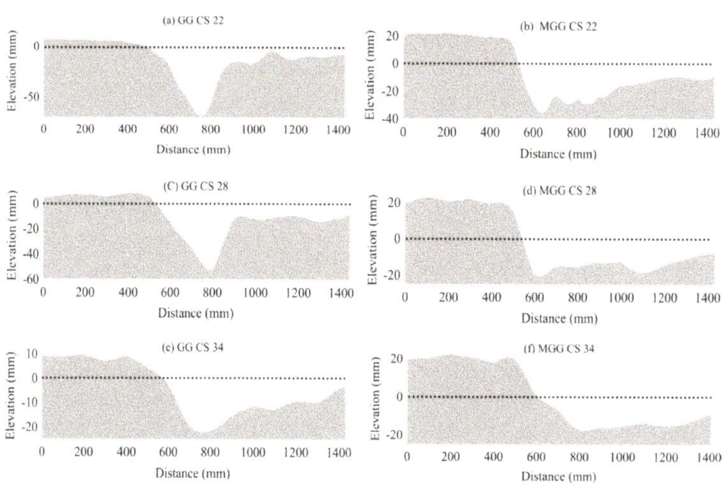

Figure 17. Elevation change profile of the longitudinal section behind the groins.

Additionally, to further explore the erosional patterns at the heads of the two different groins, cross-sectional variations at distances of 5 cm (Y = 650 mm), 15 cm (Y = 750 mm), and 25 cm (Y = 850 mm) from the groin head were plotted, as shown in Figure 18. The geotube groin showed a greater scour depth at the groin head than the mixed clay–geotube groin. The maximum scour depth for the geotube groin reached up to 100 mm, while the mixed clay–geotube groin recorded a maximum scour depth of only 60 mm. It can be inferred that the geotube groin was more susceptible to scouring at the head compared to the mixed clay–geotube groin.

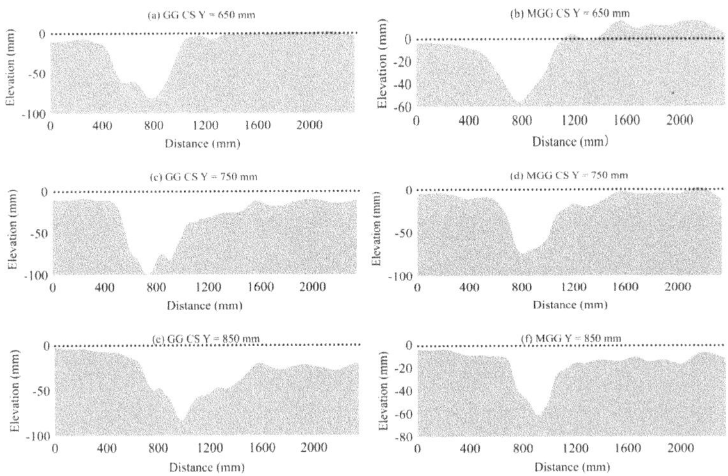

Figure 18. Topographic elevation variation diagrams of cross-sections at various distances from the groins.

During the spring tide period, the scour pit depth for the geotube groin ranged from 0 to 100 mm, as depicted in Figure 18a,c,e. Conversely, the mixed clay–geotube groin's scour pit depth ranged between 0 and 60 mm during the same phase, as shown in Figure 18b,d,f. The geotube groin's head was found to experience more severe scouring than that of the mixed clay–geotube groin. Relative to the conditions during the neap tide, the geotube groin exhibited scour pit depth ranges of 0–40 mm (see Figure 14a,c,e), while the mixed clay–geotube groin showed ranges of 0–20 mm (as shown in Figure 14b,d,f). After the spring tide phase, the scour pit depths at the heads of both groins expanded by 1–2 times compared to those during the neap tide. This indicates that the spring tide exerted a more significant influence on the scour pits at the groin heads than the neap tide.

4. Discussion

By comparing the topographical changes around the model groins under different experimental conditions, it became evident that the spring tide exerted a more significant influence on the surrounding morphology than the neap tide. This was predominantly manifested in deeper scouring pits at the groin heads, wider affected areas, and increased sediment deposition downstream of the groins. During the spring tide, the increased flow velocity could mobilize more bed sediments into the flow, resulting in a higher sediment concentration within the water column. As the sediments were transported downstream of the groins, the flow velocity diminished, and more sediments settled, leading to morphological sediment deposition.

By comparing the morphological changes downstream of groins made of different materials under the same flow conditions, we found that sediment deposition was more pronounced behind mixed clay–geotube groin. This was attributed to the faster flow velocities during spring tide, where a portion of the clay within the mixed clay–geotube groin's structure was eroded and dispersed by the water flow. Subsequently, the eroded material settled in areas with slower velocities downstream of the groin.

Figure 19 illustrates the phenomenon of head collapse in the geotube groin under the conditions of a spring tide. The geotube groin behaved approximately as a rigid body, with its base in direct contact with the original riverbed of sandy soil. Under prolonged, intense, unsteady flow, the bed sediment at the base of the geotube groin's head was progressively eroded, forming a scour hole. This resulted in an eccentric load on the groin, causing localized tensile deformation of the geotextile material. In actual engineering scenarios, when the localized tensile force exerted on the geotube exceeded its tensile strength, the geotextile material was pulled into failure, leading to sand leakage, and the groin ultimately collapsed, as shown in Figure 2.

Figure 19. Localized scour pit at the head of the geotube groin following the spring tide surge.

For the mixed clay–geotube groin, its head clay exhibited some erodibility. Our experimental findings indicated that under hydrodynamic forces, a portion of the clay from the mixed clay–geotube groin's head was eroded and fell into the scour pit, which filled the pit to a certain extent, as demonstrated in Figure 20a. Unlike the geotube groin, the clay component of the mixed clay–geotube groin exhibited some flexibility. When a scour pit formed at its head, the clay underwent deformation and fitted the edges of the pit, preventing its extension beneath the main groin structure, as seen in Figure 20b. This could reduce the eccentric load on the groin, enhancing its overall stability. Meanwhile, some of the clay carried by the flow would fill the gap between the groin and the riverbed, adhering the bottom geotube layer to the riverbed and further bolstering the groin's stability, as depicted in Figure 21.

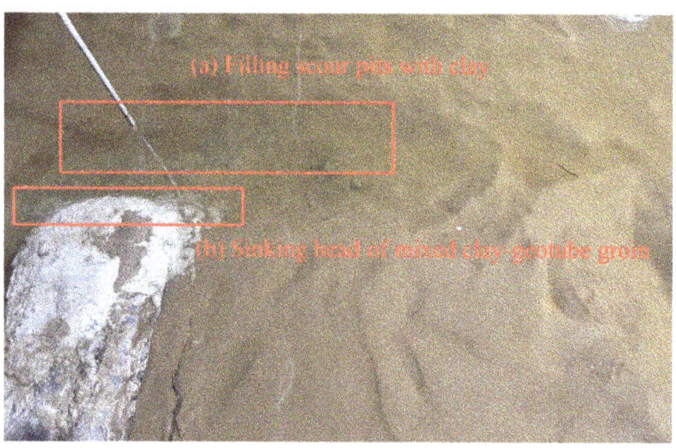

Figure 20. Scour pit at the head of the mixed clay–geotube groin.

Figure 21. Partial diagram of the contact between the geotube behind the mixed clay–geotube groin and the riverbed.

5. Conclusions

In the cofferdam construction process for the embankment toe reinforcement project at the Qiantang River Estuary, geotube groins often encountered problems of embankment head erosion, leading to the tearing of the geotextile material of geotube groins, the subsequent loss of infilled material and, finally, collapse. This significantly hindered the progress of the project. Drawing on engineering experience, adopting mixed clay–geotube groins could effectively address this issue. To provide theoretical support for this construction approach, we conducted flume model tests to study the sediment dynamics around groins of both material types under various flow conditions. The conclusions are as follows:

(1) In comparative terms, the morphological impacts on the surrounding terrain of the groins were more pronounced under conditions of spring tide inflow than under neap tide inflow conditions. Furthermore, under the same flow conditions, the scour pits at the heads of the geotube groin were conspicuously deeper than those formed around the mixed clay–geotube groin.

(2) From the perspective of sediment accretion effects behind the groins, both groins exhibited the capability to promote sediment deposition. Under two different flow conditions, the velocity of the water decreased in the region downstream of the groins, resulting in the settlement of suspended sediment, and thereby elevating the downstream terrain. It was noteworthy that during periods of spring tide, the increased flow velocity mobilized a greater quantity of bed sediment into the water column. As the flow decelerated downstream of the groins, a more substantial sediment deposition was observed. Under the same flow conditions, the mixed clay–geotube groin demonstrated a higher propensity for sediment accumulation in its downstream regions than the geotube groin. This was attributed to the fact that some of the clay within the structure of the mixed clay–geotube groin was eroded and transported downstream by the flowing water, contributing to enhanced sedimentation rates behind the groin.

(3) From the perspective of the groin stability, clay possesses certain erosive characteristics that have a unique interaction with the hydraulic environment. Under sustained water flow, some of the clay at the head of the mixed clay–geotube groin was scoured and displaced into existing scour pits, effectively serving a role in the infill and attenuation of these erosive features. Additionally, due to the inherent plasticity of clay, it had the ability to adapt its shape to conform to the contours of the scour pits, thereby bolstering the overall stability of the groin structure to a significant extent. Conversely, the geotube groin exhibited less favorable stability characteristics. Scour pits that formed at the groin head induced eccentric loading conditions, leading to deformation of the geotextile material. Once this deformation reached a critical threshold, the geotextile material experienced structural failure, causing the contained sand to leak and eventually resulting in groin collapse.

These findings highlight the benefits of mixed clay–geotube groins for their potential to enhance structural integrity and resilience against erosive forces in cofferdam construction at the Qiantang River Estuary, offering scientific support for the applicability of mixed clay–geotube groins in tidal areas.

However, it should be noted that due to the limited experimental conditions and test duration, the above conclusions were drawn under specific flow conditions, so there are certain limitations. In the future, tests under more conditions will be needed, such as setting various flow conditions, and adjusting the length and angle of the groins. Meanwhile, the durability, maintenance costs, and environmental compatibility of synthetic materials should also be investigated to protect natural water against the dispersion of artificial materials due to failures.

Author Contributions: Conceptualization, X.T. and C.X.; methodology, J.X.; validation, K.F.; formal analysis, J.X.; investigation, J.L.; resources, X.T.; writing—original draft preparation, J.X. and C.X.; writing—review and editing, J.X. and C.X.; visualization, H.G.; supervision, J.L. and C.X.; project administration, X.T. and K.F.; funding acquisition, X.P. All authors have read and agreed to the published version of the manuscript.

Funding: This research was funded by the National Natural Science Foundation of China (No. 12002310), Natural Science Foundation of Zhejiang Province (No. LZJWZ23E090010), and Science and Technology Projects of POWER CHINA HUADONG Engineering Co. Ltd. (No. KY2022-JG-03-02).

Data Availability Statement: The data presented in this study are available upon request from the corresponding author.

Acknowledgments: We would like to thank the Key Laboratory of Ocean Space Resource Management Technology at Zhejiang University of Water Resources and Electric Power for providing the experimental site. We also wish to thank the potential reviewers very much for their valuable comments and suggestions.

Conflicts of Interest: The authors declare no conflict of interest.

References

1. Pan, C.; Wang, Q.; Pan, D.; Hu, C. Characteristics of river discharge and its indirect effect on the tidal bore in the Qiantang River, China. *Int. J. Sediment Res.* **2023**, *38*, 253–264. [CrossRef]
2. Pan, C. *Special Research Report on Scouring at the Foot of the Ancient Sea Dyke on the North Shore of Qiantang River*; Zhejiang Provincial Institute of Water Resources and Estuary: Hangzhou, China, 2019. (In Chinese)
3. Wang, D. Study of local scour around submerged spur-dikes. *Adv. Hydrodyn.* **1988**, *3*, 60–69. (In Chinese)
4. Xu, H.; Li, Y.; Zhao, Z.; Wang, X.; Zhang, F. Experimental study on the local scour of submerged spur dike heads under the protection of soft mattress in plain sand-bed rivers. *Water* **2023**, *15*, 413. [CrossRef]
5. Aung, H.; Onorati, B.; Oliveto, G.; Yu, G. Riverbed morphologies induced by local scour processes at single spur dike and spur dikes in Cascade. *Water* **2023**, *15*, 1746. [CrossRef]
6. Zhang, L.; Wang, P.; Yang, W.; Zuo, W.; Gu, X.; Yang, X. Geometric characteristics of spur dike scour under clear-water scour conditions. *Water* **2018**, *10*, 680. [CrossRef]
7. Han, X.; Lin, P.; Gary, P. Influence of layout angles on river flow and local scour in grouped spur dikes field. *J. Hydrol.* **2022**, *614*, 128502. [CrossRef]
8. Reza, F.; Mahmood, S.K.; Mehdi, G.; Giuseppe, O. Temporal scour variations at permeable and angled spur dikes under steady and unsteady flows. *Water* **2022**, *14*, 3310. [CrossRef]
9. Tang, X.; Ding, X.; Chen, Z. Large eddy simulations of three-dimensional flows around a spur dike. *Tsinghua Sci. Technol.* **2006**, *11*, 117–123. [CrossRef]
10. Peng, X.; Zhao, Y.; Gao, L.; Xu, X.; Xia, J. Hydraulic and alluvial characteristics of the lower Yellow River after the sudden dam-break of permeable spur dikes. *J. Water Resour. Water Eng.* **2023**, *34*, 121–127+134. (In Chinese)
11. Zhao, W.; Yan, S.; Xuan, W.; He, C.; Lin, B. Study of erosion-lessening and siltation-promotion effect for sheet-pile groins under strong tidal estuary. *Chin. J. Hydrodyn.* **2006**, *21*, 324–330. (In Chinese)
12. Zhao, W.; Yan, S.; Xuan, W.; Lin, B. Effect of sheet-pile groins on protecting dykes and flats in tidal bore estuary. *Shuili Xubao* **2006**, *37*, 699–703. (In Chinese)
13. Zhao, W.; Yan, S.; Xuan, W.; Lin, B. Study of flat-maintenance and dyke-protection effect for sheet-pile groins. *J. Basic Sci. Eng.* **2005**, *S1*, 33–42. (In Chinese)
14. Qian, S.; Chen, G.; Jin, X.; Luo, H.; He, K. Experimental study on local scour of a spur dike in a curved river channel under the action of tidal bore. *Zhejiang Hydrotech.* **2022**, *50*, 32–35, 41. (In Chinese)

15. Zhang, Z.; Pan, C.; Zeng, J.; Chen, F.; Qin, H.; He, K.; Zhu, K.; Zhao, E. Hydrodynamics of tidal bore overflow on the spur dike and its influence on the local scour. *Ocean Eng.* **2022**, *266*, 113140. [CrossRef]
16. Pan, D.; Li, Y. Tidal bore scour around a spur dike. *J. Mar. Sci. Eng.* **2022**, *10*, 1086. [CrossRef]
17. Yu, P.; Pan, C. 1D numerical simulation of tidal bore in Qiantang River. *Chin. J. Hydrodyn.* **2010**, *25*, 669–675. (In Chinese)
18. Pan, C.; Lu, H. 2D numerical model for discontinuous shallow water flows and application to simulation of tidal bore. *J. Zhejiang Univ. (Eng. Sci.)* **2009**, *43*, 2107–2113. (In Chinese)
19. Pan, C.; Lu, H.; Zeng, J. 2D numerical simulation for sediment transport affected by tidal bore in the Qiantang River. *Shuili Xuebao* **2011**, *42*, 798–804. (In Chinese)
20. Pan, C.; Lin, B.; Mao, X. Case study: Numerical modeling of the tidal bore on the Qiantang River, China. *J. Hydraul. Eng.* **2007**, *133*, 130–138. [CrossRef]
21. Rong, G.; Wei, W.; Liu, Y.; Xu, G. Study on flow characteristics near spur dikes under tidal bore. *Shuili Xuebao* **2012**, *43*, 296–301. (In Chinese)
22. Xiao, T.; Chen, Y.; Zhang, Y.; Zhang, Z.; Zhang, Q.; Ma, L.; Li, N.; Zong, X.; Yang, J.; Yu, C. Analysis of local hydrodynamic characteristics of wading buildings under tidal action. *J. Phys. Conf. Ser.* **2021**, *2044*, 012179. [CrossRef]
23. Zhu, J.; Lin, B. Numerical study of the process of tidal bore turning over the groin. *Chin. J. Hydrodyn.* **2003**, *18*, 671–678. (In Chinese)
24. Xu, C.; Yin, M.; Pan, X. Field test and numerical simulation of tidal bore pressures on sheet-pile groin in Qiantang River. *Mar. Georesour. Geotechnol.* **2016**, *34*, 303–312. [CrossRef]
25. Cai, Y.; Cao, Z.; Wang, Y.; Guo, Z.; Chen, R. Experimental and numerical study of the tidal bore impact on a newly-developed sheet-pile groin in Qiantang river. *Appl. Ocean Res.* **2018**, *81*, 106–115. [CrossRef]
26. Xu, Z.; Xu, C.; Chen, R.; Cai, Y. Three-dimensional numerical simulation of bore against on sheet-pile groin. *J. Zhejiang Univ. (Eng. Sci.)* **2014**, *48*, 504–513. (In Chinese)
27. Zhou, M.; Huang, L.; Xiao, Z.; He, L. Stability analysis of geotube groin during construction. *Hunan Commun. Sci. Technol.* **2012**, *38*, 127–131. (In Chinese)
28. Yang, H.; Wu, W.; Wang, W.; Huang, J. Experimental study on stability of sand ribbed soft mattress of dike head for macro-tidal estuary. *J. Waterw. Harb.* **2009**, *30*, 408–412. (In Chinese)
29. Xiong, S. Design of physical sedimentation model for tidal estuaries. *Chin. J. Hydrodyn.* **1995**, *10*, 398–404. (In Chinese)
30. Xiong, S.; Hu, Y. Theory and practice of movable bed model with suspended load for tidal estuary. *J. Sediment Res.* **1999**, *1*, 2–7. (In Chinese)
31. *JTS/T 231-2021*; Technical Specification for Hydraulic Engineering Simulation Test. People's Transportation Publishing Co., Ltd.: Beijing, China, 2021. (In Chinese)

water

MDPI

Brief Report

On Internal Erosion of the Pervious Foundation of Flood Protection Dikes

Laurence Girolami [1,2,*], Stéphane Bonelli [1], Rémi Valois [3], Naïm Chaouch [1] and Jules Burgat [1]

1 RECOVER, INRAE Aix-Marseille Université, 13182 Aix-en-Provence, France
2 GéHCO, Campus Grandmont, Université de Tours, 37020 Tours, France
3 EMMAH, INRAE Université de Avignon, 84914 Avignon, France
* Correspondence: laurence.girolami@inrae.fr

Abstract: This work focuses on the mechanisms that trigger internal erosion of the pervious foundation of flood protection dikes. The origin of these permeable layers is generally attributed to the presence of a paleo-valley and paleo-channels filled with gravelly-sandy sediments beneath the river bed and dikes. These layers may extend into the protected area. Visual observations of leaks, sand boils and sinkholes in the protected area testify to internal erosion processes in the underground soil. Local geological conditions are part of the information to be sought to explain these processes: presence of permeable soils and position of interfaces. Results obtained on Agly dikes (France), using two classical geophysical methods (EMI and ERT), were analyzed using cored soils and showed that it is not enough to simply conclude to the presence of backward erosion piping. The possibility of internal erosion, such as suffusion or contact erosion, must also be considered as the cause of leaks, sand boils and sinkholes. As the results obtained are explained by the presence of a paleo-valley and paleo-channels beneath the river bed and dikes—commonly encountered in this context—the methodology presented and the results obtained are likely to be relevant for many dikes.

Keywords: river dikes; levees; paleo-valley; paleo-channels; leaks; internal erosion; sand boils; sinkholes; geophysical observations; electrical conductivity

1. Introduction

Protection dikes help keeping safe urban populations settled along rivers from risks encountered by flooding events. Underlying soils, termed as foundation, are however prone to internal erosion, which represents one of the main causes of breach failures [1,2]. This process results from the transport and migration of soil particles subjected to localized flows. Internal erosion processes include erosion along a crack or a pre-formed hole (concentrated leak erosion), migration of the fine fraction through a coarse matrix (suffusion), erosion along a contact between two soils (contact erosion), and erosion starting from a point of loss of effective stress by fluidization [3] and propagating backward towards upstream (backward erosion piping) [4–6]. Observations of surface signatures (leaks, sand boils, sinkholes and turbidity) represent the most relevant evidences of internal erosion.

The most documented signatures, termed as *sand boils*, are recognized as one of the major causes of severe damage or even failure of flood defenses worldwide [7–9]. Different mechanisms have been identified as being responsible of them [10]: detour and reappearance of a sediment-laden subsurface flow along a pre-existing fissure; earthquake-induced liquefaction; landslides; and subsurface flow under artesian conditions. The last case, discussed in this study, is driven by the imbalance of hydraulic head on both sides of the dike: the river and the protected area. As the water level rises during floods, a significant hydrostatic imbalance is formed across the foundation, increasing the pore pressure gradient and thus decreasing the effective stress acting on soil particles, which promotes their erodibility through seepage flows. Above a threshold velocity, the flow becomes strong enough to detach sand particles from the underlying soil located beneath the protected area and transport them across the cover layer of variable thickness. At this point, the hydraulic gradient quickly dissipates, depositing eroded material around the pipe head. The deposit, termed as sand boil, has a conical shape and is mostly made with fine sand. Once triggered, erosion can gradually progress towards the riverside, creating shallow pipes within the sand layer that can impede the dike stability.

Sand boils can also indicate local heave initiated at much lower hydraulic head than required for pipe progression, meaning that the process may stop at a constant hydraulic head [11]. In this case, the presence of sand boils may not lead to breach failures but can promote a late piping initiation during future periods of equivalent water levels, assuming that the duration of erosion processes is much longer than that of a flood and that pipes remain intact after flooding events. The description of such processes has become the priority of dike managers in charge of the safety of the embankment system.

Current analyses only integrate fine-scale information relating to soil and layer geometry, within a simplified description [11–14]. A broad majority of studies consider that the sand layer is horizontal within the foundation, and that such processes typically start along natural weaknesses or holes in the ground where flows are concentrated. However, such analyses cannot explain the position of sand boils around levee toes, or in their vicinity, up to 1 km [15]. While such specific hydraulic phenomena are presently well understood, the appearance of sand boils and their location relative to the dike remains difficult to predict. Local geological conditions are rarely considered, even though they have a major influence on sand boil distribution [16–18]. These local geological conditions may however explain why many kilometers of levees experience similar hydraulic gradients, but different symptoms at the surface: no signature, leaks or sand boils, and justify why few present physical and numerical models are capable of reproducing such observations made around protection dikes [19].

Another important erosion signature, widely observed around protection dikes, and described in the literature is defined as *sinkholes*. Different types of sinkholes have been reported, depending on the rocky, karstic, or granular nature of the environment [20]. The appearance of sinkholes in soils is a phenomenon that most often results from the collapse of a very low-density, mechanically unstable volume of soil. When this volume of soil is located in depth, the entire column of soil above settles. Low local density may be the result of internal erosion: suffusion [21,22] or contact erosion [23]. Settlement due to mechanical instability can occur immediately during significant erosion (when the soil is saturated), or later, when the previously eroded soil is dry and subject to saturation. Although sinkholes are observed on dikes in a non-karstic setting [7,8,24,25], the origin of these phenomena is still difficult to explain [26] and their modeling is complex [27]. Up to now, no studies have been proposed to explain their formation in foundation soils of river dikes that consider local geological conditions.

Although the observation of leaks, sand boils and sinkholes is commonly incorporated into risk analyses through expert judgement, local geological conditions are rarely considered. This work therefore aims at proposing geological observation of the foundation soils located beneath and around dikes locally affected by internal erosion, using both ElectroMagnetic Induction (EMI) and Electric Tomography (ERT) methods commonly used for river levees [28,29]. This study provides new information on the subsurface soils located beneath dikes and protected areas (nature of soils and position of interfaces), making it possible to formulate new hypotheses concerning the causes of leaks, sand boils and sinkholes. The importance of knowing local geological conditions is illustrated by the case of Agly dikes, where numerous leaks, sand boils and sinkholes have been frequently observed after floods (Figure 1), but have not yet been described [7,8,30]. Results obtained are explained by the presence of a paleo-valley and paleo-channels beneath the river bed and dikes, which are commonly encountered in this context. By using the methodology presented, the results obtained are likely to apply to numerous dikes.

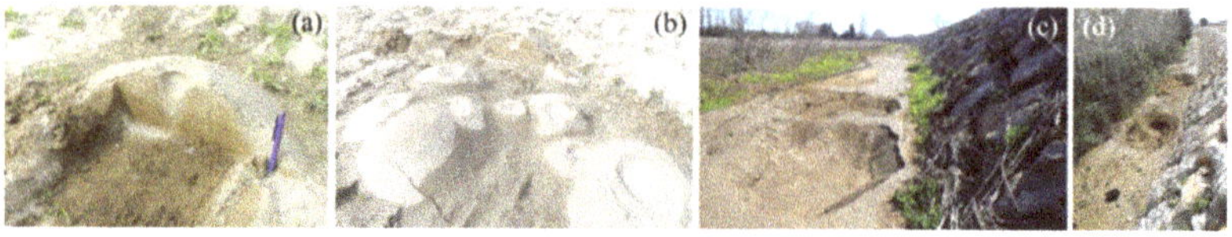

Figure 1. (**a**,**b**) Sand boils in the protected area, (**c**,**d**) Sinkholes observed around the levee toe.

2. Sand Boils and Sinkholes along the Agly Dikes

Geomorphological analyses have revealed that the Agly River is a perched river, whose minor bed is located above the natural ground of the area protected by dikes. The studied area is located downstream of a sharp widening of the major bed, leaving the possibility for the river to expand [31]. The geological context is characterized by recent Quaternary alluvial deposits of the river that may imply the probable presence of a paleo-valley and paleo-channels in the foundation soil, with a possible complex geometry. This information helps us to understand the presence of lenses of high-graded soil beneath the levees and their non-regular layouts.

Between 1969 and 1974, 13 km of dikes of 2–3 m high above the floodplain with a crest of 8 m wide, were built along the river bed of approximately 65 m wide to protect 30,000 people from risks incurred by flooding events. On the river side, the height is assessed as around 6 m, while a berm is located at mid-height of the protected area. The levee core consists of sandy loam to silty sand, while a 2 m deep drainage spur, topped with riprap, was built downstream after the flooding episode of 6 March 2013.

The diking system has been exposed to 11 floods over the period 1977–2020. Floods usually last for a few hours. The event of 1999 caused a breach on the northern bank, resulting in 35 victims, with a peak discharge of around 2110 m^3/s. Later, the 2013 flooding event caused a breach on the southern bank, with a peak flow of around 970 m^3/s. Geotechnical investigations highlighted the heterogeneous nature of the substratum and led to its simplified description: a superficial sandy silts to silty sands layer of highly variable thickness (from a few cm to 5 m), covering a roughly flat sandy layer (from fine to coarse) of up to 10 m deep above the impermeable basement, also containing highly permeable coarse materials (gravels, boulders, pebbles, cobbles), resulting from the river recalibration, and described in the borehole logs carried out after the 2013 flooding event (Figure 2b).

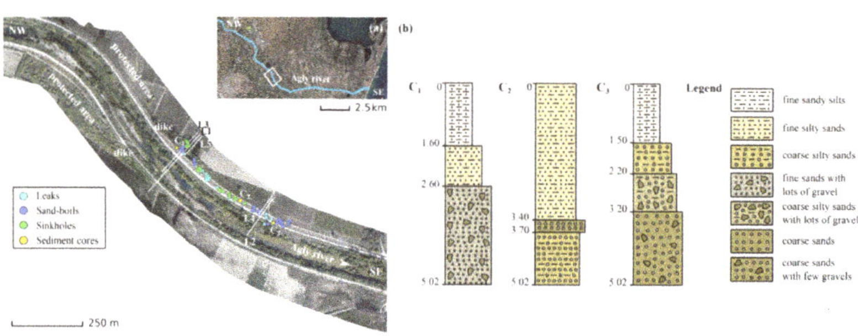

Figure 2. (**a**) Study area where both location of leaks, sand boils and sinkholes on the northern bank, as well as transverse ERT lines, have been reported. (**b**) Examples of borehole logs (C_1, C_2, C_3) performed in the area of interest and located on the map, that highlight the presence of coarse-grained soils constituting the permeable layers over a thickness of up to 5 m deep.

After each of the six floods reported from 1992 to 2014, leaks, sand boils and sinkholes were observed around dikes (Figure 1). Sand boils formed very regular cones of up to 1 m in diameter and were observed along two sections: a 700 m long section located down the northern bank; and a 500 m long section located down the southern bank, at up to around 60 m from the levee toe (Figures 1 and 2a). Additionally, sinkholes often assimilated to collapsed sand boils, were observed close to the levee toe, and exposed an irregular ellipsoidal shape of around 0.5 to 3 m long with an almost flat basement (Figure 1).

A granulometric analysis of sand boils observed in 1999, 2006 and 2013 showed that the cones were made with sandy materials, while the surrounding subsoil was composed of very little clay (a few %), sand (from 30 to 90%), and silt (from 10 to 40%) [7,8,30]. Three examples of borehole logs in the subsurface soil are given in Figure 2b.

To date, no satisfactory explanation exists to describe the appearance of leaks, sand boils and sinkholes around dikes, nor any modeling is able to reproduce these observations [7,8,30].

3. Materials and Methods

3.1. Description of Geophysical Methods

The aim of these observations is to highlight the importance of geological conditions in providing information required to explain the presence of leaks, sand boils and sinkholes. On the site studied, such information was obtained using two classical geophysical methods [28]: Electromagnetic Induction Method (EMI) and Electric Resistivity Tomography (ERT).

EMI is a well-established option for mapping the subsurface soil in a horizontally continuous manner. This method is well suited to acquire measurements of apparent electrical conductivity (in S/m; where 1 Siemens = 1 Ohm^{-1}) of the subsurface soil. The procedure consists in using coils in which an alternating current is flowing with a known frequency. Such measurements, performed by wearing the device and walking along parallel lines, provides a quick and cost-effective investigation along foundation dikes.

ERT can furthermore capture vertical information by acquiring measurements of apparent electrical resistivity (Ω.m) by injecting a current of known intensity in soils through a pair of electrodes, while measuring the potential drop which depends on the material properties through another pair of electrodes. Apparent values can thus be transformed into local real electrical resistivity after inversion processes that capture two-dimensional (2D) vertical sections of soils. The depth of investigation and resolution both depend on electrode spacing, device length, acquisition network and the investigated natural material. Resolution decreases with increasing depth of investigation.

3.2. Testing Protocols Followed on the Field

EMI was performed with a Geonics EM31-MK2. The apparent electrical conductivity of the soil was averaged over the first 6 m deep. Acquisition was carried out by walking at a velocity of around 1 m/s. A GPS was coupled to measurements. The acquisition period was taken of one second, leading to a spacing of around 1 m between each measurement. The paths were not always straight because of the vegetation, whereas the field remained approximately flat.

Three profiles were positioned owing to the repeated observations of leaks, sand boils and sinkholes made after the 2013 flooding event. One transverse profile, performed at the scale of the diked bed (L1), was made possible because of the river drought; two additional profiles were made at the scale of the levee (L2 and L3) while two more accurate profiles were made at the scale of the erosion symptom (leaks and sand boils L4; sinkholes L5). All profiles, presented here, were orthogonal to the levee (Figure 2), and assumed parallel to the underlying seepages. The characteristics of the profiles are given in Table 1. The profiles involved 64 to 120 electrodes. The inversion of apparent resistivity data was carried out using ResiPy [32].

Table 1. Characteristics of the geophysical profiles.

ERT Profiles	Equipment	Electrodes Spacing–Length	Array Type	Measurements Nb per Profile	Period of Measurements
L1	IRIS Syscal Pro	2–238 m	Wenner–Schlumberger	3254	Spring 2023
L2, L3	ABEM SAS4000	2–126 m	Wenner-α	472	Spring 2023
L4, L5	ABEM LS 2	1–63 m	Wenner-α	472	Spring 2022

4. EMI and ERT Results

To a first approximation, the lowest electrical conductivity values correspond to permeable layers containing silts, sands and gravels, while the highest values correspond to weakly permeable layers containing fine silty and clayey materials.

4.1. EMI Results

EMI measurements reveal two main conductivity zones (Figure 3):

- a zone of values of conductivity inferior to 0.01 S/m, indicating a permeable soil layer mostly made with coarse sands, silts and gravels (Figure 2b);
- a zone of values of conductivity greater than 0.02 S/m, indicating a less-permeable soil layer mostly made with fine silts [7,8,30].

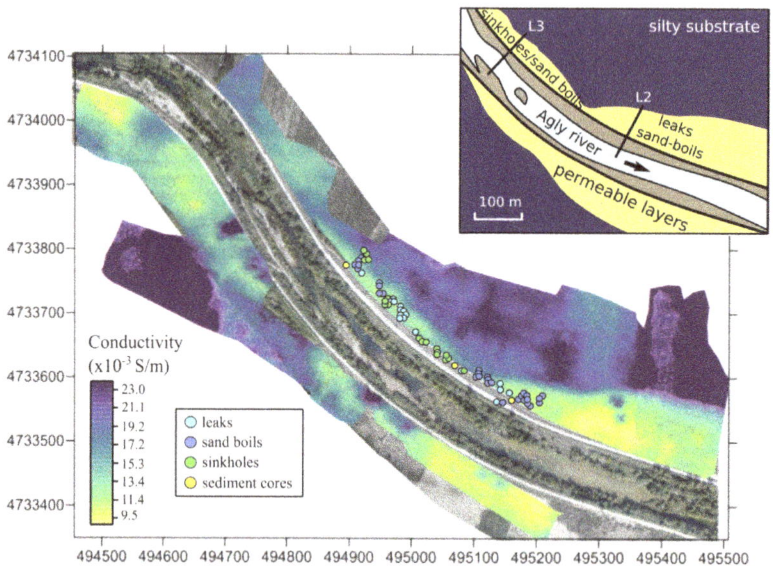

Figure 3. EMI results, map of soil conductivity averaged over the first 6 m deep. The scheme highlights two main areas where erosion signatures have been observed.

4.2. ERT Results

Values scales used in Figures 4–6 are identical to those used in Figure 3, while both conductivity and resistivity are given, so that EMI and ERT results can be compared. Figure 4 reveals two main conductivity regions:

- an area of conductivity measurements below 5×10^{-4} S/m located on the minor bed and beneath the northern and southern embankments, indicating a permeable soil layer that corresponds to old sediments that fill a paleo-valley,

- an area with measurements of conductivity above 7×10^{-3} S/m located in depth, indicating a less-permeable soil layer that corresponds to a sandy-marly substrate, which includes the groundwater at the time of measurements.

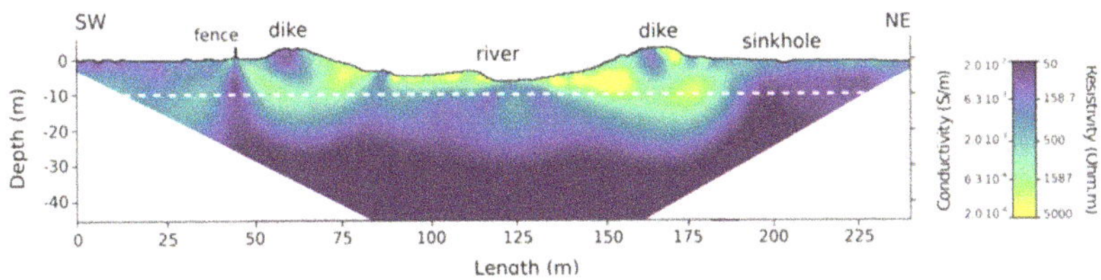

Figure 4. ERT profile along L1, exposing the soil conductivity and resistivity. The white dashed line represents the position of the groundwater. The minor bed was dried up at the time of measurement.

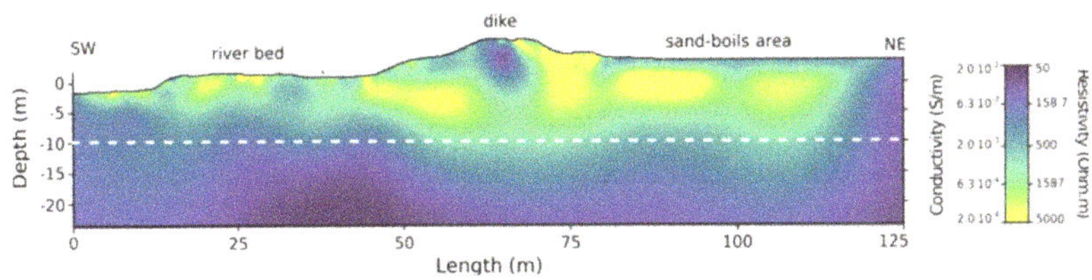

Figure 5. ERT profile along L2, exposing the soil conductivity and resistivity. The white dashed line represents the position of the groundwater. The minor bed was dried up at the time of measurement.

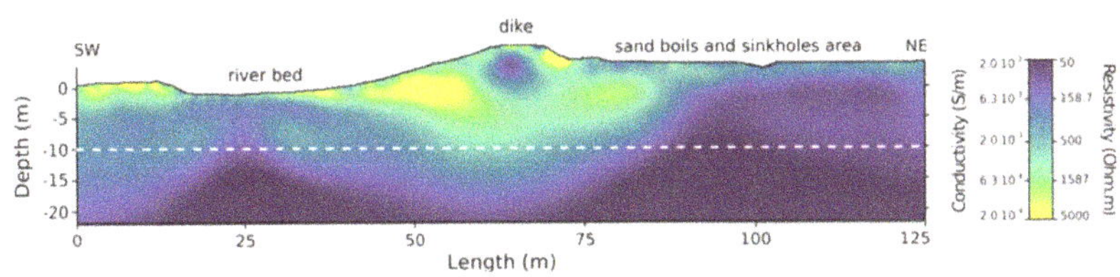

Figure 6. ERT profile along L3, exposing the soil conductivity and resistivity. The white dashed line represents the position of the groundwater. The minor bed was dried up at the time of measurement.

This result shows that paleo-sediments become saturated during a flood and form underlying seepage flows from the diked bed to the protected area. These flows travel under the dikes and circulate into the underlying soil of the protected plain. This is confirmed by profiles L2 (Figure 5) and L3 (Figure 6), which extend beyond around 60 m on each side. However, the geometry of the permeable layer in the subsoil evolves from upstream to downstream, as shown by profiles L4 and L5 (Figures 7 and 8, respectively) that cover the 60 m long area from the dike toe to the protected plain.

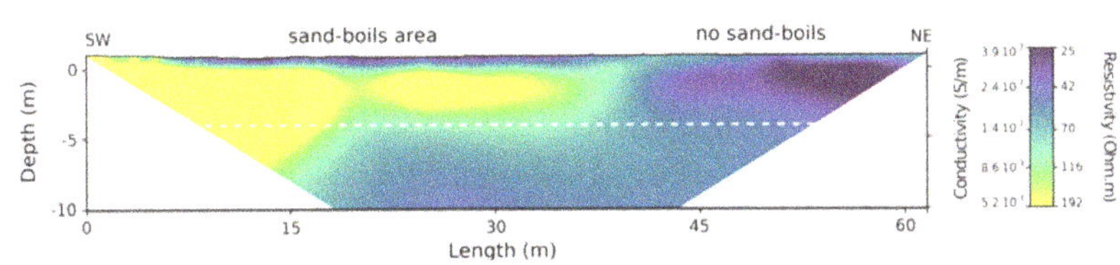

Figure 7. ERT profile along L4, exposing the soil conductivity and resistivity. The white dashed line represents the position of the groundwater. The river was flowing in its minor bed at the time of measurement.

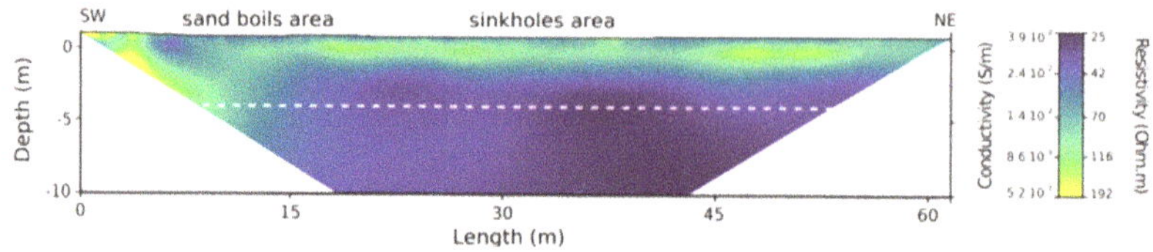

Figure 8. ERT profile along L5, exposing the soil conductivity and resistivity. The white dashed line represents the position of the groundwater. The river was flowing in its minor bed at the time of measurement.

On the site investigated, the followed methodology can be summarized as follows: (i) lateral mapping of averaged soil conductivity (based on EMI) in the protected plains. This mapping showed the presence of permeable soils and identifies vertical interfaces between high permeability and low-permeability soils. (ii) Local cross-section mapping (based on ERT). This mapping confirmed the presence of permeable soils and positions of interfaces. It was carried out at different scales to confirm the results, bearing in mind that ERT alone only provides electrical resistivity gradients, not material interfaces. The key point here was that the relationship between electrical resistivity and permeability was deduced from the analysis of cored soils.

5. Results Analysis

5.1. Analysis of EMI Results

The purpose of averaged conductivity measurements provided by EMI is to laterally delimit layers of different composition, each being roughly homogeneous over the first 6 m deep, in order to have an overview of the situation. In the study area, two different geological structures can be distinguished in the flood plain (Figure 3):

- a sandy, permeable layer (paleo-sediments) that expand from the levee toe, to up to a few tens of meters on the protected area where leaks and sand boils are observed after each flooding event.
- a sandy, permeable layer (paleo-sediments) extending from the levee toe to up to around 20 m on the plain where leaks, sand boils and sinkholes were observed after the 2013 flooding event.

5.2. Analysis of ERT Results on Leaks and Sand Boils Locations

The analysis of ERT measurements, shown in Figure 9a, exposes the geometry of this sandy permeable layer, termed a paleo-valley, filled with sediments beneath the levee. This layer is around 15–20 m thick and extends for around 60 m on both sides into the protected plain. Beyond this limit is a less-permeable silty layer, corresponding to the upwelling of a deeper material that reaches the surface, and which may constitute an impermeable barrier preventing the progression of internal flows during flooding events.

Underlying seepage flows have thus two possibilities (Figure 9a): flowing laterally, parallel to the dikes, towards the river mouth (located at around 8 km in a south-east direction), and ascending towards the surface. In any case, the flow can no longer continue in the direction transverse to the levee, since the interface between the permeable and impermeable layer is approximately vertical and extends to the surface. Near the surface, the presence of a thin layer of silty soil, less permeable, promotes the initiation of a heterogeneous fluidization through the formation of more permeable chimneys. This situation corresponds to the conditions of appearance of leaks and sand boils, which can only occur within a thick permeable layer covered by a thin heterogeneous and less permeable one, that widens towards the south-east (Figure 3b).

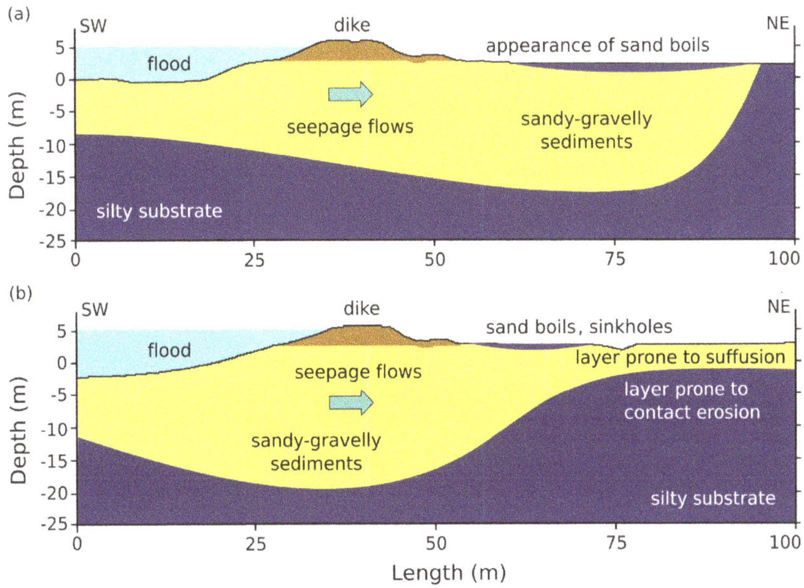

Figure 9. (**a**) ERT results analysis (L2,L4) highlighting the triggering mechanism and possible area of occurrence of leaks and sand boils. (**b**) ERT results analysis along L3, L5 highlighting the three triggering mechanisms and possible area of occurrence of leaks, sand boils and sinkholes above the thinner paleo-channel.

5.3. Analysis of ERT Results on Leaks, Sand Boils and Sinkholes Locations

Interpretation of ERT measurements in Figure 9b shows the geometry of the permeable layer beneath the levee. Its thickness decreases progressively with distance from the levee (from around 10 m to 2 m thick), then becoming approximately constant (when forming a paleo-channel). Beneath this permeable layer lies a less permeable substrate. This means that the internal flow passing beneath the levee can progress into the protected area, over 2 m deep (i.e., five times thinner) and up to 60 m long.

In this area, underlying seepage flows have three possibilities: flowing laterally, parallel to the levee; ascending towards the surface; or progressing in the protected plain until reaching the impermeable limit at a higher velocity due to the reduced thickness of the permeable layer.

Note that underlying seepage flows, parallel to the embankment, may be of less importance when the geometry of interfaces is locally assumed to be parallel to the dike. Otherwise, ascending flows are necessary to explain the appearance of leaks and sand boils at the surface during floods and may be promoted by the presence of an impermeable obstacle formed by the upwelling silty substrate that stops its lateral progression. A thinner surface soil layer, as well as a local hydraulic gradient greater than the initial one, also represent triggering factors for leaks and sand boils formation. Lateral flows, propagating towards the protected area and perpendicular to the levee, are likely to have high local velocity capable of inducing two other types of erosion (Figure 9b): (i) a selective internal erosion (suffusion) in the permeable layer that will locally increase the volume of the pore domain and give the material a possible mechanical instability. The local decompaction of the permeable layer may cause the collapse of its upper limit, as well as the above soil layer and the formation of a sinkhole at the surface. (ii) Contact erosion of the less permeable material, at the interface with the permeable material in depth. The eroded fine soil is transported by the flow into the permeable layer, which locally induces cavities, that if located superficially (i.e., about 2 m deep), can promote the soil layer collapse and the formation of sinkholes at the surface.

6. Discussion

6.1. Remarks on the Influence of Temperature and Water Content on Resistivity Variations

Although ERT and EMI measurements have been made within two different years, inherent time lapses were not limiting for a relative comparison of the geophysical properties between the different techniques and profiles. As the topography of the study area remains flat, most of the variations of the geophysical properties over time are related to seasonal variations (i.e., temperature and water content). Moreover, to correctly interpret values of electrical resistivity, in terms of soil types, it is important to know the factors influencing resistivity. For a given soil, electrical resistivity decreases with temperature, water content and density. The soils were investigated during spring, with a groundwater located from 5 to 9 m deep. Being highly permeable, the constituting material can be considered as dry, so that variations in resistivity are not attributable to water content for depths up to 5 m. The ambient temperature during measurements ranged from 20 °C to 30 °C. The temperature below 10 m can be estimated to be around 10 °C. The order of magnitude of the influence of the temperature gradient in the soil is therefore 50% on variations in resistivity measurements.

For a given soil, subjected to identical conditions of temperature and water content, values of electrical resistivity may decrease with density. ERT measurements are not well suited to investigate such density gradients as well as to detect the presence of voids; this may require a combination of Ground-Penetrating Radar (GPR) and seismic methods that will be the topic of future research. Additionally, repeating the same ERT profiles for different hydrogeological conditions can improve the knowledge of the soil layers and the location of groundwater bodies during floods or discharge events [33].

Under constant temperature and hydrogeological conditions, electrical resistivity decreases with the volume percent of clays. Analyses of sediment cores and drillings, made on the foundation soils of Agly dikes, expose a composition depleted in clays (less than few %) [7,8,30], which justifies the use of both EMI and ERT here, along with sediment sampling, to map the presence of paleo-valleys and paleo-channels and correlate them to erosion symptoms observed on protected areas.

Observations of leaks, sand boils and sinkholes also involve a combination of several physical processes in the subsurface soil, as well as a complex geometry of permeable and less permeable soil layers. Such processes probably do not consist of only one type of internal erosion, but rather several interacting mechanisms, such as suffusion and contact erosion. Erosion phenomena are described by threshold laws, which requires numerical modeling involving input data that accurately describe the investigated situation, and in particular, the geometry of soil layers and interfaces.

6.2. Numerical Modeling: Example of Results and Comments on Hydraulic Modeling Based on ERT Measurements

Two-dimensional numerical modeling using finite elements was carried out on simplified geometries deduced from the analysis of geophysical results and available soil cores. The mesh comprises 7357 linear elements. Dimensions, boundary conditions and permeabilities are given in Figure 10. The dike is assumed to be of 2 m high, while the water level is at crest on the river side and is taken to be 5 m below the natural ground on the protected area. The surface layer of sandy silt is e = 1 m thick. The two cases shown in Figure 9 were modeled. Figure 10 plots the flow vectors $\boldsymbol{q}$, and the pressure $p(x)$ beneath the surface layer of sandy silt. In both cases, $p(x)$ is very high and can exceed 1.5 m, which turns out to be a favorable condition for the appearance of a resurgence likely to cause erosion. However, in case (a) (Figure 10b), the curve $p(x)$ quickly tends towards zero after the vertical interface, pointing that erosion signatures can appear only between the dike and the vertical interface. In case (b), the values of $p(x)$ remain significant after the vertical interface, showing that erosion signatures can appear over a wider area, extending beyond the vertical interface. The flow intensity in the gravelly sand layer in the case (b) (Figure 10c) is of one order of magnitude greater than in case (a) (Figure 10b). However, in both cases, flow velocities can be locally of the same order of magnitude as those required for the initiation of suffusion (10^{-6} to 10^{-5} m/s, [34]) in this layer. Carrying out modeling to integrate the different types of erosion requires a finer description, which is beyond the scope of this paper.

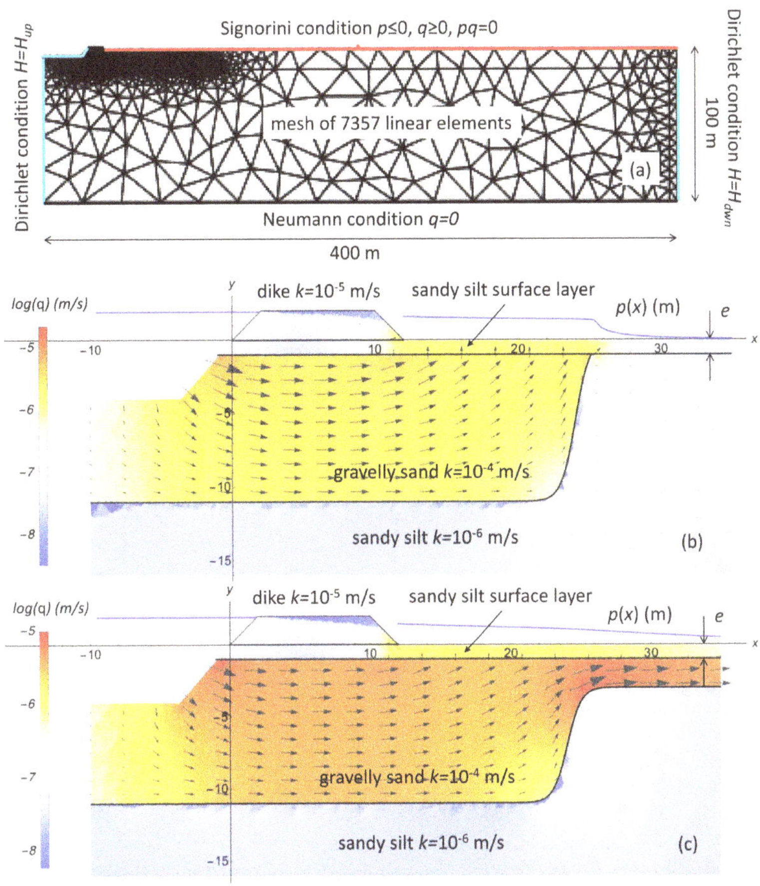

Figure 10. Finite elements modeling implying a simplified geometry. (**a**) Mesh and boundary conditions. (**b**,**c**) Results of the numerical model: flow vectors q, isocolor of $\log||q||$, and pressure $p(x) = H(x) - e$ (m) beneath the sandy silt surface layer, of thickness e, for geometries observed and reported on Figure 9a (**b**) and Figure 9b (**c**), where H is the hydraulic head, and k is the permeability.

Specific work is necessary to quantify the permeability of the soils in place, in coherence with the electrical resistivity tomograms, and allow Finite Element Modeling (FEM) of internal flows. In this context, in situ sampling by core drilling in the areas of interest is necessary in order to carry out identification and laboratory tests (particle size analysis and permeability tests). In this example, three-dimensional FEM modeling is necessary to account for the possibility of horizontal flows parallel to the levee.

The permeable zones here are correlated with the presence of a paleo-valley filled with sandy-gravelly sediments. The latter must be detected as a priority, by analyzing the geomorphological history of the river and the dike system. The combined use of EMI and

ERT methods also represents a rapid and cost-effective solution to continuously map the subsurface and effectively capture shallow lateral information [22,34]. The key outcome here is a correct description of the geometric variability of the soil layers under the dike and in the protected area. Geophysical imagery is often affected by a lack of resolution, and the geometry of interfaces is directly estimated by these methods, which inherently lack accuracy. In addition to EMI and ERT, results of geotechnical investigations, e.g., Cone Penetration Test (CPT), core sampling and borehole drilling logs, should be used to directly identify interfaces between soil layers [35].

An accurate description of the layers geometry and their interfaces is crucial to assess the reliability of the diking system, by quantifying the occurrence of the different erosion processes and their kinetics. In the case of the Agly River dikes, flood duration (lasting up to a few hours) may involve high erosion kinetics, which implies high local flow velocities. These latter require more elaborated modeling than a simple approach assuming a simplified geometry and an evaluation of the averaged hydraulic gradients, as it is often the case.

7. Conclusions

Visual observations of leaks, sand boils and sinkholes in the protected area provide evidence of internal erosion processes in the underground soil. Local geological conditions are part of the information to be sought to explain these processes: presence of permeable soils and position of interfaces. This information was obtained for the studied area by combining two classical geophysical methods (EMI and ERT). Results were interpreted using the analysis of cored soils, in order to correlate the electrical resistivity values to the nature of soils (including the orders of magnitude of permeabilities), and the position of interfaces.

Observations highlight the presence of a gravelly sand layer beneath the river bed and dikes, approximately 300 m wide and 25 m deep, which is comparable to the deposits that filled a paleo-valley resulting from the geomorphological history of the river. Important characteristics were observed: the presence of a surface layer of low permeability made with sandy silt (from 0.5 to 1 m thick), the extension of the gravelly sandy layer (of around 12 to 25 m thick) towards the protected zone, which defines a vertical interface, located at a variable distance from dikes, that preferentially guides the flows towards the surface and provide elements to explain the spatial distribution of erosion signatures during floods.

Given the nature of soils and the heterogeneity of flows, simple occurrence analyses of backward erosion piping are not sufficient to describe these erosion signatures. The possibility of internal erosion, such as suffusion or contact erosion, must also be considered. It is also necessary to consider the granulometry of the materials, and to model the initiation and evolution of erosion processes through further research.

The findings of this work highlight the importance of geological conditions to explain the formation of leaks, sand boils and sinkholes. The presence of a paleo-valley, filled with gravelly-sandy sediments beneath the river bed and dikes, turns out to be frequently encountered, suggesting that the methodology and results presented here are likely to apply to many dikes. EMI and ERT methods appear relevant for the site studied. Possible additional methods can be considered for the area of interest. Imaging results are often affected by a lack of resolution, which means that interfaces cannot be accurately located. Two conditions are necessary for success: a strong contrast between weakly and highly permeable areas, and a direct analysis from geotechnical investigations (Cone Penetration Test CPT, core sampling and borehole drilling logs).

Author Contributions: Both conception of the research work, measurements, analysis, interpretations and writing the manuscript have been made in close collaboration between all the authors. All authors have read and agreed to the published version of the manuscript.

Funding: This research was supported by INRAE (grant from Scientific bet of INRAE, Water department), Region Centre-Val de Loire (contribution of Academic Initiative Project RHEFLEXES/201900134935), and grant from Directorate General for Risk Prevention (DGPR).

Data Availability Statement: Not applicable.

Acknowledgments: We thank the managers of the Agly River dikes (SMBVA, F. Nicoleau) for having provided authorization for this work and for fruitful discussions. We thank the company Naga Geophysics for L4-L5 profiles, as part of a service order on behalf of INRAE; as well as the company Sub-C Marine for the DTM aerial lidar, as part of a service order on be-half of INRAE. We thank Laurent Peyras for his precious advices concerning the preparation of the manuscript.

Conflicts of Interest: The authors declare that they have no known competing financial interests or personal relationships that could have appeared to influence the work reported in this paper.

References

1. Foster, M.; Fell, R.; Spannagle, M. The statistics of embankment dam failures and accidents. *Can. Geotech. J.* **2000**, *37*, 1000–1024. [CrossRef]
2. Danka, J.; Zhang, L.M. Dike failure mechanisms and breaching parameters. *J. Geotech. Geoenviron. Eng.* **2015**, *141*, 04015039. [CrossRef]
3. Amin, A.; Girolami, L.; Risso, F. On the fluidization/sedimentation velocity of a homogeneous suspension in a low-inertia fluid. *Powder Technol.* **2021**, *391*, 1–10. [CrossRef]
4. Fell, R.; Fry, J.-J. (Eds.) *Internal Erosion of Dams and Their Foundations*; Taylor & Francis: London, UK, 2007.
5. Bonelli, S. (Ed.) *Erosion of Geomaterials*; John Wiley & Sons: Hoboken, NJ, USA, 2012; 371p.
6. Bonelli, S. (Ed.) *Erosion in Geomechanics Applied to Dams and Levees*; John Wiley & Sons: Hoboken, NJ, USA, 2013; 388p.
7. Zwanenburg, C.; López-Acosta, N.P.; Tourment, R.; Tarantino, A.; Pozzato, A.; Pinto, A. Lessons Learned from Dike Failures in Recent Decades. *Int. J. Geoengin. Case Hist.* **2017**, *4*, 203–229.
8. Van, M.A.; Rosenbrand, E.; Tourment, R.; Smith, P.; Zwanenburg, C. Failure paths for levees. International Society of Soil mechanics and Geotechnical Engineering (ISSMGE)—Technical Committee TC201 'Geotechnical aspects of dikes and levees'. 2022.
9. Richards, K.S.; Reddy, K.R. Critical appraisal of piping phenomena in earth dams. *Bull. Eng. Geol. Environ.* **2007**, *66*, 381–402. [CrossRef]
10. Holzer, T.L.; Clark, M.M. Sand boils without earthquakes. *Geology* **1993**, *21*, 873–876. [CrossRef]
11. van Beek, V.M. Backward Erosion Piping: Initiation and Progression. Ph.D. Thesis, Technische Universiteit Delft, Delft, The Netherlands, 2015.
12. Robbins, B.A.; van Beek, V.M. Backward erosion piping: A historical review and discussion of influential factors. In Proceedings of the ASDO Dam Safety Conference, New Orleans, LA, USA, 13–17 September 2015; pp. 1–20.
13. Luo, G.; Rice, J.D.; Peng, S.; Cao, H.; Pan, H.; Xu, G. Modelling Initiation Stage of Backward Erosion Piping through Analytical Models. *Land* **2022**, *11*, 1970. [CrossRef]
14. Pan, H.; Rice, J.D.; Peng, S.; Cao, H.; Luo, G. Analytical Modeling with Laboratory Data and Observations of the Mechanisms of Backward Erosion Piping Development. *Water* **2022**, *14*, 3420. [CrossRef]
15. DeHaan, H.; Stamper, J.; Walters, B. *Mississippi River and Tributaries System 2011 Post-Flood Report*; USACE, Mississippi Valley Division: Vicksburg, MS, USA, 2012.
16. Glynn, E.; Quinn, M.; Kuszmaul, J. Predicting piping potential along Middle Mississippi River Levees. In Proceedings of the 6th International Conference on Scour and Erosion, Paris, France, 27–31 August 2012; pp. 1473–1480.
17. Semmens, S.N.; Zhou, W. Evaluation of environmental predictors for sand boil formation: Rhine–Meuse Delta, Netherlands. *Environ. Earth Sci.* **2019**, *78*, 1–11. [CrossRef]
18. Wolff, T.F. *Performance of Levee Underseepage Controls: A Critical Review*; Rep. No. ERDC/GSL TR-02-19; USACE: Washington, DC, USA, 2002.
19. Garcia Martinez, M.F.; Tonni, L.; Marchi, M.; Tozzi, S.; Gottardi, G. Numerical Tool for Prediction of Sand Boil Reactivations near River Embankments. *J. Geotech. Geoenviron. Eng.* **2020**, *146*, 06020023. [CrossRef]
20. Gutierrez, F. Sinkhole Hazards. In *Oxford Research Encyclopedia of Natural Hazard Science*; Oxford University Press: Oxford, UK, 2016; pp. 1–92.
21. Nguyen, T.K.; Benahmed, N.; Hicher, P.Y.; Nicolas, M. The Influence of Fines Content on the Onset of Instability and Critical State Line of Silty Sand. In *Bifurcation and Degradation of Geomaterials in the New Millennium, IWBDG 2014*; Chau, K.T., Zhao, J., Eds.; Springer Series in Geomechanics and Geoengineering; Springer: Berlin/Heidelberg, Germany, 2015.
22. Valois, R.; Camerlynck, C.; Dhemaied, A.; Guerin, R.; Hovhannissian, G.; Plagnes, V.; Rejiba, F.; Robain, H. Assessment of doline geometry using geophysics on the Quercy plateau karst (South France). *Earth Surf. Process. Landf.* **2011**, *36*, 1183–1192. [CrossRef]
23. Philippe, P.; Beguin, R.; Faure, Y.H. Contact erosion. In *Erosion in Geomechanics Applied to Dams and Levees*; John Wiley & Sons: Hoboken, NJ, USA, 2013; pp. 101–192.

24. Bianchi, E.; Borgatti, L.; Vittuari, L. The Medium- to Long-Term Effects of Soil Liquefaction in the Po Plain (Italy). In *Engineering Geology for Society and Territory*; Lollino, G., Ed.; Springer International Publishing: Cham, Switzerland, 2015; Volume 6, pp. 421–425.
25. Morton, L.W.; Olson, K.R. Sinkholes and sand boils during 2011 record flooding in Cairo (Illinois). *J. Soil Water Conserv.* **2015**, *70*, 49A–54A. [CrossRef]
26. Garner, S.J.; Fannin, R.J. Understanding internal erosion: A decade of research following a sinkhole event. *Hydropower Dams Int.* **2010**, *17*, 93–98.
27. Yin, Z.-Y.; Yang, J.; Laouafa, F.; Hicher, P.-Y. A framework for coupled hydro-mechanical continuous modelling of gap-graded granular soils subjected to suffusion. *Eur. J. Environ. Civ. Eng.* **2023**, *27*, 2678–2699. [CrossRef]
28. Dezert, T.; Fargier, Y.; Palma Lopes, S.; Cote, P. Geophysical and Geotechnical methods for fluvial levee investigation: A review. In *Engineering Geology*; Elsevier: Amsterdam, The Netherlands, 2019; 18p.
29. Karim, M.Z.; Tucker-Kulesza, S.E.; Rutherford, C.J.; Bernhardt-Barry, M. Geophysical Engineering to Identify Seepage Channels in the Hager Slough Levee. In *Eighth International Conference on Case Histories in Geotechnical Engineering*; Geotechnical Special Publications GSP 311; American Society of Civil Engineers: Reston, VA, USA, 2019.
30. Tourment, R.; Benahmed, N.; Nicaise, S.; Meriaux, P.; Salmi, A.; Rougé, M. Lessons learned on the damaged on the levees of the Agly River, analysis of the sand-boils phenomena, Q. 103 R.21. In Proceedings of the 26th ICOLD Congress, Vienna, Austria, 1–7 July 2018.
31. Puig, C.; Carozza, J.-M. Les changements de tracés des cours d'eau d'après les sources historiques et géomorphologiques dans la plaine du Roussillon depuis le XIIe siècle: Approche théorique et premiers résultats. In *Les Plaines Littorales en Méditerranée Nord-Occidentale, Regards Croisés D'histoire, D'archéologie et de Géographie, de la Protohistoire au Moyen Age*; Ropiot, V., Puig, C., Mazière, F., Eds.; Monique Mergoil: Paris, France; pp. 297–312.
32. Blanchy, G.; Saneiyan, S.; Boyd, J.; McLachlan, P.; Binley, A. ResIPy, an intuitive open source software for complex geoelectrical inversion/modeling. *Comput. Geosci.* **2020**, *137*, 104423. [CrossRef]
33. Côté, S. Internal Stability Criteria for Low-Plasticity Materials Subjected to Water Flow. Master's Thesis, University of Laval, Québec, QC, Canada, 2010; 237p. (In French).
34. Valois, R.; Galibert, P.Y.; Guerin, R.; Plagnes, V. Application of combined time-lapse seismic refraction and electrical resistivity tomography to the analysis of infiltration and dissolution processes in the epikarst of the Causse du Larzac (France). *Near Surf. Geophys.* **2016**, *14*, 13–22. [CrossRef]
35. Chavez Olalla, J.; Winkels, T.G.; Ngan-Tillard, D.J.M.; Heimovaara, T.J. Geophysical tomography as a tool to estimate the geometry of soil layers: Relevance for the reliability assessment of dikes. *Georisk Assess. Manag. Risk Eng. Syst. Geohazards* **2022**, *16*, 678–698. [CrossRef]

water

MDPI

Article

Flooding Depth and Flooding Duration with the Zonation of Riparian Plant Communities in the Three Gorges Reservoir of China

Xuemei Yi [1], Yuanyang Huang [2], Yi Jiang [1], Maohua Ma [1], Qiao Chen [1,*] and Shengjun Wu [1]

1 Chongqing Institute of Green and Intelligent Technology, Chinese Academy of Sciences, Chongqing 400714, China; yixuemei@cigit.ac.cn (X.Y.)
2 Chongqing Water Resources and Electric Engineering College, Chongqing 402160, China
* Correspondence: chenqiao@cigit.ac.cn; Tel.: +86-13301856629

Abstract: The hydraulics of flows, especially the flooding process, influence the patterns of riparian plant zonation. Different characteristics of the flooding process should be analyzed to correlate plant zonation with flooding due to their different effect modes. The effects of flooding characteristics on riparian plants have yet to be studied, especially in the field. Thus, two elements of the flow regime, flooding duration and depth, were analyzed in relation to the riparian plants of the Three Gorges Reservoir. The taxonomic indices and the functional diversity of the riparian plants in three seasons in 2019 and the corresponding inundation character were surveyed. Our results showed that the riparian plant diversity and functional diversity varied by season. A significant negative relationship between plant diversity and flooding depth was observed, while flooding duration was not a significant predictor in different seasons. The greater explanatory capacity of flooding depth than that of flooding duration suggests that flooding depth could be a better indicator of the zonation of the riparian vegetation in this area. Concerning the vital component of flow hydraulics, growing opportunities to study flooding depth and strategies that consider both flooding time and flooding depth in a reservoir should be offered, as they will assist in refining process-based river restoration.

Keywords: flooding depth; flooding duration; riparian zone; plant zonation; Three Gorges Reservoir

1. Introduction

Riparian zones serve as a framework for understanding how the organization, diversity, and dynamics of communities associated with fluvial ecosystems represent an unusually diverse mosaic of landforms, communities, and environments within a larger landscape [1,2]. The water regime of a riparian zone is a significant determinant of plant zonation patterns and is essential to river life, but it can also be a source of stress. Flow spatial- temporal patterns from local to regional scales exert direct and indirect control over plant communities [3]. River regulation affects riparian plants by modifying water level fluctuations via altering flow regimes [4,5], especially for the large dams.

With intensive human intervention and occupation, natural river regimes are gradually being replaced by regulated flow, thereby dramatically affecting riparian zones and stoking international concern [6–8]. The impacts caused by dam construction include changing river sediment transportation and scouring–silting patterns and effects on biological activities, downstream fluvial geomorphology, and sedimentary environments [9,10]. The Three Gorges Dam (TGD) has entered its twentieth year after its first turbine operation in June 2003. The dam, which has 20 times the power generation capacity of the Hoover Dam in the United States, has been hailed as a key component of solving China's energy crisis [11]. However, major concerns about this large-scale engineering project's social consequences and catastrophic environmental repercussions, especially with respect to its impact on the nearby riparian ecosystem, have been expressed despite its benefits.

After a dam's construction, the habitat-changing characteristics of the neighboring riparian zone are significantly related to the magnitude and frequency of flow duration [12]. The Three Gorges Dam, the biggest dam in China, caused the Three Gorges Reservoir (TGR) to experience a seasonal reversal of its hydrological regime [13]. The riparian plants of the TGR then faced inundation for up to several months [14], which resulted in a typical degree of plant zonation. Plant species zonation is a characteristic feature of water depth gradients in wetland environments and lake shorelines due to the wide availability of water for growth in these areas [15]. It has been reported that plant diversity and species richness are negatively affected by frequent flooding [16]. With the severe stress of a riparian zone, riparian vegetation species must complete their life cycles under a limited growth period, which can be as short as three months [17]. Riparian plants appear to have more distinctive preferences and distribution patterns after twenty years of reversed hydrological regime adjustment.

Different life-history strategies are needed to allow riparian plants to adapt to a varying environment. Flow hydraulics influence the vegetation distribution and zonation in riparian zones. This is especially the case for flooding processes, which bring about scour and deposition [18,19]. Flooding and water level fluctuations disturb riparian landscapes by determining the amount of exposed land and affect riverine vegetation by imposing physiological constraints and precipitating losses in cover and species diversity [20]. How hydrological characteristics interact with and affect riparian zones is complex, particularly with respect to seasonal variations in flow and alternating wet and dry cycles [21]. Flow regulation can lead to changes in the responses of riparian plant species guilds [20,22,23], it has motivated research on the link between hydrologic alterations and biota [3,24,25]. Different life-history strategies are needed for riparian plants to adapt to varying environments.

With intensifying human demands for water and the continued alteration of rivers by humans, there is a growing need to predict vegetation responses to flow alteration [26], including responses related to the flooding process. The different aspects of flooding dynamics should be analyzed separately to correlate riparian plant zonation with flooding [27]. Flooding duration has attracted more attention than other factors in riparian plant research because of the unnatural long-term flooding of the TGR, where high-level water fluctuations result in an extremely high flooding depth. Such flooding depths change the physical and chemical composition [28] of a riparian zone's soil, which affects the area's riparian plants. Research linking flood factors with riparian vegetation have shown the importance of flooding depth [29,30]. Current research focuses on the effects of long-term flooding [16,31] on riparian plants and the effects of flooding depth on plant traits in the laboratory [32,33], but the plant community's response to flooding depth lacks attention in the field, especially regarding the extreme flooding caused by big dams. Inundation with seasonal variation presents a different impact compared to flooding events over a long period [34]. Whole-year data, including on seasons, appear to be more explanatory than growing season data [27].

Despite the importance of flooding depth for the riparian plants in the TGR, little information is available about the effects of flooding depth on this plant community as assessed in the field, particularly when considering seasonal variation. Therefore, a study of inundation characteristics and the corresponding plant community was conducted in the riparian zone of the TGR. The taxonomic indices and the functional diversity of the riparian plant communities surveyed in different seasons in 2019 were analyzed. The relationship between the inundation characteristics and the plant community in the riparian zone affected by the dam was studied. This study explores the correlation between the inundation factor and riparian plants to provide greater insight into riparian ecosystem management with flow regulation.

2. Materials and Methods

2.1. Study Sites

The study was conducted at the TGR of the Yangtze River in China (Figure 1). This area has a humid, subtropical monsoonal climate [35]. The area's average annual temperature is approximately 18 °C, and its annual precipitation level is about 1100 mm [36]. Monthly precipitation levels vary between 18.72 cm and 197.32 cm, with the maximum in July and the minimum in January (influenced by monsoons), presenting strong seasonality and significant inter-annual changes. Relative humidity ranges between 75.98% and 81.66%, and wind velocity ranges between 0.9 m/s and 1.24 m/s without a clear seasonal pattern [37]. The average annual frost-free period is 268 days, accounting for 73% of the total days in the year.

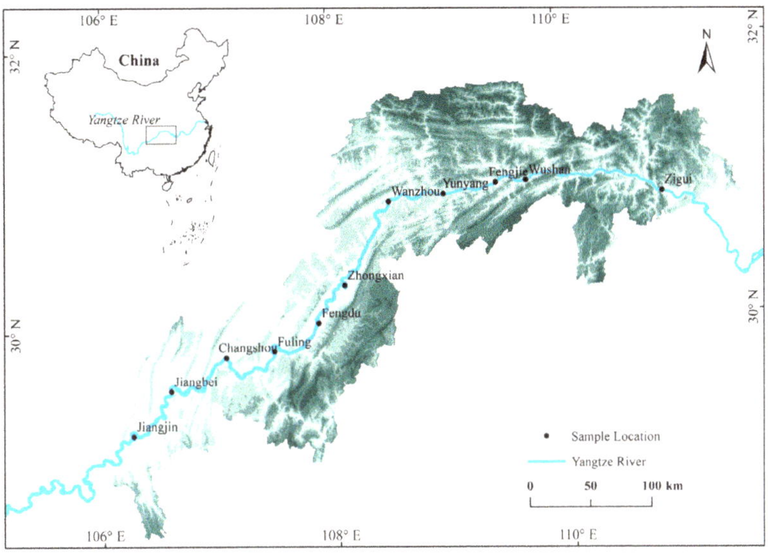

Figure 1. Location of the study site.

2.2. Vegetation Investigation

Twelve sites in the TGR ranging from Jiangjin (Chongqing) to Zigui (Hubei) were chosen to include a wide range of hydrologic and geomorphic conditions and vegetation types. At each site, three transects (145–155 m, 155–165 m, and 165–175 m for the sites from Changshou to Zigui and 155–165 m, 165–175 m, and 175–185 m for Jiangjin and Jiangbei) were delineated along an elevation gradient from the river to the band. Three quadrats measuring 1 m×1 m were established in each transect. The elevation of each quadrat was determined by applying their GPS positions to digital elevation maps using ArcGIS software 10.5 (ESRI Inc., Redlands, CA, USA). Detailed observation and sampling were conducted in three different seasons of 2019 based on the period when the riparian zone emerged from the flood (April, the early period; July, the middle period; and September, the end period). All plant species in the quadrats were identified at each quadrat. The number and coverage of each plant species were recorded. The plant species richness and plant species abundance were estimated. To avoid bias, the same botanist performed all of these vegetation assessments during the project. The site's topography, soil type, and ground cover were recorded along with these vegetation assessments.

Herbaceous plants, including *Cynodon dactylon*, *Cyperus rotundus*, *Echinochloa crusgalli*, *Bidens pilosa*, *Setaria viridis*, *Xanthium strumarium*, etc., are the dominant plants in the Three Gorges Reservoir riparian zone. Only a few arborvitae seedlings were found in the upper part of the riparian zone. They were not counted due to the predictably short period of their emergence. Functional traits such as dispersal type, growth form, life cycle, shoot height, and flowering phenology [38] (Table S2), which are considered to determine a plant's role in ecosystems and associated ecosystem services [39], were selected. The analyzed plant species' life forms and functional traits were recorded from the "Flora of China" (http://www.iplant.cn/, accessed on 7 September 2022).

2.3. Inundation Characteristics

The water levels of the TGR varied as a result of anti-seasonal impoundment (the water level was highest (175 m) in the winter and lowest (145 m) in the summer), thus producing a Water Level Fluctuation Zone (WLFZ) of approximately 350 km^2 between the minimum and maximum water level lines. Located at the end part of the reservoir, the water levels of Jiangjin and Jiangbei range from 160 m to 190 m. The water level in front of the TGD is based on the Wusong Elevation System. The water level at the closest hydrological station to the study sites was used as the site's river water level (Table S1). The flooding times were calculated using the total number of flooded days in one year (i.e., 365 days, ranging from the investigated date to the same date of the last year). The average flooding depth was defined as the average inundation depth in meters, which was averaged over the different inundation events within a year (from the investigated day to the same day of the previous year).

2.4. Data Analysis

Species richness, total coverage, and Shannon–Wiener Index H were used to analyze the plant community. Species richness is the number of different plant species recorded in the quadrats. Total coverage is the sum of the species coverage, which is the percentage of the quadrat area that is covered by one species. The Shannon–Wiener Index is a measure of the diversity of species in a community. The higher the value of H, the greater the species diversity in a given community. This index is calculated as follows:

$$\mathrm{H} = -\sum p_i * lnp_i \tag{1}$$

where H is the Shannon–Wiener diversity Index, and p_i is the relative proportion of individuals belonging to one of the species found. The lower the value of H, the lower the diversity. A value of H = 0 indicates a community with only one species.

The Rao's quadratic diversity index Q [40,41] accounts for the trait differences between species pairs. This index is defined as the expected dissimilarity between two individuals of a given species assemblage selected at random with replacement. The dissimilarity ranges from 0 to 1 and is based on specified functional traits or phylogenetic dissimilarity:

$$\mathrm{Q} = \sum_{ij}^{S} d_{ij} p_i p_j \tag{2}$$

where p_j is the relative abundance of species j (j = 1,2,...,S), while d_{ij} is the dissimilarity between species i and j ($d_{ij} \geq 0$, $d_{ii} = 0$).

The relationship between plant species richness and functional diversity was tested using three models [42], namely, linear, exponential, and sigmoid logistic models, to understand the consequences of disturbing the functioning of a given ecosystem. Model selection was performed according to Akaike's Information Criterion (AIC). The overall responses to flooding duration and flooding depth for riparian plant species richness and total coverage were examined via regression analysis and by using a set of linear models. Regression coefficients describe the relationship between a predictor variable and a response. In linear regression, coefficients are values that multiply the predictor values.

The data were analyzed using SPSS for Windows, Version 12 (SPSS. Inc., Chicago, IL, USA). Raw data of all the variables were checked for normal distribution rates using the one-sample Kolmogorov–Smirnov test as well as for homogeneity of the variances using Levene's test. A t-test and Analysis Of Variance (ANOVA) followed by Tukey's post hoc tests were used to compare the means between the different seasons. Pearson's correlation test was used to determine the associations between variables. Data were analyzed using R 4.1.2 (The R Foundation for Statistical Computing, Vienna, Austria), with statistical significance determined at $\alpha = 0.05$.

3. Results

The riparian plant community varied with the seasons (Figure 2). The diversity index was lowest in April and highest in September. Species richness in July and September was significantly higher than that in April. The lowest Shannon–Wiener Index value was observed in April, and it was significantly lower than that in September. The functional diversity was highest in September, and July had a significantly higher index than April. The total coverage in April was significantly lower than that in July and September. The plant communities presented the lowest species richness and total coverage in April; diversity and the functional diversity were also lower in this month.

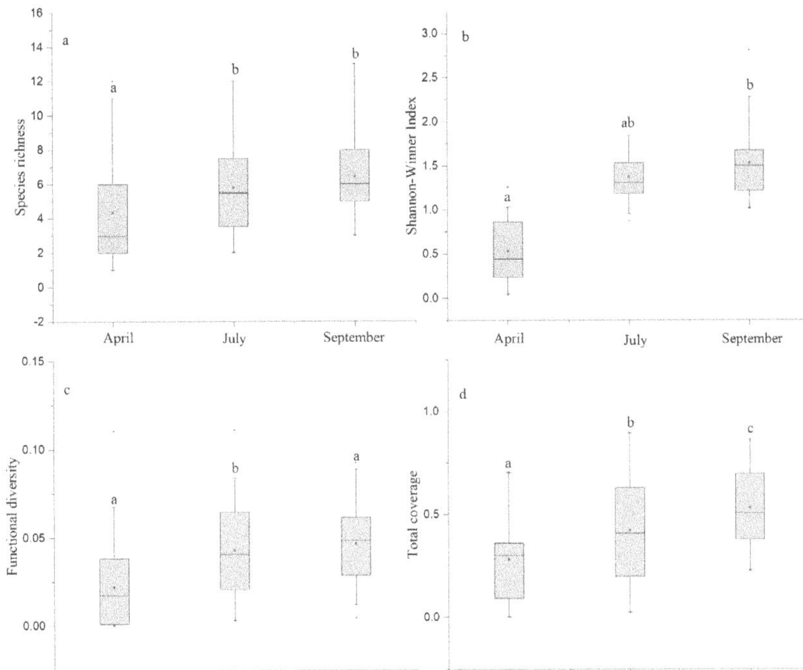

Figure 2. Species richness (**a**), Shannon Winner Index (**b**), Functional diversity (**c**) and total coverage (**d**) of riparian plants in different seasons (bars with different letters indicate statistically significant differences, bars with the same letters indicate differences that are not significant, and bars with "ab" indicate that there are no significant differences from the bars of both a and b ($p \leq 0.05$)).

The test showed a sigmoidal relationship between plant species richness and functional diversity in April, while a linear relationship was shown in July and September (Figure 3). The sigmoidal relationship in April showed a functional redundancy at low levels of species richness, followed by a rapid increase at intermediate levels until functional diversity reached an asymptote at a relatively high level. This linear relationship indicated that functional diversity increased as the species richness increased with a relatively steady functional redundancy.

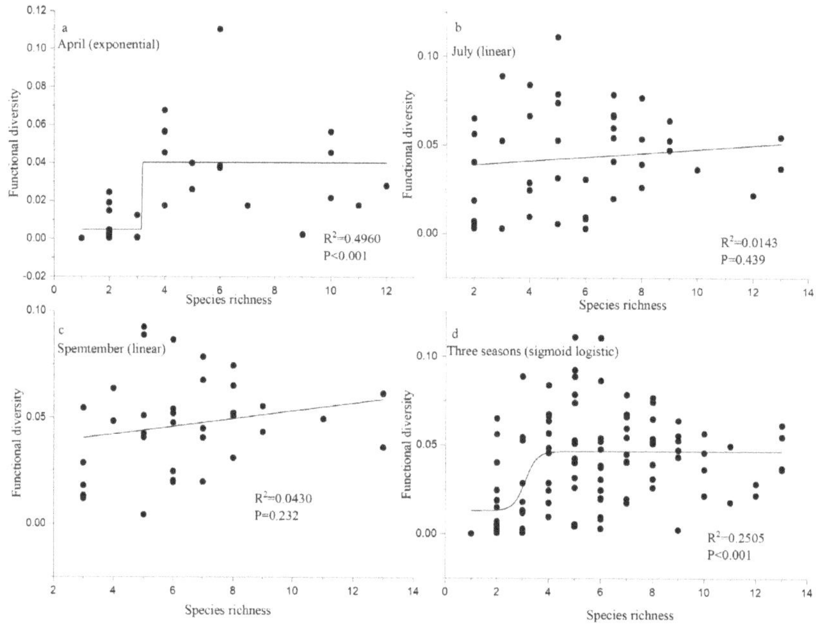

Figure 3. Relationship between species richness and functional diversity in different seasons ((**a**)—April, (**b**)—July, (**c**)—September, (**d**)—three seasons).

A significant negative relationship between the plant community and inundation characteristics was observed, as revealed via linear regression (Figure 4). Species richness and the total coverage of riparian plants decreased with increasing flooding days and depth. Aside from July, the correlation coefficient was higher between species richness and flooding time than total coverage. The total coverage of April was more correlated, i.e., had a higher correlation coefficient, with flooding depth than species richness. Flooding depth had a higher correlation coefficient in relation to plant communities than flooding duration, except for species richness in April and total coverage in July. The correlation coefficient was highest between flooding depth and species richness in April ($R^2 = 0.4636$, $p < 0.001$) and lowest between flooding duration and total coverage in September ($R^2 = 0.1128$, $p < 0.05$).

The relationship between the Shannon–Wiener Index and inundation characteristics revealed that the Shannon–Wiener Index decreased with increasing inundation characteristics (Figure 5). Flooding times in April had the highest correlation coefficient in relation to the Shannon–Wiener Index, while the lowest was in July. The correlation coefficient for the relationship between flooding time and the Shannon–Wiener Index in three seasons was higher than that in July and September but lower than that in April. Flooding depth showed a similar trend with respect to flooding time. The correlation coefficient for the relationship between flooding depth and the Shannon—Wiener Index was higher in September ($R^2 = 0.2036$, $p < 0.05$) and three seasons ($R^2 = 0.2075$, $p < 0.001$).

A regression model comparing flooding time and flooding depth with species richness and total coverage of riparian plants was developed to determine the significance of inundation character with respect to riparian plants' diversity. The regression model showed that flooding depth explained a more significant proportion of the overall variation in the riparian plant community than that of flooding duration in July and September (Table 1). It can be noticed that the flooding depth in July had the highest significant coefficient value ($p < 0.05$) in relation to explaining the species richness and total coverage of riparian plants, while the flooding time in July had the lowest value in this regard. Aside from flooding depth in April, Negative coefficient values were observed for inundation character in three months.

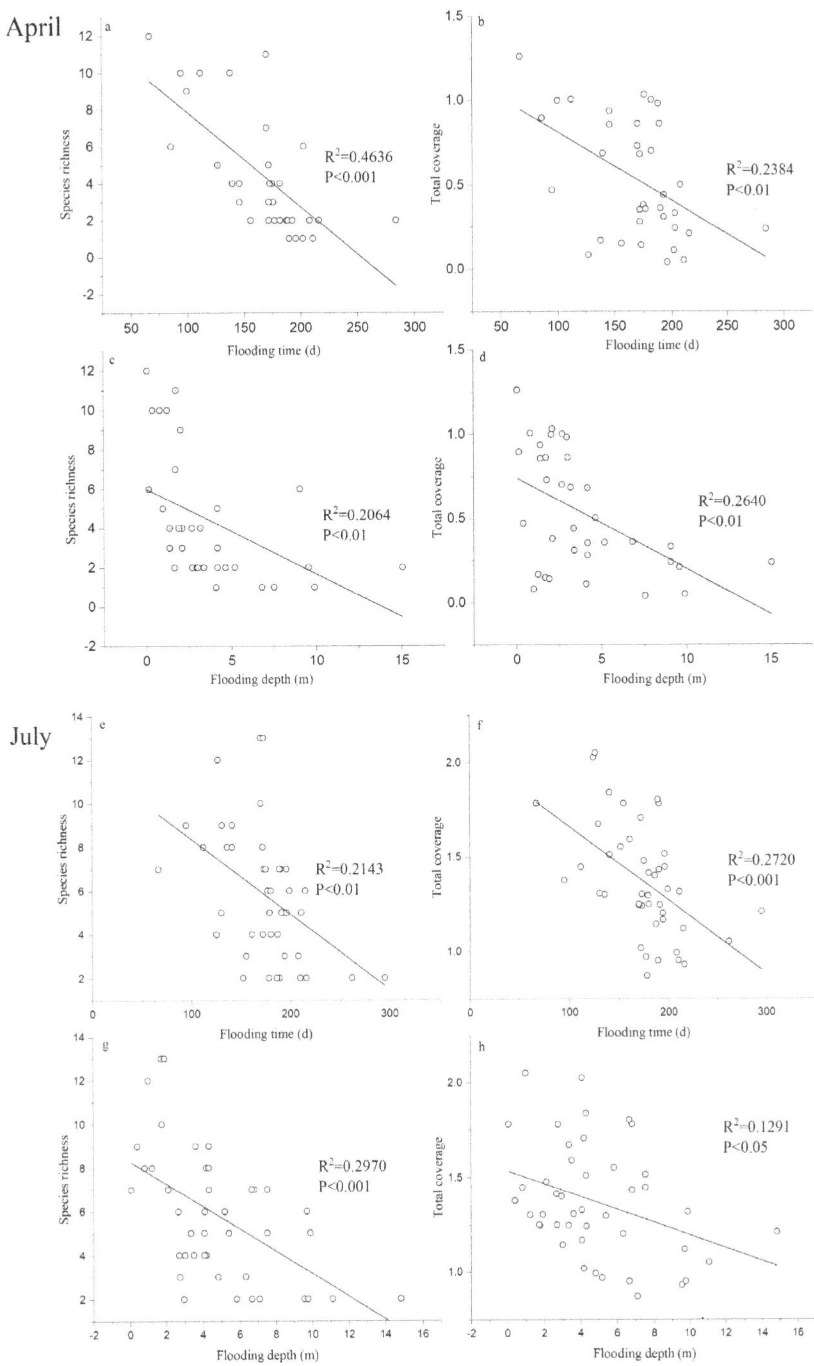

Figure 4. *Cont.*

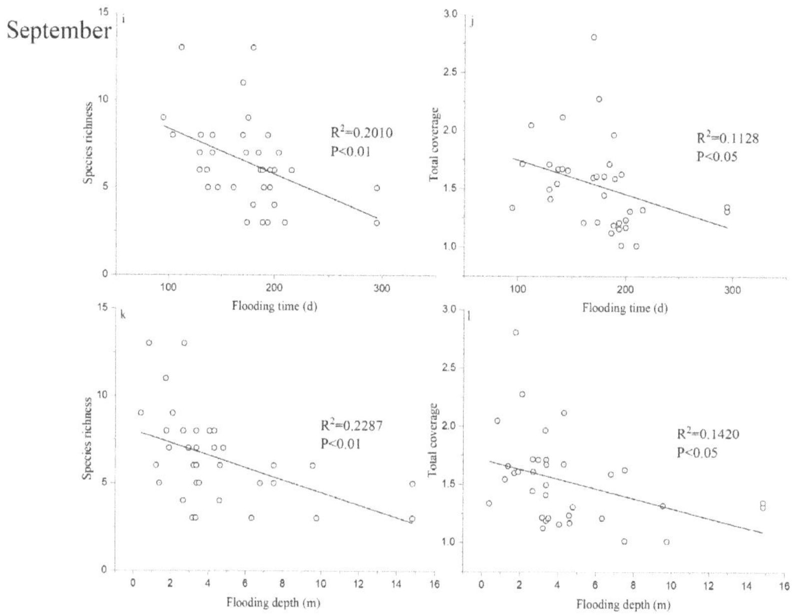

Figure 4. Species richness (**a**,**c**,**e**,**g**,**i**,**k**) and total coverage (**b**,**d**,**f**,**h**,**j**,**l**) versus inundation characteristics in different seasons.

Table 1. Regression coefficient of inundation character for riparian plants in different seasons.

		Estimate	Std. Error	*p*-Value
April	Intercept	15.587	2.202	$p < 0.001$
	flooding time	−0.068	0.016	0.000206 ***
	flooding depth	0.201	0.206	0.336
	regression equation	$Y = 15.587 - 0.068X_1 + 0.201X_2$, $p < 0.001$		
July	Intercept	11.064	2.222	$p < 0.001$
	flooding time	−0.010	0.017	0.560
	flooding depth	−0.447	0.213	0.042 *
	regression equation	$Y = 11.064 - 0.010X_1 - 0.447X_2$, $p < 0.001$		
September	Intercept	11.080	2.246	$p < 0.001$
	flooding time	−0.011	0.017	0.531
	flooding depth	−0.286	0.213	0.189
	regression equation	$Y = 11.080 - 0.011X_1 - 0.286X_2$, $p < 0.001$		

Notes: *** $p < 0.001$ and * $p < 0.05$.

Here, Y, denoting riparian plants, includes species richness and total coverage; X_1 denotes flooding time; and X_2 denotes flooding depth.

Std. Error is the standard error of the estimate, calculated as $S/\sqrt{n}$, where S is the sample standard deviation and n is the sample size.

A *p*-value is a standard notation used to denote probability. If the *p*-value is less than a certain significance level (e.g., $\alpha = 0.05$), then the predictor variable is said to have a statistically significant relationship with the response variable in the model.

The flooding times were calculated according to the total number of flooded days in one year (constituting the 365 days ranging from the investigated date to the same date of the previous year). Flooding depth was defined as the average inundation depth in meters, which was averaged over the different inundation events within a year (ranging from the investigated day to the same day in the previous year). The flooding time and depth were analyzed using R after a normality test.

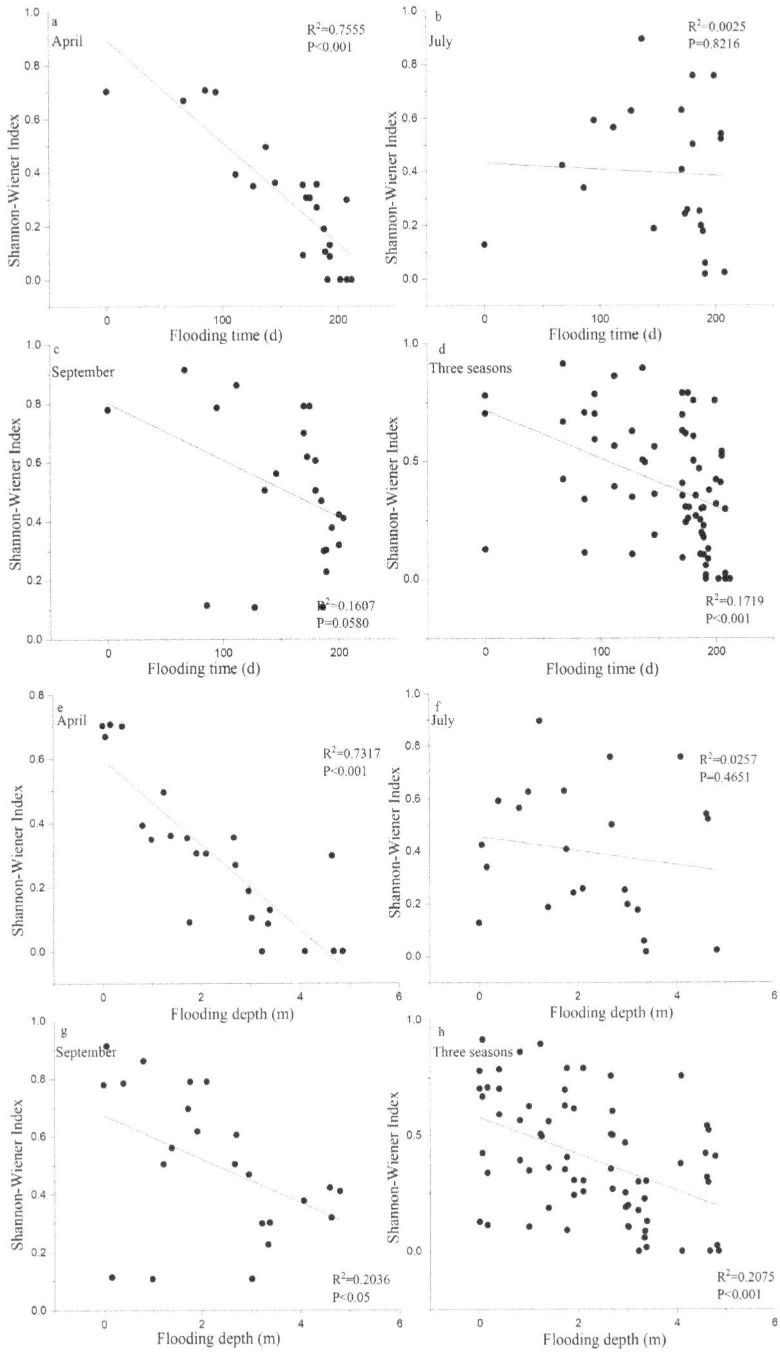

Figure 5. Shannon–Wiener Index versus inundation characteristics in different seasons ((**a**,**e**)—April, (**b**,**f**)—July, (**c**,**g**)—September,(**d**,**h**)—three seasons).

4. Discussion

Hydrological regimes are considered the main factor in determining the assembly of riparian plant communities [43,44]. When flow regulation alters flow regimes, biotic and abiotic pathways are triggered. As a result of flow regulation, riparian vegetation in the TGR area has changed in terms of its species composition, biomass, height, cover, and diversity [14]. Different vegetation types have various adaptive capacities with respect to water level fluctuations, and in general, perennial herbaceous plants are more tolerant of anti-seasonal inundation than trees [45]. The response of riparian vegetation also depends on the magnitude, frequency, and duration of water level fluctuations [46].

4.1. Plant Community Varied throughout the Seasons in the Riparian Zone

The analyzed riparian zone is subject to unnaturally prolonged flooding of up to 30 m caused by the Three Gorges Dam on submerging days. Hydrological regimes significantly impact riparian plants [21] in terms of five key flow characteristics, namely, timing, duration, frequency, rate of change, and magnitude, particularly regarding seasonal changes in flow and alternating wet and dry conditions. The length of a growing season is influenced by flow inundation characteristics, which determine the amount of exposed land during the summer or when plants can grow and reproduce. The climatic seasonal differences of a riparian zone can also affect plant growth and survival, leading to changes in species in the riparian zone. The analyzed riparian zone is drier in April, which can decrease the number of plant species that can survive in the area. The riparian zone may become wetter in July, which can increase the number of plant species that can survive in the area. The changes in plant species can have a significant impact on the ecosystem of a riparian zone and its surrounding areas. These factors can vary significantly within and between years, further increasing the disturbance of riparian ecosystems.

Flooding and water level fluctuations cause landscape disturbances that create a variety of habitats with different plant adaptations. In the emerging period of the growing season, the riparian zone is affected by dam regulation and natural seasonal flooding. The lower riparian zone is faced with a vigorous flooding intensity with a long duration, while the upper zone is faced with a relatively weaker intensity. Significant differences were observed in the riparian plant communities throughout the seasons. The highest index of plant community diversity was found in September; the most significant variation was observed in April, and the smallest variation was in July. The variability of this index shows that the variation coefficients in April were all greater than those of the other seasons. The plants' responses to the inundation varied because of their growth and survival sensitivity to timing and duration [47,48]. When emerging from inundation, few plants adapted to the flooding tended to appear in April, leading to low species richness and functional diversity. As the emerging time increased, more plant species appeared, and the diversity and functional diversity increased. The difference in the timing and duration of inundation made a difference in the riparian plant community.

4.2. Different Plant Community Constructions in Different Seasons of the Riparian Zone

Functional diversity was found to be sigmoidal related to species richness in April, while it was linear in July and September. Coupling flooding depth with duration, different construction processes were identified in the riparian zone. Submerged from the months of flooding, the severe stress caused by habitat change led to a subset of species having a limited number of shared functional traits of riparian plants in April. A relatively high turnover rate, a low recovery rate, and low stabilization to changes in taxonomic diversity were observed in the riparian plant community in April. Unique functions were added, with species added at a higher diversity level. This two-phase functional redundancy indicates a contrasting state caused by disturbance-resistant traits [42] appearing in April. Affected by the extremely long-term flooding, the lower riparian zone was relatively harsher for the plants than the higher riparian zone.

As the emerging time increased, more plant species with relatively unique traits entered this area, resulting in a positive linear relationship between species and functional diversity in July and September. This relationship indicates that low functional redundancy and changes in species diversity will lead to changes in functional diversity [42,49]. Located on the hillside of the reservoir side, this riparian zone tends to experience a period of drought during the exposure period. This can lead to a decrease in the number of plant species that can survive in the area. These changes in plant species can have a significant impact on the ecosystem of a riparian zone and its surrounding areas. However, it is essential to note that riparian vegetation is unique to riparian zones, and its success depends on planting at the proper elevations and seasons. The characteristics of riparian vegetation vary substantially and correspond to geographic variations in climate, hydrologic regimes, and associated geomorphology. The interval of flooding from July to September immersed this zone in a period of alternating flooding and drought. Species with unique traits were more adapted to the riparian habitat, but few differences among species resulted in low functional redundancy.

4.3. Flooding Depth: A Better Indicator for the Zonation of the Riparian Vegetation

Flooding and fluctuating water availability strongly constrain riparian plant ecological strategies, and the relationship between the two may be generalizable to diverse biomes. Human activities, such as dam construction and management for hydropower and water storage and diversion, have severely modified the natural flow regimes in many river systems. In particular, flow regimes are expected to shift under even the most conservative of climate change scenarios [50]. Extreme floods cause significant geomorphic [51] and chemical changes along the water depth gradient [28]. This changing pattern varies with flooding depth, resulting in a variety of plants in a riparian zone.

Due to the dam's regulation, the riparian plants in the Three Gorges Reservoir experienced extreme flooding, which lasted more than half a year and had a depth of 30 m. Research has shown that even slight modifications to the historic natural flow regime have significant consequences for the structure of riparian plant networks [5]. Flooding time is often used as a simple indicator of a riparian plant's water stress. However, in this study, flooding time alone did not capture the full extent of the stress, as it did not account for the depth of the water that covered the plant. When the water level rose, the plants were submerged deeper, which significantly impacted the soil properties and physiology and limited the plants' growth and survival in the flooded environment. The flooding depth contained a message: the depth to which a riparian plant submerges can affect the character of the soil and restrict plant growth.

Research has also shown that the inundation depth and frequency primarily determine riparian plant species composition [27]. A significant negative relationship between plants and flooding depth was observed in our study, while flooding duration was not retained as a significant predictor. Compared to flooding time, flooding depth presented a more proper reflection of the riparian stress experienced by the plants. The greater correlation between the flooding depth and community structure compared to that of duration suggests the better adaptability of management and restoration based on flooding depth rather than duration in environment-changing circumstances. In our study, the seasonal data showed that flooding depth could be a powerful indicator of plant zonation in the riparian zone.

5. Conclusions

River regulation, which involves managing and controlling water flowing through dams, can positively and negatively impact riparian plant communities. Riparian plants are important influencers of the flow regime of a river reservoir and are sensitive indicators of changes in the hydrological environment. Hydrological changes affect the distribution patterns, growth patterns, reproduction modes, and survival strategies of plants in riparian zones. In our study, the depth and duration of floods were analyzed in comparison with the riparian plant community's species richness and total coverage. The higher correlation between flooding depth and the riparian plant community compared to that of flooding duration suggests that flooding depth is a better indicator of the zonation of riparian vegetation affected by dam regulation. Our study emphasizes the importance of flooding depth as a determinant of ecological strategies in riparian plant communities. Regarding the key component of ecological flow, further opportunities to experiment with flooding depth should be offered, as these will assist in refining process-based river restoration. More work is needed on the interaction and influence of vegetation and local habitats on the flooding depth in the Three Gorges Reservoir.

Supplementary Materials: The following supporting information can be downloaded at: https://www.mdpi.com/article/10.3390/w15183228/s1, Table S1: Location and the hydrological station; Table S2: Plant functional traits and ecosystem processes and service. References [52–54] are cited in the Supplementary Materials.

Author Contributions: Conceptualization, X.Y.; methodology, X.Y.; software, X.Y., Y.H. and Y.J.; validation, X.Y. and M.M.; formal analysis, X.Y. and Y.H.; investigation, X.Y. and Y.J.; resources, S.W.; data curation, X.Y.; writing—original draft preparation, X.Y.; writing—review and editing, X.Y., M.M. and Q.C.; visualization, Q.C.; supervision, Q.C.; project administration, S.W.; funding acquisition, S.W. All authors have read and agreed to the published version of the manuscript.

Funding: This study was supported by the National Natural Science Foundation of China [51709250; 51779241; 41601042] and the Three Gorges' follow-up scientific research project from Chongqing Municipal Bureau of Water Resources (No. 5000002021BF40001).

Data Availability Statement: The data presented in this study are available on request from the corresponding author. The data are not publicly available due to restrictions.

Acknowledgments: We thank Xiaolei Su and Songlin Zhang for their constructive comments on earlier drafts. Yiguo Ran and Xiaoxiang Sun provided invaluable help in the data analysis and field work. The author would also like to thank the very valuable comments from the reviewers and editors that greatly improved the quality of the paper.

Conflicts of Interest: The authors declare no conflict of interest.

References

1. Gregory, S.V.; Swanson, F.J.; McKee, W.A.; Cummins, K.W. An Ecosystem Perspective of Riparian Zones: Focus on links between land and water. *BioScience* **1991**, *41*, 540–551. [CrossRef]
2. Naiman, R.J.; Décamps, H. The Ecology of Interfaces: Riparian Zones. *Annu. Rev. Ecol. Syst.* **1997**, *28*, 621–658. [CrossRef]
3. Palmer, M.A.; Ruhi, A. Linkages between flow regime, biota, and ecosystem processes: Implications for river restoration. *Science* **2019**, *365*, eaaw2087. [CrossRef] [PubMed]
4. Nilsson, C.; Berggren, K. Alterations of riparian ecosystems caused by river regulation. *BioScience* **2000**, *50*, 783–792. [CrossRef]
5. Tonkin, J.D.; Merritt, D.M.; Olden, J.D.; Reynolds, L.V.; Lytle, D.A. Flow regime alteration degrades ecological networks in riparian ecosystems. *Nat. Ecol. Evol.* **2018**, *2*, 86–93. [CrossRef] [PubMed]
6. Lou, Y.; Mei, X.; Dai, Z.; Wang, J.; Wei, W. Evolution of the mid-channel bars in the middle and lower reaches of the Changjiang (Yangtze) River from 1989 to 2014 based on the Landsat satellite images: Impact of the Three Gorges Dam. *Environ. Earth Sci.* **2018**, *77*, 394. [CrossRef]
7. Kiss, T.; Balogh, M. Characteristics of point-bar development under the influence of a dam: Case study on the Dráva River at Sigetec, Croatia. *J. Environ. Geogr.* **2015**, *8*, 23–30. [CrossRef]
8. Skalak, K.J.; Benthem, A.J.; Schenk, E.R.; Hupp, C.R.; Galloway, J.M.; Nustad, R.A.; Wiche, G.J. Large dams and alluvial rivers in the Anthropocene: The impacts of the Garrison and Oahe Dams on the Upper Missouri River. *Anthropocene* **2013**, *2*, 51–64. [CrossRef]
9. Gu, Z.K.; Fan, H.; Wang, Y.H. Dynamic characteristics of sandbar evolution in the lower Lancang-Mekong River between 1993 and 2012 in the context of hydropower development. *Estuar. Coast. Shelf Sci.* **2020**, *237*, 11. [CrossRef]
10. Fan, H.; He, D.M.; Wang, H.L. Environmental consequences of damming the mainstream Lancang-Mekong River: A review. *Earth-Sci. Rev.* **2015**, *146*, 77–91. [CrossRef]
11. Yang, X.; Lu, X.X. Ten years of the Three Gorges Dam: A call for policy overhaul. *Environ. Res. Lett.* **2013**, *8*, 041006. [CrossRef]
12. Li, D.; Lu, X.X.; Chen, L.; Wasson, R. Downstream geomorphic impact of the Three Gorges Dam: With special reference to the channel bars in the Middle Yangtze River. *Earth Surf. Process. Landf.* **2019**, *44*, 2660–2670. [CrossRef]
13. Willison, J.; Li, R.; Yuan, X. Conservation and ecofriendly utilization of wetlands associated with the Three Gorges Reservoir. *Environ. Sci. Pollut. Res.* **2013**, *20*, 6907–6916. [CrossRef] [PubMed]

14. Chen, Z.; Yuan, X.; Roß-Nickoll, M.; Hollert, H.; Schäffer, A. Moderate inundation stimulates plant community assembly in the drawdown zone of China's Three Gorges Reservoir. *Environ. Sci. Eur.* **2020**, *32*, 79. [CrossRef]
15. Strayer, D.L.; Findlay, S. Ecology of freshwater shore zones. *Aquat. Sci.* **2010**, *72*, 127–163. [CrossRef]
16. Ye, C.; Butler, O.M.; Chen, C.; Liu, W.; Du, M.; Zhang, Q. Shifts in characteristics of the plant-soil system associated with flooding and revegetation in the riparian zone of Three Gorges Reservoir, China. *Geoderma* **2020**, *361*, 114015. [CrossRef]
17. Lin, J.; Zhou, S.; Liu, D.; Zhang, S.; Yu, Z.; Yang, X. Relative contribution of environmental and nutritional variables to net primary production of Cynodon dactylon (Linn.) Pers in the riparian zone of a Three Gorges tributary. *Ecol. Evol.* **2020**, *10*, 7073–7081. [CrossRef]
18. Hupp, C.R.; Rinaldi, M. Riparian Vegetation Patterns in Relation to Fluvial Landforms and Channel Evolution Along Selected Rivers of Tuscany (Central Italy). *Ann. Assoc. Am. Geogr.* **2007**, *97*, 12–30. [CrossRef]
19. Sandercock, P.J.; Hooke, J.M.; Sandercock, P.J.; Hooke, J.M. Assessment of vegetation effects on hydraulics and of feedbacks on plant survival and zonation in ephemeral channels. *Hydrol. Process.* **2010**, *24*, 695–713. [CrossRef]
20. Bejarano, M.D.; Jansson, R.; Nilsson, C. The effects of hydropeaking on riverine plants: A review. *Biol. Rev.* **2018**, *93*, 658–673. [CrossRef]
21. Toner, M.; Keddy, P. River hydrology and riparian werlands: A predictive model for ecological assembly. *Ecol. Appl.* **1997**, *7*, 236–246. [CrossRef]
22. Aguiar, F.C.; Segurado, P.; Martins, M.J.; Bejarano, M.D.; Nilsson, C.; Portela, M.M.; Merritt, D.M. The abundance and distribution of guilds of riparian woody plants change in response to land use and flow regulation. *J. Appl. Ecol.* **2018**, *55*, 2227–2240. [CrossRef]
23. Butterfield, B.J.; Palmquist, E.; Ralston, B.E. Hydrological regime and climate interactively shape riparian vegetation composition along the Colorado River, Grand Canyon. *Appl. Veg. Sci.* **2018**, *21*, 572–583. [CrossRef]
24. Jansson, R.; Strom, L.; Nilsson, C. Smaller future floods imply less habitat for riparian plants along a boreal river. *Ecol. Appl.* **2019**, *29*, e01977. [CrossRef]
25. Su, X.; Bejarano, M.D.; Yi, X.; Lin, F.; Ayi, Q.; Zeng, B. Unnatural flooding alters the functional diversity of riparian vegetation of the Three Gorges Reservoir. *Freshw. Biol.* **2020**, *20*, 1585–1595. [CrossRef]
26. Lytle, D.A.; Merritt, D.M.; Tonkin, J.D.; Olden, J.D.; Reynolds, L.V. Linking river flow regimes to riparian plant guilds: A community-wide modeling approach. *Ecol. Appl.* **2017**, *27*, 1338–1350. [CrossRef]
27. Vreugdenhil, S.J.; Kramer, K.; Pelsma, T. Effects of flooding duration, -frequency and -depth on the presence of saplings of six woody species in north-west Europe. *For. Ecol. Manag.* **2006**, *236*, 47–55. [CrossRef]
28. Engloner, A.I.; Németh, K.; Gere, D.; Stefán, D.; Óvári, M. Effects of water depth and water level fluctuation on the total and bio-available element concentrations in riverine reed stands. *Ecol. Indic.* **2020**, *114*, 106328. [CrossRef]
29. Zhang, Y.; Li, Z.; Ge, W.; Chen, X.; Wang, T. Impact of extreme floods on plants considering various influencing factors downstream of Luhun Reservoir, China. *Sci. Total Environ.* **2021**, *768*, 145312. [CrossRef]
30. Kramer, K.; Vreugdenhil, S.J.; Werf, D. Effects of flooding on the recruitment, damage and mortality of riparian tree species: A field and simulation study on the Rhine floodplain. *For. Ecol. Manag.* **2008**, *255*, 3893–3903. [CrossRef]
31. Garssen, A.G.; Baattrup-Pedersen, A.; Voesenek, L.A.C.J.; Verhoeven, J.T.A.; Soons, M.B. Riparian plant community responses to increased flooding: A meta-analysis. *Glob. Chang. Biol.* **2015**, *21*, 2881–2890. [CrossRef] [PubMed]
32. Chen, H.; Zamorano, M.F.; Ivanoff, D. Effect of Flooding Depth on Growth, Biomass, Photosynthesis, and Chlorophyll Fluorescence of Typha domingensis. *Wetlands* **2010**, *30*, 957–965. [CrossRef]
33. Sorrell, B.K.; Tanner, C.C.; Brix, H. Regression analysis of growth responses to water depth in three wetland plant species. *AoB Plants* **2012**, *2012*, pls043. [CrossRef] [PubMed]
34. New, T.; Xie, Z. Impacts of large dams on riparian vegetation: Applying global experience to the case of China's Three Gorges Dam. *Biodivers. Conserv.* **2008**, *17*, 3149–3163. [CrossRef]
35. Chen, J.; Li, G.; Xiao, B.; Wen, Z.; Lv, M.; Chen, C.; Jiang, Y.; Wang, X.; Wu, S. Assessing the transferability of support vector machine model for estimation of global solar radiation from air temperature. *Energy Convers. Manag.* **2015**, *89*, 318–329. [CrossRef]
36. Lv, M.; Chen, J.; Mirza, Z.A.; Chen, C.; Wen, Z.; Jiang, Y.; Ma, M.; Wu, S. Spatial distribution and temporal variation of reference evapotranspiration in the Three Gorges Reservoir area during 1960–2013. *Int. J. Climatol.* **2016**, *36*, 4497–4511. [CrossRef]
37. Chen, J.; He, L.; Yang, H.; Ma, M.; Chen, Q.; Wu, S.; Xiao, Z. Empirical models for estimating monthly global solar radiation: A most comprehensive review and comparative case study in China. *Renew. Sustain. Energy Rev.* **2019**, *108*, 91–111. [CrossRef]
38. Yi, X.; Huang, Y.; Ma, M.; Wen, Z.; Wu, S. Plant trait-based analysis reveals greater focus needed for mid-channel bar downstream from the Three Gorges Dam of the Yangtze River. *Ecol. Indic.* **2020**, *111*, 105950. [CrossRef]
39. DÍAz, S.; Lavorel, S.; McIntyre, S.U.E.; Falczuk, V.; Casanoves, F.; Milchunas, D.G.; Skarpe, C.; Rusch, G.; Sternberg, M.; Noy-Meir, I.; et al. Plant trait responses to grazing—A global synthesis. *Glob. Chang. Biol.* **2006**, *13*, 313–341. [CrossRef]
40. Rao, C.R. Diversity and dissimilarity coefficients: A unified approach. *Theor. Popul. Biol.* **1982**, *21*, 24–43. [CrossRef]
41. Botta-Dukát, Z. Rao's quadratic entropy as a measure of functional diversity based on multiple traits. *J. Veg. Sci.* **2005**, *16*, 533–540. [CrossRef]
42. Sasaki, T.; Okubo, S.; Okayasu, T.; Jamsran, U.; Ohkuro, T.; Takeuchi, K. Two-phase functional redundancy in plant communities along a grazing gradient in Mongolian rangelands. *Ecology* **2009**, *90*, 2598–2608. [CrossRef] [PubMed]
43. Moor, H.; Rydin, H.; Hylander, K.; Nilsson, M.B.; Lindborg, R.; Norberg, J. Towards a trait-based ecology of wetland vegetation. *J. Ecol.* **2017**, *105*, 1623–1635. [CrossRef]

44. Grygoruk, M.; Kochanek, K.; Mirosław-Świątek, D. Analysis of long-term changes in inundation characteristics of near-natural temperate riparian habitats in the Lower Basin of the Biebrza Valley, Poland. *J. Hydrol. Reg. Stud.* **2021**, *36*, 100844. [CrossRef]
45. Jian, Z.; Ma, F.; Guo, Q.; Qin, A.; Xiao, W. Long-term responses of riparian plants' composition to water level fluctuation in China's Three Gorges Reservoir. *PLoS ONE* **2018**, *13*, e0207689. [CrossRef]
46. Hoppenreijs, J.H.T.; Eckstein, R.L.; Lind, L. Pressures on Boreal Riparian Vegetation: A Literature Review. *Front. Ecol. Evol.* **2022**, *9*, 806130. [CrossRef]
47. Eck, W.; Lenssen, J.; Steeg, H.; Blom, C.; Kroon, H.D.J.H. Seasonal Dependent Effects of Flooding on Plant Species Survival and Zonation: A Comparative Study of 10 Terrestrial Grassland Species. *Hydrobiologia* **2006**, *565*, 59–69.
48. Mccoy-Sulentic, M.E.; Kolb, T.E.; Merritt, D.M.; Palmquist, E.; Ralston, B.E.; Sarr, D.A.; Shafroth, P.B. Changes in Community-Level Riparian Plant Traits over Inundation Gradients, Colorado River, Grand Canyon. *Wetlands* **2017**, *37*, 635–646. [CrossRef]
49. Mayfield, M.M.; Boni, M.F.; Daily, G.C.; Ackerly, D. Species and Functional Diversity of Native and Human-Dominated Plant Communities. *Ecology* **2005**, *86*, 2365–2372. [CrossRef]
50. Rood, S.B.; Pan, J.; Gill, K.M.; Franks, C.G.; Samuelson, G.M.; Shepherd, A. Declining summer flows of Rocky Mountain rivers: Changing seasonal hydrology and probable impacts on floodplain forests. *J. Hydrol.* **2008**, *349*, 397–410. [CrossRef]
51. Li, Z.; Zhang, Y.; Wang, J.; Ge, W.; Jiao, Y. Impact evaluation of geomorphic changes caused by extreme floods on inundation area considering geomorphic variations and land use types. *Sci. Total Environ.* **2021**, *754*, 142424. [CrossRef] [PubMed]
52. Van der Pijl, L. Ecological dispersal classes, established on the basis of the dispersing agents. In *Principles of Dispersal in Higher Plants*; Springer: Berlin/Heidelberg, Germany, 1982; pp. 22–90.
53. Hodgson, J.; Grime, J.; Hunt, R.; Thompson, K. *The Electronic Comparative Plant Ecology*; Springer: Berlin/Heidelberg, Germany, 1995.
54. Mabry, C.M. *Floristic Analysis of Central Iowa Woodlands, and Comparison of Reproduction and Regeneration in Common and Restricted Herbaceous Species*; Iowa State University: Ames, IA, USA, 2000.

water

MDPI

Article

The Investigation of Local Scour around Bridge Piers with the Protection of a Quasi-Stumps Group

Yisheng Zhang [1,2], Jiangfei Wang [3], Qi Zhou [3], Yingchun Cai [1,2] and Wei Tang [1,2,*]

1 School of Water Conservancy and Transportation, Zhengzhou University, Zhengzhou 450001, China; yishengzhang@zzu.edu.cn (Y.Z.); yccai@zzu.edu.cn (Y.C.)
2 Yellow River Laboratory, Zhengzhou University, Zhengzhou 450001, China
3 Henan Puze Expressway Co., Ltd., Zhengzhou 450046, China; tqycfv@ksdud.com (J.W.); egawxd@ksdud.com (Q.Z.)
* Correspondence: tangwei251231@163.com; Tel.: +86-0371-6778-3172

Abstract: In this study, a quasi-stumps group structure was proposed and placed upstream of the bridge piers to mitigate the scour of the waterflow on the riverbed. Both experiment and numerical simulations using FLOW 3D were employed to study the protective effect of this structure. The numerical results were in good agreement with the experimental findings. It was found that the quasi-stumps group can effectively reduce the flow velocities around the bridge piers, thereby promoting the deposition of suspended sediment. As a result, there was no erosion around the piers, and instead, siltation was formed, which contributed to the stability of the piers. The deposition height around the piers increased as the L (the horizontal distance between the quasi-stumps group and the piers) decreased and both the P (the height of the quasi-stumps group) and S (the ratio of the area of a single leaf on the quasi-stumps group to the cross-sectional area of a single pier) increased. As the L, P, and S increased, the quantity of suspended sediment deposition over the entire riverbed increased. The optimal combination of the quasi-stumps group's protective effect was determined to be L = D (pier diameter), P = H (water depth), and S = 0.148.

Keywords: bridge piers; local scour; quasi-stumps group; numerical simulation; FLOW 3D

1. Introduction

The transportation system serves as the cornerstone of national development, and bridges are an integral component of this system. As society progresses, there has been a significant increase in demand for travel and cargo transportation, leading to higher requirements for bridge quantities and capacities. Bridges spanning rivers are particularly crucial, as they not only endure the air corrosion on piers but also the destructive impact of waterflow on pier foundations. Of the two, the latter poses a more significant threat, making river-crossing bridges more susceptible to damage. When water flows through bridge piers, the velocity increases, and the turbulence is intensified near the piers, resulting in backwater upstream of the piers. This creates a pressure gradient from top to bottom, which generates a downward flow that impacts the riverbed. At the same time, the flow on both sides of the piers separates significantly, forming a strong vortex system and a horseshoe vortex with the downward flow. The intense horseshoe vortex generates significant shear stress on the riverbed, causing severe scouring around the bridge piers, resulting in local scour holes. This phenomenon is almost inevitable in alluvial riverbeds, and it affects the stability and safety of bridge piers. Therefore, appropriate measures must be taken during the design stage to prevent scouring. Suitable protective measures can significantly reduce local scouring around bridge piers and improve the safety of bridge structures.

Smith [1] counted and analyzed the causes of 143 bridges that had major accidents from 1847 to 1975 and found that the damage to 70 bridges was caused by the displacement of piers during flooding. The investigation report by Lagasse et al. [2] showed that about

60% of bridge accidents were caused by scours. The scours on riverbeds around piers can also pose threats to people's lives. For example, in April 1987, pier NO. 3 of the bridge crossing the Schoharie River near Fort Hunter, New York, was damaged due to excessive local scour caused by a major flood, resulting in the collapse of the bridge's main girder and the death of 10 people in the accident. In 1989, a bridge on Highway 51 in Tennessee collapsed, and eight people fell into the water and died due to excessive local scour. Although local scours around bridge piers occur all the time, they are particularly strong during floods. During a flood in 1956, the maximum scour depth around pier NO. 7 of the Zhengzhou Yellow River Bridge reached 8 m, and the maximum scour depth around pier NO. 11 reached 14 m, with obvious tilting of the piers. Ultimately, the bridge was destroyed during the catastrophic flood of the Yellow River in 1958. The investigation of the US Federal Highway Administration showed that local scours are one of the primary issues to be considered in bridge design and protection [3]. Therefore, in the process of designing bridge foundations, it is necessary to take some measures to protect the riverbed around the piers, so as to ensure the safety and stability of the piers.

Based on the protective mechanism of a scour, Chiew [4] classified the protective measures of pier foundations into the following two categories: (1) reducing the scouring capacity of a submerged flow and horseshoe vortex in the incoming flow and (2) enhancing the scour resistance of the riverbed materials around bridge piers. Wang et al. [5] referred to these two types of protective measures as active protection and passive protection. Active protective measures include changing the pier shape, collar, slot, sacrificial pile in front of the pier, threaded pier, a splitter plate downstream pier, etc. Jalal and Hassan [6] used FLOW 3D software to study the influence of different pier shapes (circular, rectangular, octagonal, oval, and lenticular) on local scours. The results showed that the scour depth of the lenticular pier was the minimum, while that of the rectangular pier was the maximum. The scour depth of lenticular piers is 40% lower than that of other piers, on average. Jahangirzadeh et al. [7] investigated the effect of different sizes and shapes of collars on reducing the local scour depths around bridge piers. The study found that a collar size W = 3.0D~3.5D can best and most economically reduce the scour depth, where W is the side length of a square collar or the diameter of a circular collar, and D is the pier diameter. Additionally, the square collar provided better protective effects than the circular collar. Nazari-Sharabian et al. [8] used FLOW 3D software to simulate the anti-scouring effect of sacrificial piles with different numbers and different distribution patterns. It was found that, when a single pile is installed in front of a pier, installing a single pile at a distance of five times the pier diameter upstream of the pier can more effectively reduce the scour depth. When three and five piles are installed, the optimal distance for the three piles is six and four times of the pier diameter, and the optimal distance for the five piles is four, six, and eight times the pier diameter. Dey et al. [9] spirally wrapped helical wires and cables on a pier to form a threaded pile. Through experimental studies, it was found that, when the thread angle was 15° and the diameter ratio of the cable to pier was 0.1, the maximum scour depth of the triple-threaded pile could be reduced by 46.3%. The threaded pile could weaken the vortex shedding, which was instrumental in inducing a scour. Furthermore, many researchers have studied the protective effects of active protective measures against scouring a riverbed [10–12]. Passive protection measures include riprap countermeasures, foundation expansion, a tetrahedron frame group, etc. Chiew [13] studied the stability of riprap around piers and identified three different failure modes of riprap: shear failure; settlement failure, and edge failure; a semi-empirical method was proposed to determine the size of the rocks to protect the riprap pile. The experimental data showed that, in the absence of a filter layer, a thicker riprap layer could prevent settlement, and two empirical relationships were proposed for the influence of riprap layer thickness and coverage on the stability of the riprap layer. Parola et al. [14] found that the best protective effect was achieved when the top surface of the expanded foundation was level with the riverbed surface. As the top surface of the foundation lowered or raised, the protective effect weakened accordingly. Li et al. [15] found that the tetrahedron frame group could reduce the flow velocities around it and dissipate the energy of the waterflow, thus achieving the effect of preventing erosion and promoting sediment deposition, as well as protecting the bedload and suppressing the local scour. In addition, many researchers have studied the passive protective measures [16–18]. The above are the protective measures for the local scour of bridge piers that have been studied more at present.

The above measures are effective in preventing scouring of the riverbed when the flow is clear, but when the water carries silt, these measures cannot effectively cause the suspended sediment in the water to deposit. Therefore, in the case of muddy water, it is necessary to reduce the erosion of the riverbed and also try to make the suspended sediment settle near the bridge piers to fill in the scour holes. At present, most studies on the local scour of piers focus on clean water, while only a few scholars have studied the influence of muddy water on the local scour of piers. For a clear water scour, the sediment is removed from the scour hole but not supplemented by the approach flow. The maximum scouring depth is gradually approached when the flow is no longer capable of removing the bed sediment from the hole. However, for a muddy water scour, the scour hole is continuously provided with sediment by the approach flow. The maximum scouring depth is reached when the average amount of sediment transported into the scour hole by the approach flow is equal to the average amount of sediment removed from the hole due to the local scour. Therefore, the maximum scouring depth fluctuates periodically about a mean value [19].

For erosion in muddy water, the current methods to reduce the scour depth include installing plants and thin cylinders, etc. upstream of piers, and the main research methods are experiments and numerical simulations. Abt et al. [20] studied the sediment deposition and retention capacity in vegetated riverbeds. The results showed that the presence of vegetation not only promotes the deposition of suspended sediment but also causes some suspended sediment to adhere to the vegetation. Ultimately, vegetation can retain 30~70% of suspended sediment, and it was found that the amount of sediment siltation is mainly related to the flow and the length of the plant leaves. Sun et al. [21] studied the effect of tall and short submerged plants on suspended sediment deposition in a turbid water tank. It was found that the siltation was seven times greater when tall and short plants were mixed than when no plants were present, the effect of promoting sediment deposition was the next most effective when only tall plants were present, and the effect gradually increased with the decrease in plant spacing. Nabaei et al. [22] studied the influence of vegetation on the flow structure and turbulence anisotropy around a semi-elliptical abutment. They found that the presence of vegetation in the channel significantly reduced the primary vortex. The tangential and radial velocities decreased with the vegetation in the channel bed, and the Reynolds shear stress in the channel decreased obviously. Amir et al. [23] studied the impact of vegetation coverage on local scours around bridge piers. The results indicated that the depth of the local scour around bridge piers in a water tank with vegetation was only half of that without vegetation, and the scour equilibrium time was reduced by 42% compared to that without vegetation. Wang et al. [24] studied the influence of the arrangement density, water depth, and sediment particle size on the sediment starting velocity by installing flexible rubber cylinders parallel in a water tank. The experimental results showed that the starting velocity of the sediment increased with the decrease in the arrangement density and the increase in the water depth and sediment particle size, and the corresponding equation of the starting velocity of the sediment was derived. Tang et al. [25] set up stainless steel fine cylinders with different arrangement densities and diameters in still water and measured the sediment settling velocities of particles with diameters ranging from 0.1 cm to 0.45 cm. The study found that, within the experimental particle size range, the settling velocity of the sediment in the presence of slender cylinders decreased with the increase of the cylinder density and sediment particle size. They also derived a sediment settling velocity formula that took into account the density and diameter of the slender cylinders. Zhao at al. [26] investigated the influence of the submerged vegetation density λ on the turbulent flow characteristics of an open channel.

The results showed that, for $0.04 < \lambda < 0.1$, the turbulent statistical profiles exhibited similar characteristics to the bed shear flow and free shear flow. When $\lambda = 1.44$, the turbulent statistics above the vegetation top exhibited features of a boundary shear flow. Xiong et al. [27] proposed and verified a three-dimensional simulation model for a live bed pier scour considering suspended sediment loads, which could simulate the dynamic balance of inputs and outputs of sediment around the piers in a reliable and computationally efficient manner. Yu at al. [28] employed six different RANS turbulence models to simulate the flow structure in an open channel with a developed vegetation patch and compared the results with the experimental data. They concluded that CFD is an effective research method and also observed that different turbulence models exhibited varying simulation performances. Chen et al. [29] used FLOW 3D software to conduct numerical simulations of the hydraulic characteristics of rigid submerged plants. They modified the Shields number from the perspective of turbulent kinetic energy and improved the sediment module, so that the FLOW 3D software could simulate the flow field and sediment movement more accurately, and the results of the numerical simulation were in good agreement with the experimental data. Compared with a single pier, a group of piers are increasingly used in bridge constructions. A group of piers is shown in Figure 1. Therefore, many scholars have studied the local scours around groups of piers in riverbeds through experiments and numerical simulations. They have found that the waterflow between groups of piers is obviously discontinuous and nonlinear, and the scouring process in riverbeds has also become more complex [30–34]. Studying the sediment movement around groups of piers in rivers has more reference value for practical engineering.

Figure 1. A group of bridge piers.

Based on the above content, a quasi-stumps group structure placed upstream of bridge piers was proposed in this study. This structure can effectively lower the flow velocities around bridge piers, impede the transportation of bedload sediment, and intercept suspended sediment, so as to achieve the purpose of reducing the local scour. Additionally, it can provide protection against floating objects upstream colliding with bridge piers. The group of piers used in this study was obtained by proportionally reducing the actual engineering and consisted of six circular piers. In this paper, the protective effect of the quasi-stumps group on the riverbed around the bridge piers was studied by experiment and numerical simulations. The effects of three factors (the distance between the quasi-stumps group and the bridge pier, the height of the quasi-stumps group, and the leaves area of the quasi-stumps group) of this structure on the protective effect were also analyzed. Based on the analysis results, the combination of a quasi-stumps group with the best protective effect was proposed.

2. Materials and Methods

2.1. Structure of Quasi-Stumps Group

The quasi-stumps group structure placed upstream of the piers was proposed to reduce the local scour in the riverbed around the piers. The quasi-stumps group is composed of nine quasi-stumps. A single quasi-stump of 8 cm high (one-third the water depth) contains twelve leaves. The top view and side view of the quasi-stumps group are shown in Figure 2 (D is the pier diameter). The structure not only reduces the flow velocities reaching the piers, thus reducing the erosion of the riverbed by the flow, but also promotes the suspended sediment in the water to be deposited around the piers to replenish the scour holes.

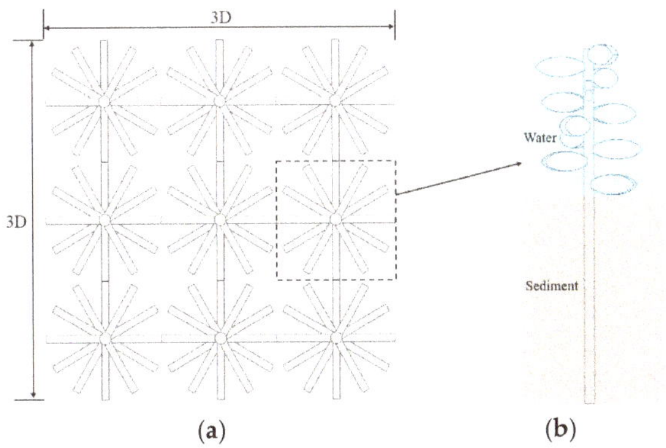

Figure 2. The structure of the quasi-stumps group. (**a**) Top view. (**b**) Side view.

2.2. Experimental Setup

The experiments were carried out in the hydraulic laboratory at Zhengzhou University's School of Water Conservancy and Civil Engineering. The experimental platform consists of a rectangular channel made of glass with dimensions of 72 cm (width) × 50 cm (height) and a length of 12 m, two water tanks, a pumping station, a circulating pipe, an electromagnetic flowmeter, a valve, a sand mixing pool, the group of bridge piers, and the quasi-stumps group installed in the sink, as shown in Figure 3. In order to achieve flow stability, a channel measuring 4.3 m in length was positioned at the forefront of the group of piers. The water in the experiment was a silt-carrying flow with a sediment content of 2.7 kg/m^3, meaning that each cubic meter of water and sand mixture contained 2.7 kg of sediment. The median grain size of the sediment in the inlet flow was 0.063 mm. In the experiment, the median grain size (d_{50}) of the sediment in the riverbed was 0.34 mm, and the sediment particle size ranged from 0.15 mm to 0.41 mm in diameter. The median grain size of the sediment in the riverbed used in this study was referenced from the experiments of Jahangirzadeh et al. [35] and Obied et al. [36]. The median particle size and sediment content of the suspended sediment in the inlet flow were from the study of Debnath and Chaudhuri [37]. The thickness of the sediment was 10 cm (in the preliminary experiment, the scour depth of the riverbed was much smaller than 10 cm). The density of the sediment was 2650 kg/m^3. The underwater angles of repose for the sediment in the riverbed and inlet flow were 32° and 30°, respectively. Each experiment lasted for the duration of one hour, which matched the duration of the numerical simulation. Prior to each experiment, the flume was gradually filled with water to ensure saturation of the riverbed material. Measurements of the water depth, scour depth, deposit depth, and velocities were taken using a needle water level gauge and a flow velocity meter, respectively. To minimize the wall effects on the scour rate, the flume width was maintained at a minimum of 10 times the diameter of the pier [38], which was set to 6 cm based on the dimensions of the experimental flume.

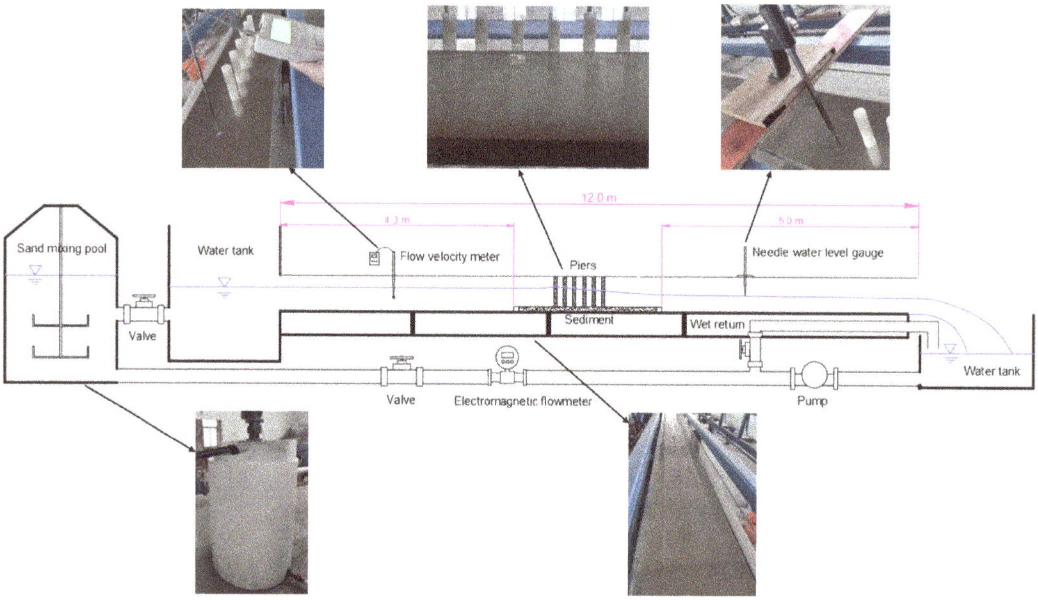

Figure 3. The experimental facilities.

The group of piers in the experiment consists of six piers, and the piers are numbered as shown in Figure 4. The model of the group of piers is obtained by proportionally reducing the piers of the Yellow River Super Major Bridge of Yangxin Expressway from Puyang City to Hubei Province, and the scale is 100:3. In the experiment, the flow velocity upstream of the piers is controlled at 0.48 m/s, which is obtained by converting the flow velocity in the actual project to the model experiment. The scale of the flow velocity is $\sqrt{100/3}$.

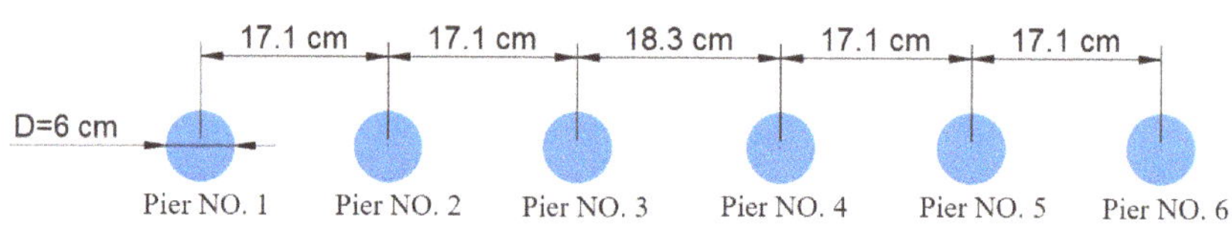

Figure 4. The dimensions, spacing, and numbers of the piers group.

2.3. Numerical Simulation

FLOW 3D uses the RNG k−ε turbulence model to solve Reynolds-averaged Navier–Stokes equations. It is an improved version of the k-ε turbulence model, where RNG stands for "renormalization group". It is a method that utilizes the theory of a renormalization group from statistical physics to model turbulence. k is the turbulent kinetic energy, and it refers to the intensity or energy of the vortical motion in a turbulent flow field. In turbulence, the fluid contains vortices of various scales that continuously interact and exchange energy. Turbulent kinetic energy quantifies the strength of these vortices and is described in the turbulent model by an equation governing its evolution. ε is the turbulent dissipation rate, and it represents the rate at which turbulent kinetic energy is transformed or dissipated. Turbulence is a process of continuous energy conversion, where kinetic energy is converted into internal energy, and there is also a process of internal energy returning to kinetic energy. The turbulent dissipation rate describes the rate of this energy conversion or dissipation and is described in the model by an equation governing its evolution.

The hydrodynamics solver employed in this study is fully coupled with the sediment transport module, enabling the combined use of semi-empirical formulas and hydraulic principles to predict the riverbed evolution. The equilibrium states of the scour are determined by analyzing changes in the riverbed and calculating the flow field.

The turbulent flow in this study is governed by the continuity equation, which can be mathematically expressed as follows:

$$V_F\frac{\partial\rho}{\partial t}+\frac{\partial}{\partial x}(\rho uA_x)+R\frac{\partial}{\partial y}(\rho vA_y)+\frac{\partial}{\partial z}(\rho wA_z)+\xi\frac{\rho uA_x}{x}=R_{DIF}+R_{SOR} \quad (1)$$

where V_F is the fractional volume open to flow; ρ is the fluid density; (u, v, and w) are the velocity components in the (x, y, and z) directions, respectively; (A_x, A_y, and A_z) are the portion of area open to flow in the (x, y, and z) directions, respectively; R_{DIF} is the turbulent diffusion term; R_{SOR} is the mass source; R is a coefficient that depends on the chosen coordinate system; and ξ is related to the compressibility of the fluid. In this numerical simulation, water is an incompressible fluid, and the Cartesian coordinate system is used, so $R = 1$ and $\xi = 0$. x is the coordinate value along the x-axis.

The equations of motion for the fluid velocity components (u, v, and w) in the three coordinate directions are Navier–Stokes equations with some additional terms:

$$\frac{\partial u}{\partial t}+\frac{1}{V_F}\left(uA_x\frac{\partial u}{\partial x}+vA_y\frac{\partial u}{\partial y}+wA_z\frac{\partial u}{\partial z}\right)=-\frac{1}{\rho}\frac{\partial p}{\partial x}+G_x+f_x-b_x-\frac{R_{SOR}}{\rho V_F}(u-u_w-\delta u_s) \quad (2)$$

$$\frac{\partial v}{\partial t}+\frac{1}{V_F}\left(uA_x\frac{\partial v}{\partial x}+vA_y\frac{\partial v}{\partial y}+wA_z\frac{\partial v}{\partial z}\right)=-\frac{1}{\rho}\frac{\partial p}{\partial y}+G_y+f_y-b_y-\frac{R_{SOR}}{\rho V_F}(v-v_w-\delta v_s) \quad (3)$$

$$\frac{\partial w}{\partial t}+\frac{1}{V_F}\left(uA_x\frac{\partial w}{\partial x}+vA_y\frac{\partial w}{\partial y}+wA_z\frac{\partial w}{\partial z}\right)=-\frac{1}{\rho}\frac{\partial p}{\partial z}+G_z+f_z-b_z-\frac{R_{SOR}}{\rho V_F}(w-w_w-\delta w_s) \quad (4)$$

where (G_x, G_y, and G_z) are the body accelerations in the (x, y, and z) directions; (b_x, b_y, and b_z) are the flow losses through porous media in the (x, y, and z) directions; and (f_x, f_y, and f_z) are the viscous accelerations in the (x, y, and z) directions, respectively. Viscous acceleration describes the acceleration experienced by a moving object in fluid due to the influence of viscous forces. When an object moving in fluid changes its velocity or direction, the surrounding fluid's viscous forces cause the object to undergo an additional acceleration exerted by the fluid, and this acceleration is known as the viscous acceleration. (u_w, v_w, and w_w) are the speeds of the source components, they are used in simulations to represent the velocities of the source component of an external input or perturbation. The velocities can be used to simulate injection, withdrawal, agitation, or other external influences in the fluid. t is the time, and p is the pressure. The determination of the δ value is related to the type of the source. When the source is the stagnation pressure type, δ = 0. When the source is the static pressure type, δ = 1. δ = 1 in this study. (u_s, v_s, and w_s) are the velocities of the fluid relative to the source at its surface, and they represent the relative relationships between the velocities of the fluid on the surface of the source and the movements of the source itself. In FLOW 3D, to accurately simulate sediment erosion, it is necessary to consider the relative motion of the flow and sediment particles, as well as their shape and velocity distribution, in order to better simulate the interaction between the flow and sediment. In this numerical simulation, the sources are the waterflow and the sediment in the inlet flow.

The properties of various non-cohesive sediments can be set in the sediment scour model, including the grain size, mass density, angle of repose, critical shear stress, and parameters for entrainment and transport. In FLOW 3D, sediment has two states of existence: suspended and packed sediment. The sediment movement is estimated through the prediction of the sediment erosion, advection, and deposition processes.

Due to the impracticality of computing the flow dynamics for individual sediment grains, an empirical model is employed. In FLOW 3D, the model utilized is based on the work of Mastbergen and van den Berg [39]. The critical Shields parameter is calculated using the Soulsby–Whitehouse equation [40]. The first step in calculating the critical Shields parameters is to compute the dimensionless parameters $d_{*,i}$:

$$d_{*,i} = d_i \left[\frac{\rho_f \left(\rho_i - \rho_f \right) \|\mathbf{g}\|}{\mu_f^2} \right]^{1/3} \tag{5}$$

where i represents the ith type of sediment, as multiple sediments with different characteristics can be defined in FLOW 3D; d_i is the sediment diameter; ρ_f is the fluid density; ρ_i is the sediment density; $\|\mathbf{g}\|$ is the magnitude of the acceleration of gravity $\mathbf{g}$; and μ_f is the dynamic viscosity of the fluid. The dimensionless critical Shields parameter $\theta_{cr,i}$ is computed using the Soulsby–Whitehouse equation [40]:

$$\theta_{cr,i} = \frac{0.3}{1 + 1.2d_{*,i}} + 0.055[1 - \exp(-0.02d_{*,i})] \tag{6}$$

The modification is applied to the dimensionless critical Shields parameter of sediment on the slope $\theta'_{cr,i}$ [40]:

$$\theta'_{cr,i} = \theta_{cr,i} \frac{\cos\alpha sin\beta + \sqrt{\cos^2\beta \tan^2\varphi - \sin^2\alpha \sin^2\beta}}{\tan\varphi} \tag{7}$$

where α represents the angle between the upslope direction and flow direction, while β denotes the angle of the bed slope, and φ indicates the user-defined angle of repose of the sediment.

The calculation of the local Shields parameter θ_i is as follows:

$$\theta_i = \frac{\tau}{\|\mathbf{g}\| d_i \left(\rho_i - \rho_f \right)} \tag{8}$$

where τ is local bed shear stress.

The calculation for the entrainment lift velocity $\mathbf{u}_{lift,i}$ is as follows [39]:

$$\mathbf{u}_{lift,i} = \alpha_i \mathbf{n_s} d_*^{0.3} \left(\theta_i - \theta'_{cr,i} \right)^{1.5} \sqrt{\frac{\|\mathbf{g}\| d_i \left(\rho_i - \rho_f \right)}{\rho_f}} \tag{9}$$

where α_i represents the entrainment parameter, while $\mathbf{n_s}$ denotes the outward-pointing normal line at the packed bed interface.

The equation proposed by Soulsby [40] to calculate the settling velocity of sediment $u_{settling,i}$ is used:

$$u_{settling,i} = \frac{v_f}{d_i} \left[\left(10.36^2 + 1.049 d_*^3 \right)^{0.5} - 10.36 \right] \tag{10}$$

where v_f is the kinematic viscosity of fluids. The assumption is made that the settling motion occurs aligned with the force of gravity:

$$\mathbf{u}_{settling,i} = u_{settling,i} \frac{\mathbf{g}}{\|\mathbf{g}\|} \tag{11}$$

To describe the interactions between particles, the settling velocity is determined using the Richardson-Zaki correlation:

$$\overset{*}{\mathbf{u}}_{settling} = \mathbf{u}_{settling,i} [1 - min(0.5, c_s)]^{\zeta} \tag{12}$$

where c_s represents the volumetric fraction that accounts for the total amount of suspended sediment. ζ is computed as

$$\zeta = \zeta_{user}\zeta_0 \tag{13}$$

where ζ_{user} is the Richardson-Zaki coefficient multiplier, and ζ_0 represents the Richardson-Zaki coefficient.

By utilizing Equation (14), the dimensionless bedload transport rate Φ_i [41] was determined, while Equation (15) was employed to calculate the volumetric bedload transport rate $q_{b,i}$:

$$\Phi_i = \beta_i\left(\theta_i - \theta'_{cr,i}\right)^{1.5} C_{b,i} \tag{14}$$

$$q_{b,i} = \Phi_i\left[\|\mathbf{g}\|\left(\frac{\rho_i - \rho_f}{\rho_f}\right)d_i^3\right]^{1/2} \tag{15}$$

where the coefficient β_i typically equals 8.0, while $C_{b,i}$ represents the volume fraction of the sediment.

2.4. Computational Domain

Based on the experimental conditions, the dimensions of the simulation region were 270 cm in length, 72 cm in width (12D), and 44 cm in height. The distance between pier NO. 1 and the inlet was 76 cm (12.6D), and the distance between pier NO. 6 and the outlet was 108 cm (18D). The initial riverbed had a sediment thickness of 10 cm, while each individual pier had a diameter of 6 cm. The median particle size of the sediment was 0.34 mm. The water depth was 24 cm (4D). Moreover, in order to prevent the initial scouring, baffles with a height matching that of the riverbed were positioned at both the inlet and outlet locations, as shown in Figure 5a. For the boundary conditions, the inlet was specified as a velocity inlet, while the outlet was set as the outflow. The boundaries on both sides were defined as walls. The top mesh was assigned a symmetric boundary condition, and there was an 8 cm high air space between the water surface and the top mesh, as shown in Figure 5b. A structured orthogonal grid was implemented across all the computing domains. The main body utilized a grid size of 0.012 m × 0.012 m × 0.012 m, while the grid size in the area of the quasi-stumps group was 0.0036 m × 0.0036 m × 0.0036 m. The sediment concentration of the inlet flow was 2.7 kg/m^3, and the median particle size of the sediment was 0.063 mm. The sediment in the riverbed had a critical Shields number of 0.061, an entrainment coefficient of 0.018, and a bedload coefficient of 8. Regarding the suspended sediment, its critical Shields number was 0.043, the entrainment coefficient was 0.02, and the bedload coefficient was 9.3. The RNG k-ε turbulence model was used in this numerical simulation, and the free surface of the flow was tracked using the Tru VOF method.

The grid is the basis of the numerical simulation. The degree of discretization of the equations is determined by the grid partition, while the accuracy and stability of the calculations depend on the size of the grid partition. Therefore, a sensitivity analysis of the grid size is required, as shown in Table 1. Taking the scouring of the piers group without the quasi-stumps group as an example, d_m is the maximum scouring depth, and d_a is the average scouring depth in Table 1. Ten points are selected in the scour pit, and their scour depths are measured, and the average scour depth is obtained by calculating their average value. The sensitivity analysis shows that, while the grid size decreased to 0.012 m, the mean error tended to be stable and less than 5%; when size decreased to 0.010 m, the mean error only decreased by 0.3%; however the number of grids increased by 364,185, which made the calculation time twice as long as the original one. Therefore, taking into account the error and computational time, a grid size of 0.012 m was determined as the optimal choice for the numerical simulation.

Prior to conducting the actual experiments, the pre-experiment and its corresponding numerical simulations of the scour in the riverbed around the bridge piers without the quasi-stumps group were performed. According to the criterion of Chiew [4], the experiment's duration was determined based on the point at which the change in the scour depth remained below 1 mm for a continuous period of eight hours, indicating that the scour reached its equilibrium state and maximum depth. In the pre-experiment, it took 62 h to attain scour equilibrium. The scouring area is the sum of the areas where the scour depth exceeds 1 mm. The variation in the maximum scouring depth and scouring area with time are shown in Figure 6. After one hour, the increase in the maximum scouring depth and scouring area became very small. The maximum scour depth and scour area of the simulation reached 97.1% and 96.0% of the maximum value, respectively, and the results of the experiment attained 96.4% and 95.8% of the maximum value, respectively. Therefore, the experimental duration and simulated time were both configured for a period of 3600 s.

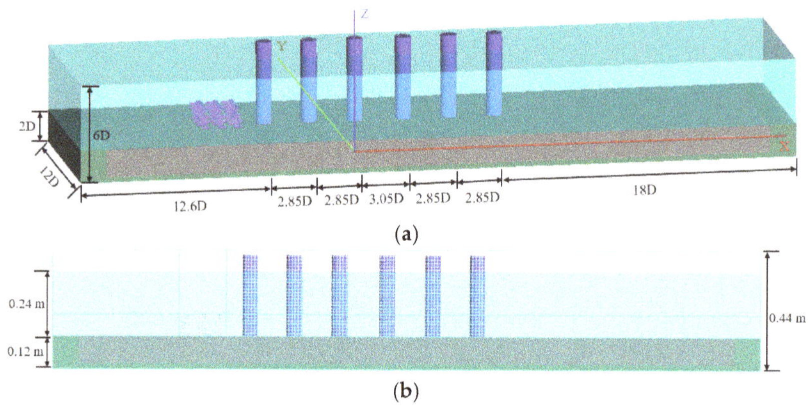

Figure 5. The model of the numerical calculations. (**a**) Overview and dimensions of the model. (**b**) Side view of the grids.

Table 1. The sensitivity analysis of the mesh size.

Test Number	Mesh Size (m)	Mesh Number	Numerical Simulation		Laboratory Experiment		Mean Error
			d_m	d_a	d_m	d_a	
1	0.016	216,384	0.0178	0.0153	0.0159	0.0134	13.1%
2	0.014	305,133	0.0171	0.0145			7.9%
3	0.012	491,175	0.0166	0.0141			4.8%
4	0.010	855,360	0.0166	0.0140			4.5%

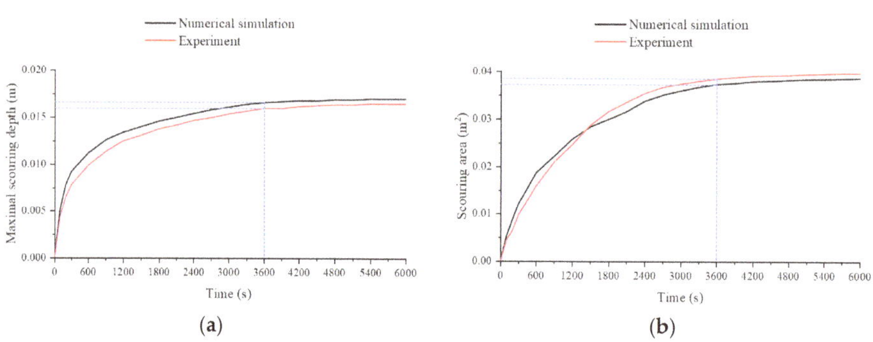

Figure 6. The contrast in the experimental (red) and simulated (black) variations of the maximum scouring depth. (**a**) and scouring area (**b**) with time. (**a**) Maximum scouring depth. (**b**) Scouring area.

2.5. Experimental Variables and Working Conditions

A total of 12 groups of experiments were set up in this study, among which the experiment NO. 0 did not include the quasi-stumps group. It was used to compare with the experiments involving the quasi-stumps group in order to analyze the protective effect of the quasi-stumps group on the riverbed scour around the bridge piers. In addition, the influence of different quasi-stumps groups on the protective effect was also studied. Three factors were set for the quasi-stumps group: the distance between the quasi-stumps group and bridge piers, the height of the quasi-stumps group, and the area of leaves on the quasi-stump. There were five different values for the distance and leaves area and three different values for the height, which were used to analyze the degree of influence of different factors on the protective effect. The experimental working conditions were set up as shown in Table 2. In Table 2, V is the average flow velocity upstream of pier NO. 1, H is the water depth, S is the ratio of the area of a single leaf to the cross-sectional area of a single pier, P is the height of the stumps group in the water, and L is the horizontal distance between the downstream edge of the stumps group and the upstream edge of pier NO. 1. The meanings of L, P, and S are shown in Figure 7. The meanings of these symbols are added to Appendix A.

Table 2. The experimental working conditions.

Experimental Number	V (m/s)	H (m)	S	P	L
0	0.48	0.24	Without quasi-stumps group		
1			0.088	H/3	D
2			0.088	H/3	2D
3			0.088	H/3	3D
4			0.088	H/3	4D
5			0.088	H/3	5D
6			0.088	2H/3	D
7			0.088	H	D
8			0.103	H/3	D
9			0.118	H/3	D
10			0.133	H/3	D
11			0.148	H/3	D

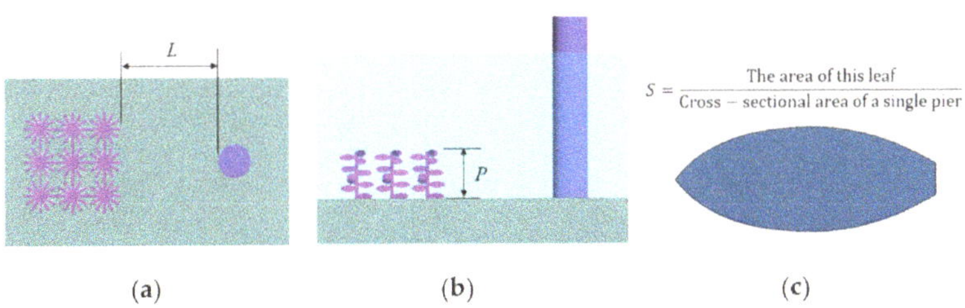

Figure 7. The meanings of L (**a**), P (**b**), and S (**c**).

2.6. Grid and User-Defined Parameters

The mesh is a structured finite-difference mesh featuring rectangular prism-shaped cells, which ensures an efficient computational performance and few cumulative numerical errors. In contrast to the typical body-fitting mesh, the mesh of FLOW 3D is less accurate in describing complex geometric shapes. Therefore, in order to ensure that all structures are identified, it is necessary to perform local mesh encryption in areas with small geometry dimensions. It is important to maintain a size ratio of no more than 5 between neighboring grids and also that the particle size of the sediment is less than one-tenth of the grid size. It is very important to find a suitable mesh size; if the size is too big, it will not be able to accurately capture the outline of the piers and quasi-stumps group, which will affect the accuracy of the calculations; if it is too small, it will increase the time required for the calculations. Consequently, it is crucial to strike a balance between precision and computational resources when determining an appropriate mesh size. Figure 8 presents the top-down perspective of the computed meshes surrounding the quasi-stumps group and piers. Figure 9 illustrates the outline of the quasi-stumps group and piers captured by the grid. The most suitable grid size for this numerical simulation was found by trying.

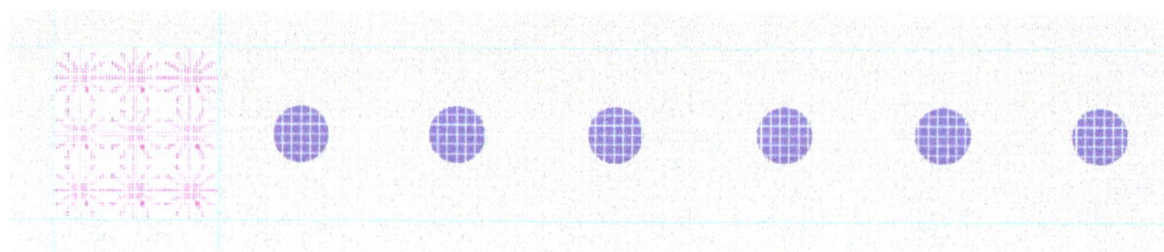

Figure 8. The top-down view of the grids for the quasi-stumps group and piers.

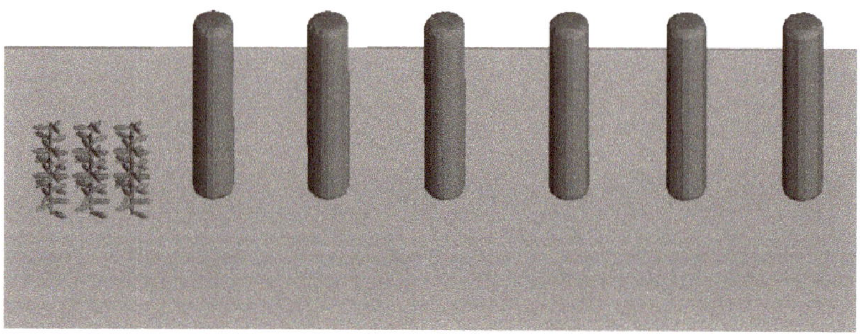

Figure 9. The outline of the quasi-stumps group and piers captured by the grids.

The effect of user-defined parameters on sediment movement was studied before conducting the actual experiment. The user-defined parameters include the underwater angle of repose φ, maximum packing fraction C_V, entrainment coefficient α, and bedload coefficient β. The maximum packing factor C_V of the sediment is the highest density of sediment particles packed on the riverbed. When the maximum packing factor is reached, the gap between sediment particles is the smallest, and the most compact state of accumulation is reached. (1) The underwater angle of repose φ of the sediment in the numerical simulation is the same as that of the sediment used in the experiment. (2) As the maximum packing fraction C_V increases within the range of (0,1), there is a small reduction in the diffusion height of the suspended sediment. It is observed that the maximum scouring depth at $C_V = 0.1$ is only 5.8% higher compared to $C_V = 0.9$. Therefore, the variation of C_V has little influence on the depths of the scour holes. C_V is set to 0.64. (3) The depths of the scour holes are significantly influenced by the entrainment coefficient α. A higher value of α leads to a greater amount of sediment being carried away by the flow, resulting in accelerated erosion of the riverbed. At $\alpha = 0.018$ and $\alpha = 0.4$, the depth of the scour hole is 2.5 and 5.1 times greater, respectively, compared to $\alpha = 0.001$. The α of the bedload sediment and suspended sediment are set to 0.018 and 0.02, respectively. (4) The magnitude of sediment transport along the riverbed increases with an increase in the bedload coefficient β. The scouring depth is 1.5 times higher at $\beta = 13$ compared to

β = 0.058, indicating that the effect of the β value on the scour depth is small. The β values of the bedload sediment and suspended sediment are set to 8.0 and 9.3, respectively.

3. Results

3.1. Validation

In experiment NO. 3, the experimental and simulated siltation are shown in Figure 10. It is evident that the simulated outcomes are in close agreement with the experimental results. The regions that cause siltation are approximately the same in the experiment and simulation. The relative error between the simulated and experimental results of the maximum siltation height was only 5.7% by measurement.

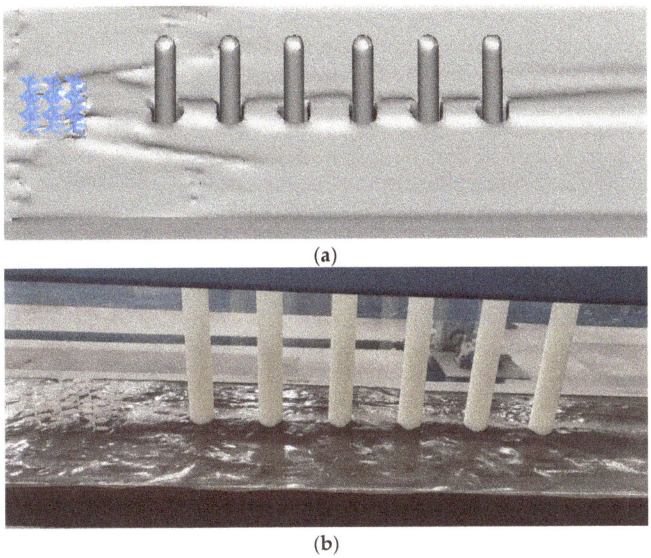

Figure 10. The comparison of the simulated (**a**) and experimental (**b**) results of the silted riverbed for experiment NO. 3. (**a**) The results of the numerical simulation. (**b**) The results of the experiment.

The flow velocity is the primary factor causing erosion and deposition in riverbeds, so before verifying the three-dimensional topography of the sedimentation, it is necessary to compare the agreement between the experimental and numerical simulation flow fields. Five measurement points were selected around the bridge piers, as shown in Figure 11.

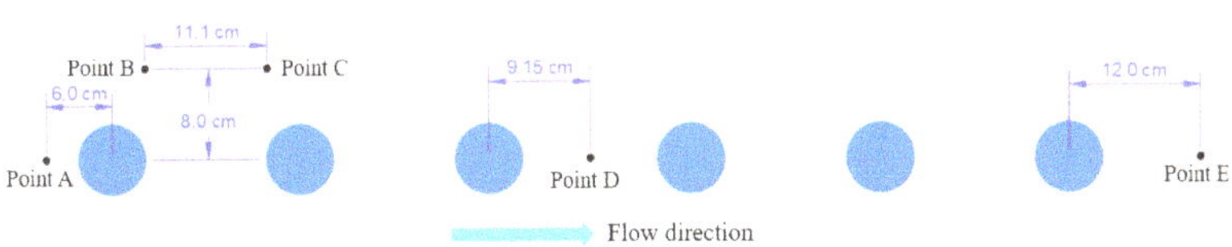

Figure 11. The locations of points A, B, C, D, and E.

Validation was conducted on experiments NO. 0, NO. 5, and NO. 10, with all points located at one-fourth of the water depth. The flow velocities and relative errors for each point in the experiment and numerical simulation are presented in Table 3. The relative error is calculated as follows:

$$Relativ\ eerror = \frac{Simulated\ value - Experimental\ value}{Experimental\ value} \times 100\% \quad (16)$$

Table 3. Relative errors between the numerical simulated and experimental flow velocity (m/s) at each point for experiments NO. 0, 5, and 10.

Relative Error / Points	Experiment NO. 0		Relative Error	Experiment NO. 5		Relative Error	Experiment NO. 10		Relative Error
	Experiment	Numerical Simulation		Experiment	Numerical Simulation		Experiment	Numerical Simulation	
Point A	0.341	0.332	−2.64%	0.233	0.253	8.58%	0.044	0.045	2.27%
Point B	0.447	0.423	−5.37%	0.338	0.354	4.73%	0.221	0.219	−0.90%
Point C	0.438	0.411	−6.16%	0.334	0.347	3.89%	0.229	0.225	−1.75%
Point D	0.147	0.144	−2.04%	0.121	0.123	1.65%	0.082	0.085	3.66%
Point E	0.253	0.254	0.40%	0.205	0.206	0.49%	0.147	0.148	0.68%

From Table 3, it can be observed that the relative errors between the numerical simulation and experiment flow velocities ranged from −6.16% to 8.58%, with the highest relative error occurring at point A of experiment NO. 5. This indicates a good agreement between the numerical simulated and experimental flow fields.

The simulated results of the dimensionless maximum deposition height d_h/D and deposition area A_h/A_D are compared with the experimental results to validate the accuracy of the simulated results obtained using FLOW 3D, where d_h is the maximum deposition height, A_h is the deposition area on the riverbed, and A_D represents the cross-sectional area of a single pier: $A_D = \pi D^2/4$. The meanings of these symbols are added to Appendix A. Figure 12 shows a comparison between the experimental and simulated results of d_h/D and A_h/A_D at 60 min. The data in Figure 12 are from experiments NO. 1 to 11. Formula (16) was used to calculate the relative errors. The boundary lines in Figure 12 are the straight lines passing through the origin point and the point of the maximum relative error. The relative error of the dimensionless deposition height (Figure 12a) ranged from −9.44% to 4.87%, respectively, and the dimensionless deposition area (Figure 12b) ranged from −7.76% to 12.33%, respectively. Under various conditions, the simulated results demonstrated a maximum relative error of 12.33% in comparison to the experimental findings, which indicated that the numerical simulation and the experiment were in good agreement. Therefore, FLOW 3D software can be used to study the erosion and sedimentation around bridge piers with a quasi-stumps group.

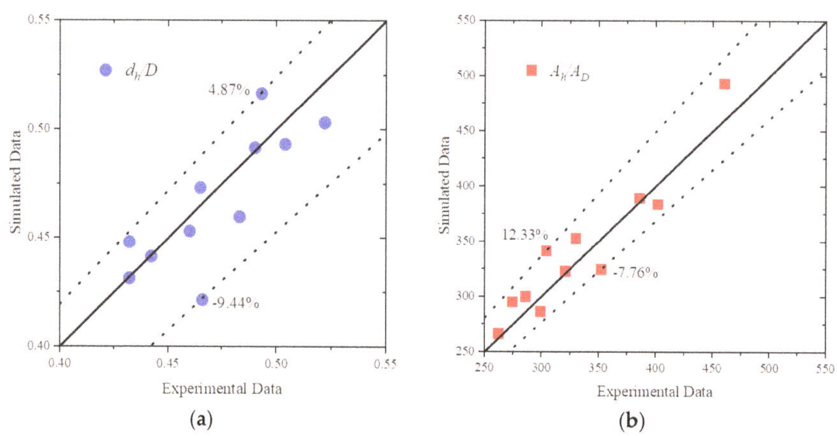

Figure 12. Comparison of the simulated and experimental data for the maximum deposition height (**a**) and deposition area (**b**). (**a**) Maximum deposition height. (**b**) Deposition area.

Although there is good agreement between the numerical simulated and experimental results regarding the maximum deposition height and deposition area, this does not necessarily imply that the deposition patterns in the numerical simulations and experiments are identical. Therefore, it is also necessary to compare the location generating the maximum deposition height and the distribution of the deposition areas. By selecting experiments NO. 2, NO. 4, NO. 6, NO. 8, and NO. 10, the locations generating the maximum deposition height in both the experiments and simulations were compared. Their spatial distribution is illustrated in Figure 13a. Through measurements and calculations, it was found that the average distance between the highest deposition points in the experiments and simulations was 6.34 cm (1.06D). This indicates that the positions generating the maximum deposition in the experiments and simulations were relatively close. Experiments NO. 3, NO. 5, and NO. 9 were selected to compare the deposition distributions between the experiments and simulations. The comparison for experiment NO. 9 is illustrated in Figure 13b. Through measurements and calculations, it was found that the overlapping deposition area accounted for 82.25% of the total experimental area on average. Therefore, the distribution of deposition areas in the simulations closely approximated that of the experiments.

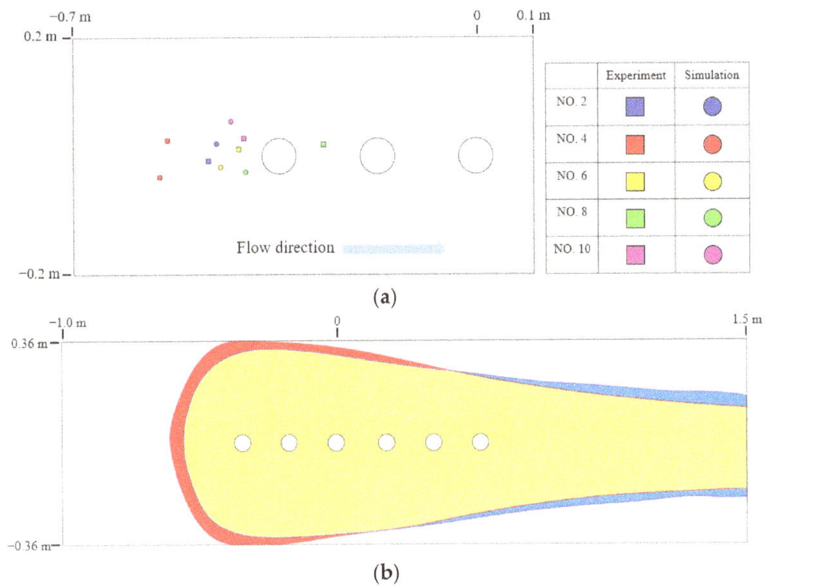

Figure 13. A comparison between the locations generating the maximum deposition height (**a**) and the distribution of the deposition areas (**b**) in the experiments and simulations. (**a**) The locations generating the maximum deposition height in the experiments and numerical simulations. (**b**) The distribution of the deposition areas in the numerical simulation and experiment for experiment NO. 9. The first column of the table indicates the number of the experiment, for example, NO. 2 represents experiment No. 2. Yellow represents the overlapping deposition areas between the simulation and experiment, red represents the deposition areas unique to the numerical simulation, and blue represents the deposition areas unique to the experiment.

3.2. Protective Effect of the Quasi-Stumps Group

The three-dimensional riverbed topography after being scoured and silted at 3600 s without (experiment NO. 0) and with (experiment NO. 3) the quasi-stumps group is shown in Figure 14. In the absence of the quasi-stumps group, the riverbed around piers NO. 1 and NO. 2 was seriously scoured, which threatened the stability of the piers. In the presence of the quasi-stumps group, the riverbed around the piers was not scoured; instead, a large amount of sediment was deposited upstream of pier NO. 1, and the height and extent of siltation downstream of pier NO. 6 also increased significantly. These findings indicate that the presence of the quasi-stumps group was highly beneficial for the stability of the bridge piers. The main reason why the quasi-stumps group could change the erosion and deposition patterns around the piers was that this structure could effectively reduce its downstream flow velocities, so that the suspended sediment originally moving with the flow could be deposited on the riverbed. Meanwhile, the shear stress of the flow on the riverbed was decreased due to the decrease in the flow velocities, thus weakening the scouring effect on the riverbed. Therefore, under the dual effect of promoting siltation and weakening scouring, the riverbed around the piers gradually formed siltation.

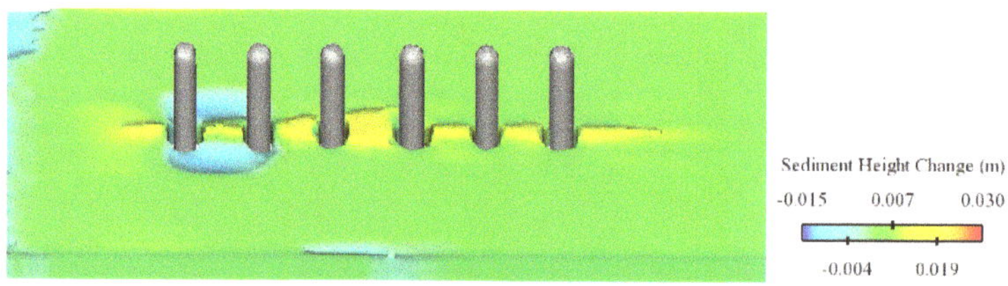

(**a**)

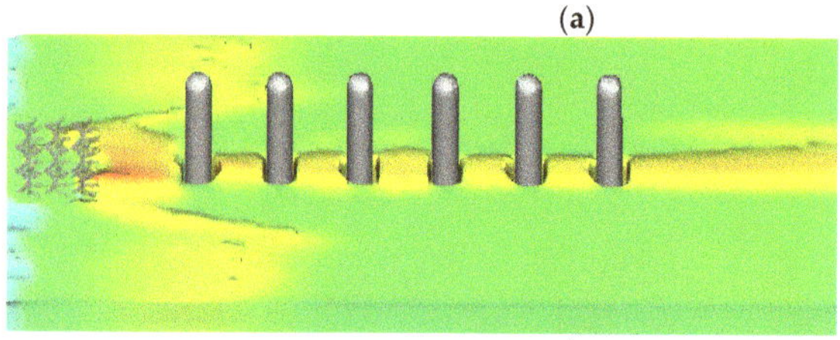

(**b**)

Figure 14. Comparison of the three-dimensional riverbed topography around the piers after erosion and sedimentation without (**a**) and with (**b**) the quasi-stumps group. (**a**) Without the protection of the quasi-stumps group. (**b**) With the protection of the quasi-stumps group.

During the whole experiment, the quantities of suspended sediment deposition in experiment NO. 0 (without quasi-stumps group) and experiment NO. 3 (with quasi-stumps group) were 3.69 kg and 9.16 kg, respectively. The latter was 2.48 times more than the former, indicating that the quasi-stumps group effectively promoted the deposition of suspended sediment. In experiments NO. 0 and NO. 3, the quantities of sediment scoured from the riverbed were 8.25 kg and 1.75 kg, respectively. It was observed that the former was 4.71 times greater than the latter. This indicates that the quasi-stumps group effectively weakens the erosion on the riverbed.

The increase in suspended sediment deposition and the decrease in sediment scoured from the riverbed were mainly attributed to the quasi-stumps group that could effectively lower the flow velocities around piers. Measurements of the flow velocities around the piers were taken, and the locations of the measurement points are shown in Figure 11. The flow velocity at each point both without (experiment NO. 0) and with (experiment NO. 3) the quasi-stumps group is shown in Figure 15, and all the points are located at one-fourth the water depth. Based on Figure 15, it is evident that the flow velocities around the piers with the quasi-stumps group were significantly lower than those without the quasi-stumps group.

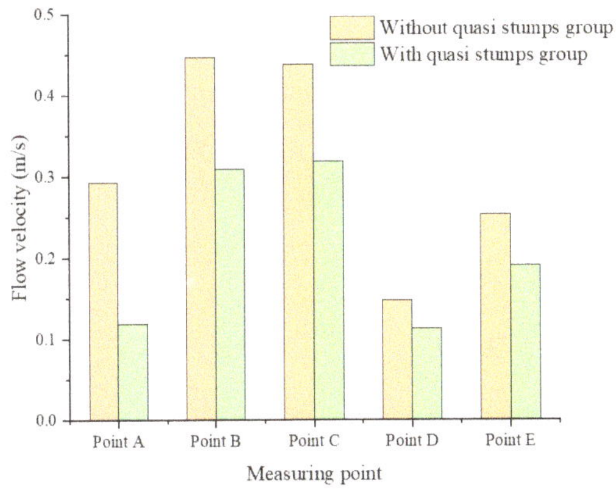

Figure 15. Comparison of the flow velocity at each point around the piers with and without the quasi-stumps group.

3.3. The effect of Horizontal Distance L on Siltation Characteristics

The distance between the quasi-stumps group and the piers significantly impacted the sediment deposition around the piers. After each experiment, the bed topographic elevation along the centerline of the piers was measured using a needle water level gauge. Figure 16 shows the longitudinal section of the riverbed topography along the centerline (along the flow direction) of the piers group at different distances *L*. At the initial time, the height of the riverbed was 0.12 m. As the value of *L* increased, the siltation height of both the upstream and downstream areas of the piers decreased gradually. Additionally, the reduction in flow velocities around the piers became less significant as the quasi-stumps group moved farther away from the piers. As a result, the deposition of the suspended sediment decreased. There was no erosion around the piers. Starting from pier NO. 1, the deposition height gradually decreased with the increasing *X*. The average deposition height for *L* = 5D, 4D, 3D, and 2D were 0.824, 0.878, 0.920, and 0.966 times that for *L* = D, respectively.

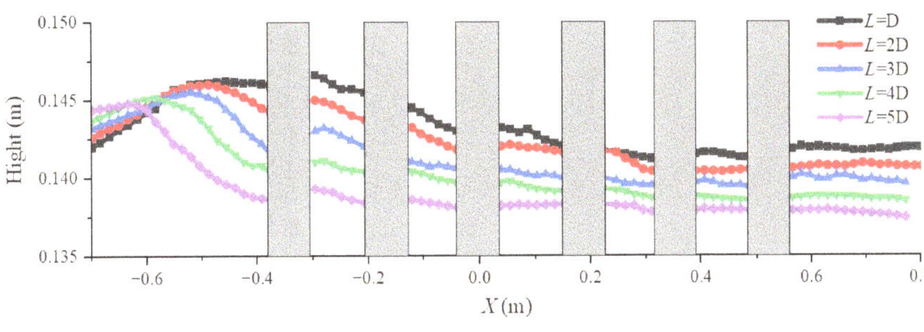

Figure 16. The longitudinal section of the riverbed along the centerline of the piers group at different *L*.

Figure 17 shows the siltation on the left side along the centerline of pier NO. 1 (perpendicular to the flow direction) at different distances *L*. As the value of *L* increases, the siltation height on both sides of pier NO. 1 gradually decreases, and the degree of change near the pier (*Y* = 0.03–0.18 m) is greater than that away from the pier (*Y* = 0.18–0.36 m). At the same *L*, the deposition height away from the pier is greater than that near the pier. This is due to the velocities and fluctuations of the flow near the pier that are higher than those farther away from the pier. From Figure 17, it can be observed that there are irregular variations in the sedimentation height in the areas far from the bridge piers. This is mainly because the quasi-stumps group has less influence on the flow velocities and sediment movements in the area away from the pier compared to the area near the pier.

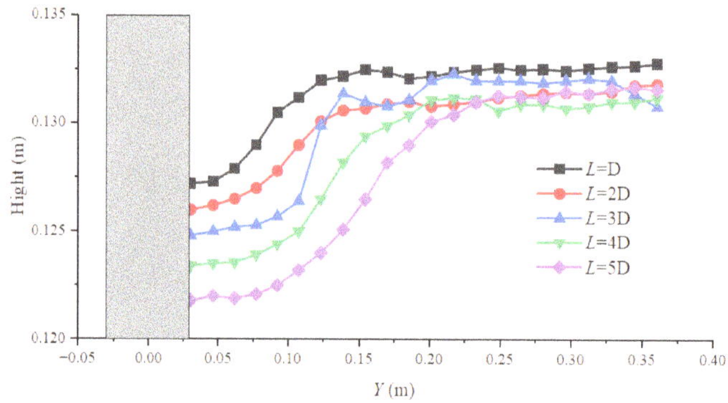

Figure 17. A cross-section of the riverbed along the centerline of pier NO. 1 on the left side at different L.

Figure 18 shows the variations in the dimensionless maximum deposition height d_h/D and deposition area A_h/A_D with the distance L. With the increase of L, the maximum deposition height decreases slowly, and the deposition area increases gradually. The average growth rates of the maximum deposition height and deposition area are −1.79% and 6.43%, respectively. The influence of L on the maximum deposition height is insignificant. The closer the quasi-stumps group is situated to the inlet, the greater the range of downstream velocity reduced by it, resulting in an increased deposition of the suspended sediment in the downstream area. Consequently, the deposition area also increases, albeit within a relatively small range of change.

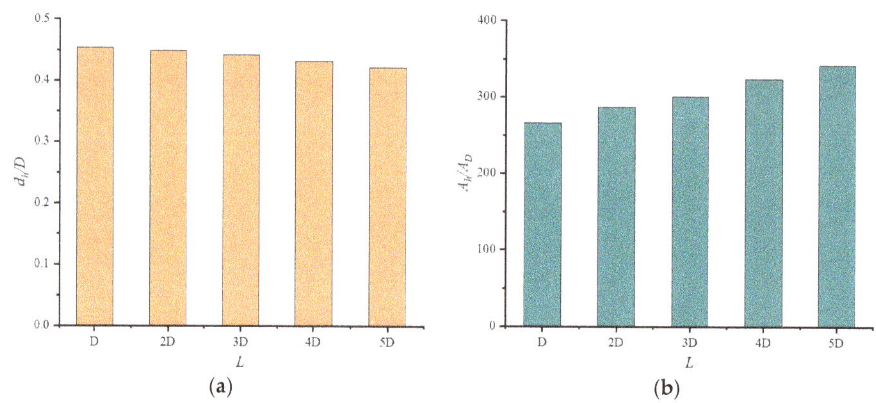

Figure 18. The change in the dimensionless maximum siltation height (**a**) and siltation area (**b**) with L. (**a**) Maximum siltation height. (**b**) Siltation area.

The quantity of sediment scoured from the riverbed and the quantity of suspended sediment deposition during the experiment vary with the distance L, as shown in Figure 19, and the duration of the experiment is one hour. From Figure 19, it is evident that the quantity of suspended sediment deposition is significantly greater than the quantity of sediment scoured from the riverbed. As the distance L increases, the quantity of suspended sediment deposition increases gradually, the quantity of sediment scoured from the riverbed decreases gradually, and their average growth rates are 9.69% and −4.01%, respectively. The total sediment in the riverbed increases with an increase in L, which is consistent with the previously observed variation of siltation area with L. On average, the quantity of suspended sediment deposition is 5.25 times greater than the quantity of sediment scoured from the riverbed.

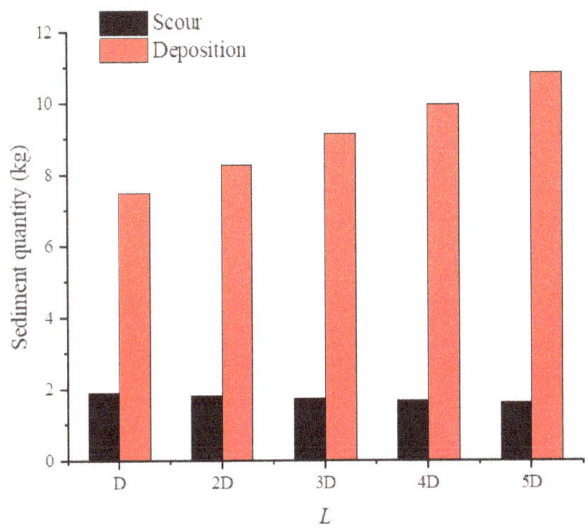

Figure 19. The variation in sediment scoured from the riverbed, and the deposition of suspended sediment with L.

The primary cause for alterations in the deposition height, deposition area, and sediment quantity is attributed to the changes in the flow velocities around the piers. Therefore, the flow velocities at various points around the piers under different distances L were measured. All points are situated at one-fourth the water depth. The changes in the flow velocity at each point concerning L are shown in Figure 20. Based on Figure 20, it is evident that the flow velocity of each point increases with an increase in L, and the change at point A is the largest. Points B and C have similar velocities, which are the highest among the five points. Points D and E are less affected by the changes in L. From Figures 19–21, it can be observed that, as L increases, the flow velocities at points A, B, C, D, and E gradually increase. However, at the same time, the deposition area and the deposited sediment also increase. This is mainly due to the change in location of the quasi-stumps group. The sedimentation is more significantly influenced by the position of the quasi-stumps group. As L increases, the closer the quasi-stumps group is to the flow inlet, and the larger the area in the flume where the flow velocities are reduced due to the presence of the quasi-stumps group; therefore, more sediment is deposited. Although the flow velocity at each point increases with the increase in L, all the flow velocities are reduced relative to the inlet flow velocity of 0.48 m/s.

3.4. The Effect of the Height P of the Quasi-Stumps Group on Siltation Characteristics

The movement of the sediment around the piers will be influenced by the height P of the quasi-stumps group. Figure 21 shows the longitudinal section of the riverbed topography at different heights P along the centerline of the piers group. With the increase in height P, the siltation height gradually increases. This is due to the enhanced ability of the quasi-stumps group to reduce the flow velocities and intercept the suspended sediment, leading to more suspended sediment being deposited downstream. The maximum siltation height occurs in the region between the quasi-stumps group and pier NO. 1. The average siltation height for P = 2H/3 and P = H is 1.081 and 1.166 times higher than that for P = H/3, respectively.

Figure 22 shows the siltation height along the left side of the centerline of pier NO. 1 at different heights P. The siltation height gradually increases with the increase in P, and the changes in the area (Y = 0.03–0.12 m) near the pier are greater.

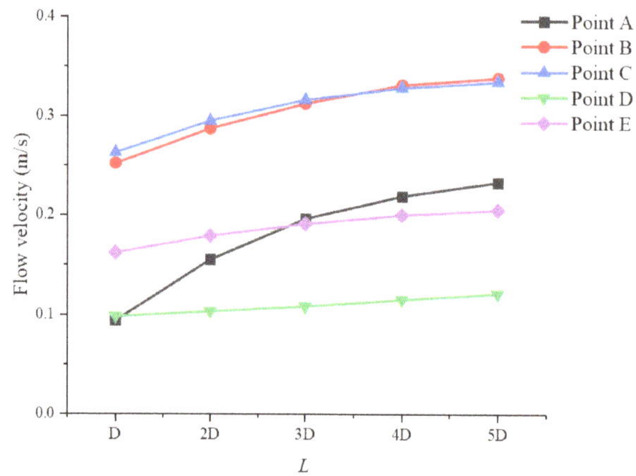

Figure 20. The variation of the flow velocities at five points around the piers with *L*.

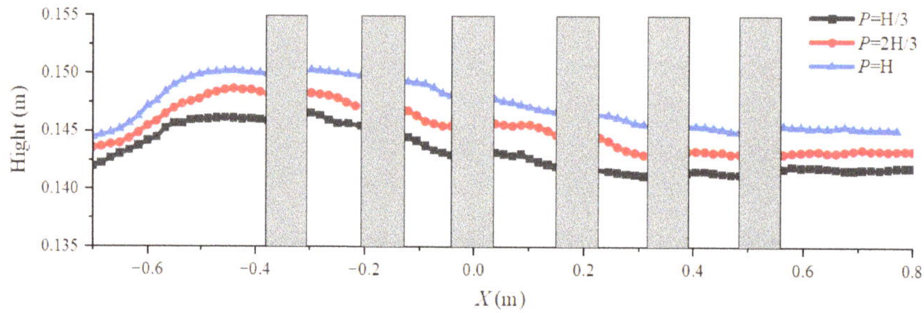

Figure 21. The longitudinal section of the riverbed along the centerline of the piers group at different *P*.

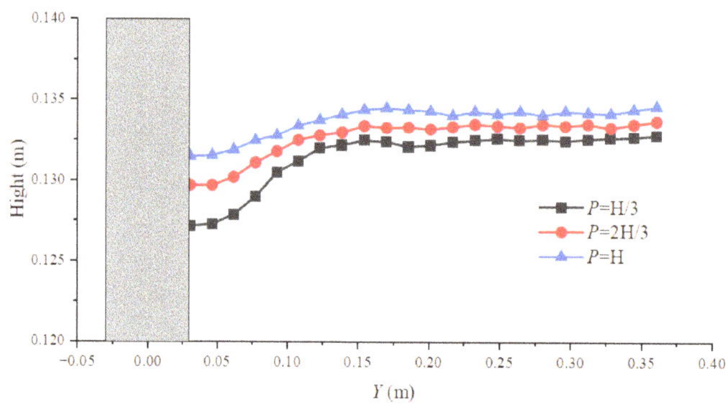

Figure 22. The cross-section of the riverbed along the centerline of pier NO. 1 on the left side at different *P*.

The dimensionless maximum siltation height d_h/D and siltation area A_h/A_D vary with the height *P*, as shown in Figure 23. It is observed that both the maximum siltation height and siltation area gradually increase with the increase in *P*, and their average growth rates are 6.77% and 18.55%, respectively. Notably, the siltation area appears to be more significantly influenced by the height of the quasi-stumps group than the maximum siltation height.

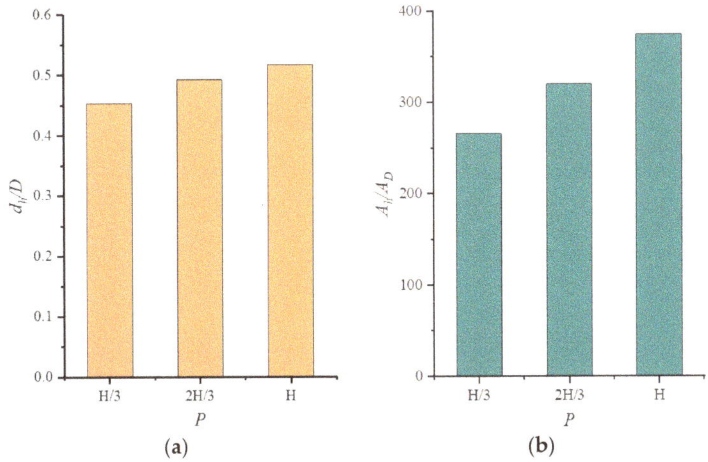

Figure 23. The change in the dimensionless maximum siltation height (**a**) and siltation area (**b**) with the *P*. (**a**) Maximum siltation height. (**b**) Siltation area.

During the experiment, the quantity of sediment scoured from the riverbed and the quantity of suspended sediment deposition vary with the height *P*, as shown in Figure 24. With the increase in height *P*, the quantity of sediment scoured from the riverbed gradually decreases, the quantity of suspended sediment deposition gradually increases, and their average growth rates are −6.18% and 24.59%, respectively. The height *P* has a significant impact on both, especially the deposition of the suspended sediment. On average, the quantity of the suspended sediment deposition is 5.39 times greater than the quantity of sediment scoured from the riverbed. Therefore, the ability of the quasi-stumps group to promote the deposition of the suspended sediment is significantly impacted by its height.

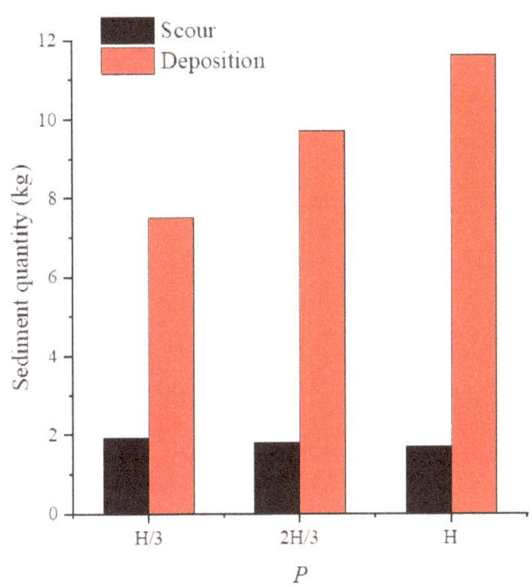

Figure 24. The variations in sediment scoured from the riverbed, and the deposition of the suspended sediment with the *P*.

The flow velocities of points around the bridge piers under different heights P were measured, and all points were located at half the depth of the water. The selection of half of the water depth was to better demonstrate the influence of the height of the quasi-stumps group on the flow velocities, as the heights of the quasi-stumps group were one-third of the water depth, two-thirds of the water depth, and equal to the water depth. The variations in the flow velocity at each point with L are shown in Figure 25. The flow velocity at each point decreases with the increase in height P, but the degree in decrease is different. Among them, the changes of points D and E are the smallest, and the flow velocities of points B and C are similar. When P is increased from H/3 to 2H/3, there is a significant decrease in the flow velocities observed at point A. This is due to the fact that the height of the quasi-stumps group changes from being below point A to being above it during this process. As the height P increases, the flow velocity decreases at various points around the piers, leading to a gradual increase in the maximum siltation height, siltation area, and quantity of the sediment deposition. This outcome is rational and in accordance with expected trends.

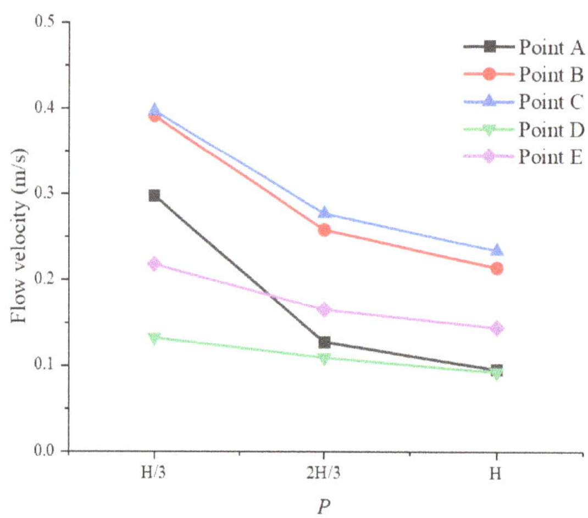

Figure 25. The variations in the flow velocities at five points around the piers with the P.

3.5. The Effect of the Ratio S between the Single Leaf Area and the Cross-Sectional Area of Single Pier on the Siltation Characteristics

The siltation around the bridge piers is influenced by the area of leaves in the quasi-stumps group. S represents the ratio of the area of a single leaf to the cross-sectional area of a single pier. Figure 26 shows the riverbed topography along the centerline of the piers group in a longitudinal section at different S. The siltation heights both upstream and downstream of the piers gradually increase with the increase in S. This indicates that, the larger the single leaf area, the stronger the ability of this structure to promote sediment deposition. The locations of the highest siltation points are nearly identical for different values of S. The average deposition heights for S = 0.148, 0.133, 0.118, and 0.103 are 1.156, 1.124, 1.079, and 1.034 times higher than S = 0.088, respectively.

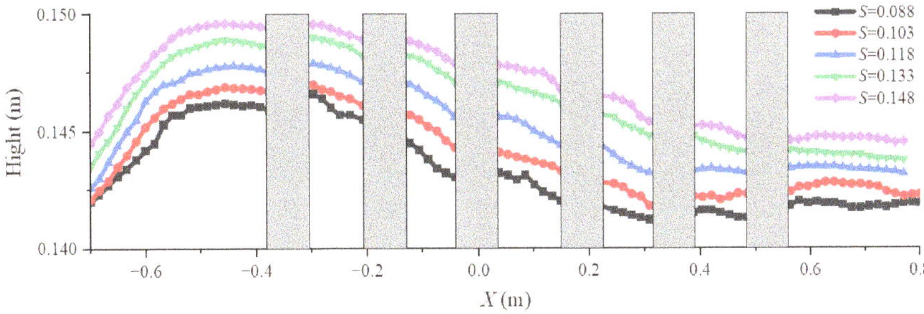

Figure 26. The longitudinal section of the riverbed along the centerline of the piers group at different *S*.

The data in Figure 27 demonstrate the height of siltation along the left side of the centerline of pier NO. 1 at different ratios *S*. The siltation height gradually increases with the increase in *S*. The deposition height of the riverbed close to the pier (Y = 0.03–0.15 m) is comparatively lower than that of the riverbed farther away from the pier (Y = 0.15–0.36 m). When S = 0.148, the siltation height at each point along the left side of the pier is approximately equal.

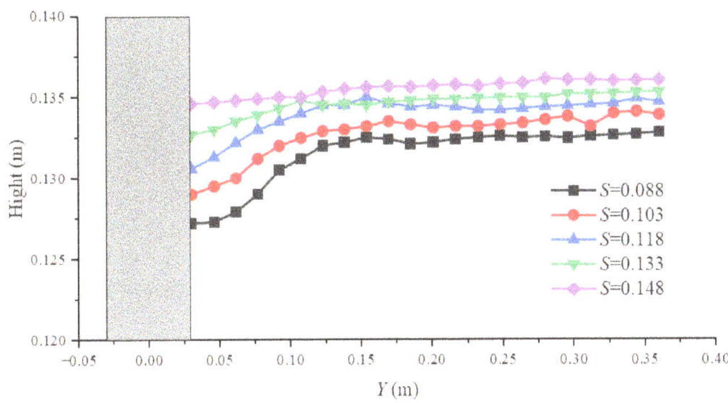

Figure 27. The cross-section of the riverbed along the centerline of pier NO. 1 on the left side at different *S*.

Figure 28 shows the variations of the dimensionless maximum siltation height d_h/D and siltation area A_h/A_D with the *S*. The results indicate that both the maximum siltation height and siltation area gradually increase with the increase in *S*. Their average growth rates are 2.66% and 10.0%, respectively. The effect of the single leaf area on the maximum siltation height is very small.

The quantity of sediment scoured from the riverbed and the quantity of suspended sediment deposition during the experiment vary with the ratio *S*, as shown in Figure 29. As the value of *S* increases, the quantity of sediment scoured from the riverbed decreases gradually, and the quantity of the suspended sediment deposition increases gradually. Their average growth rates are −7.58% and 10.22%, respectively. Furthermore, the mass of the suspended sediment deposition is, on average, 5.77 times greater than the mass of sediment scoured from the riverbed.

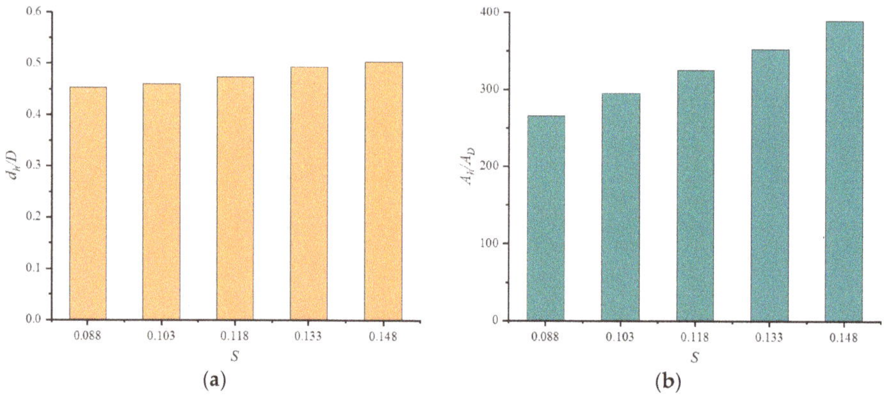

Figure 28. The changes in the dimensionless maximum siltation height (**a**) and siltation area (**b**) with the *S*. (**a**) Maximum siltation height. (**b**) Siltation area.

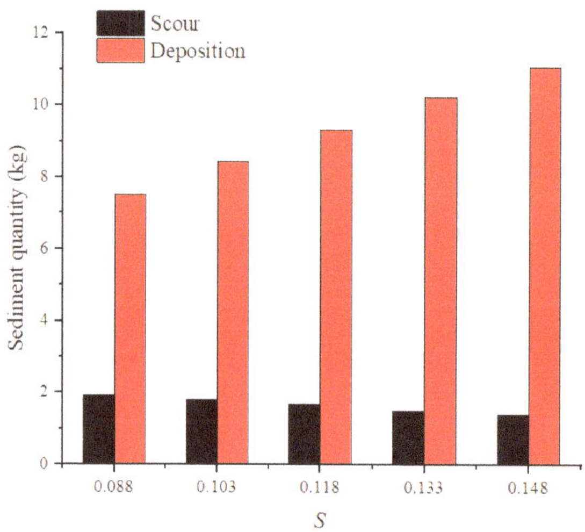

Figure 29. The variations in the sediment scoured from the riverbed and the deposition of the suspended sediment with the *S*.

During the experiment, the flow velocities around the piers were measured, and the measurement point locations are illustrated in Figure 11, where each point is posited at one-fourth the water depth. Figure 30 illustrates the variations in the the flow velocity at each point with the *S*. As the *S* increases, the flow velocities of all the points decrease. The most significant change is observed at point A, and the smallest change is noted at points D and E. These findings are consistent with the results of the variations in the siltation height, siltation area, and mass of the sediment on the riverbed with the *S*.

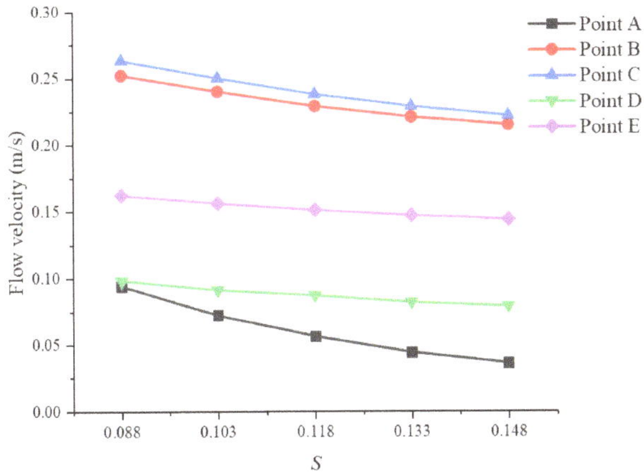

Figure 30. The variations in the flow velocities at five points around the piers with the S.

4. Discussion

The shapes of the scour holes around the piers are shown in Figure 14a, and they are different from the shapes of the scour holes obtained by Daneshfaraz et al. [42] and Kim et al. [43] through experiments, who believed that the riverbed upstream of the first pier should be scoured. However, as can be seen from Figure 14a, the riverbed upstream of the first pier was not scoured, but a small amount of sediment siltation was formed. This was mainly due to the RNG k-ε turbulence model used in the numerical simulation of this study, and the inlet flow was the carrying flow with a sediment content of 2.7 kg/m^3. The cross-section of the riverbed at the $Y = 0$ plane is shown in Figure 31, where the solid black line represents the original riverbed surface. It can be observed from the figure that the riverbed upstream of pier NO. 1 was not all silted up, and the scour pit was formed within a range 2.3 cm upstream of this pier. The scour pit is shown in the red dotted box in Figure 31.

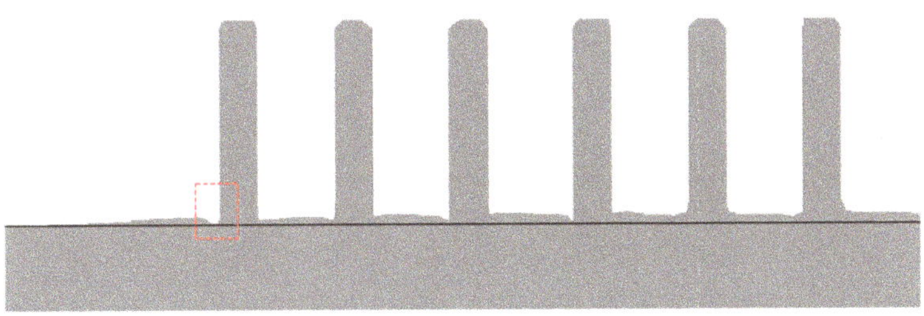

Figure 31. The height of the riverbed on the $Y = 0$ plane for experiment NO. 0.

The Reynolds-averaged Navier–Stokes equation model (RANS) cannot effectively capture the high-energy turbulent eddies at the junction between the upstream face of the bridge pier and the riverbed, which will cause the riverbed erosion upstream of the pier to be less than the actual situation [44]. The RNG k-ε turbulence model used in this study, as a kind of Reynolds-averaged Navier–Stokes model, also has this problem. Many simulation results based on RANS models exhibit this same phenomenon as well. For instance, Wang et al. [45] calculated the scour holes around a pier using the RNG k-ε turbulence model, as shown in Figure 32a. Li et al. [46] obtained the scour holes at a flow velocity of 0.4 m/s using the k-ω turbulence model, as shown in Figure 32b. ω is also the turbulence dissipation rate, which represents the rate of energy dissipation in a turbulent flow. In a turbulent flow, vortices are continuously generated and dissipated, leading to energy conversion and dissipation during this process. Compared to the k-ε turbulence model, the k-ω turbulence model solves for a specific rate ω of kinetic energy dissipation, and it is a low Reynolds number model. Its nonlinearity is greater, making it more difficult to converge, and it is quite sensitive to the initial guess of the solution. The k-ε turbulence model performs well in many common turbulent flow fields, while the k-ω turbulence model is more suitable for simulating rotational flows. Wang et al. [44] obtained the scour hole using the realizable k-ε turbulence model, as shown in Figure 32c.

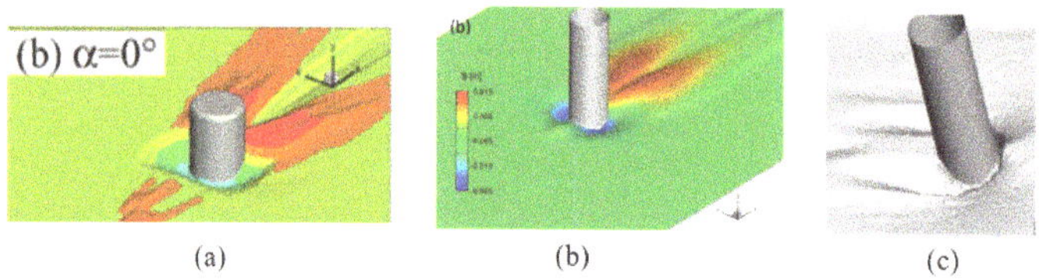

Figure 32. The scour pits obtained by other researchers: (**a**) Ref. [45]; (**b**) Ref. [46]; (**c**) Ref. [44].

The streamlines around the first and second piers of experiment NO. 0 are shown in Figure 33, which are similar to the streamlines of the longitudinal section obtained by Li et al. [47]. It can be observed from the figure that the streamlines upstream of the first pier are smoother, so the erosion effect of the flow on the riverbed is small. Figure 34 shows the velocity nephogram on the plane of Z = 0.15 m. The flow velocities in the area upstream of the first pier are reduced, as shown in the red box in Figure 34, which allows some of the suspended sediment located at the bottom of the flow to be deposited there.

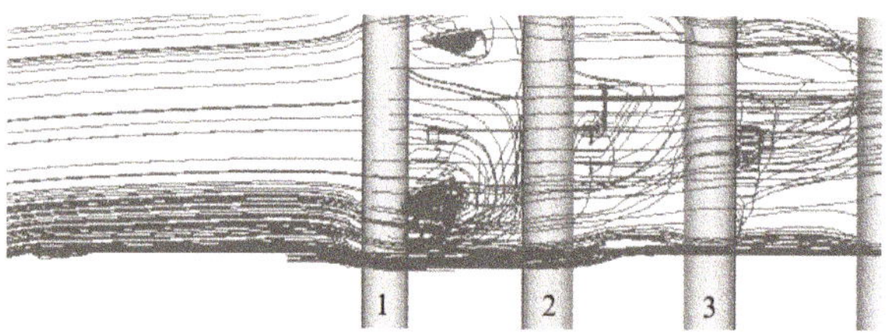

Figure 33. The streamlines around the piers in experiment NO. 0.

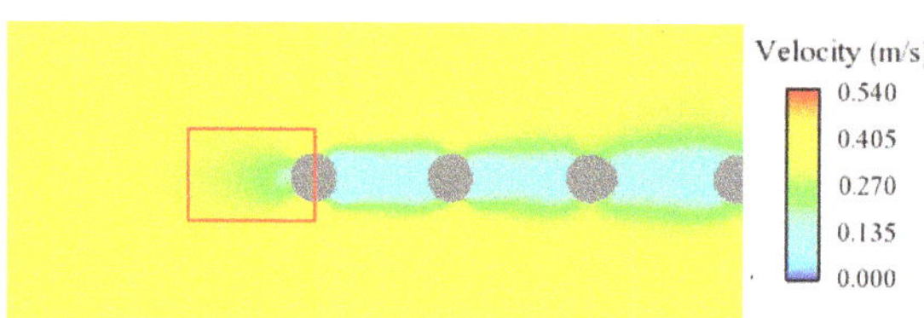

Figure 34. The velocity nephogram of experiment NO. 0 on the plane of Z = 0.15 m.

For experiments NO. 1 to 11 with the protection of the quasi-stumps group, the main reason for the significant accumulation of sediment around the piers is that the quasi-stumps group can effectively reduce the downstream flow velocities, thus weakening the shear stress of the flow on the riverbed and promoting the deposition of more suspended sediment around the piers. Additionally, the sediment concentration in the water downstream of the quasi-stumps group is higher than that of the inlet flow due to the decrease in the flow velocities. The presence of suspended sediment in the flow will reduce the shear stress on the riverbed and diminish the turbulence within the flow [48] and further reduce the erosion of the riverbed by the flow.

The flow field and scour around the group of bridge piers are very complex and influenced by numerous factors, such as the flow velocity, sediment particle size and gradation, distance between piers, angle between the piers group and the flow direction, water depth, and more. The current study in this paper is very limited and only investigates the effects of the quasi-stumps group and its L, P, and S on the siltation characteristics around the piers group. Many of the initial conditions in this study are predefined and remain unchanged. In future research, the initial conditions can be modified, and the effects of these initial conditions on the scour and siltation can be analyzed to better understand the scour mechanism of a riverbed around bridge piers.

5. Conclusions

In this study, both experiments and numerical simulations using FLOW 3D were employed to examine the protective effect of a quasi-stumps group on the local scour of bridge piers. Specifically, the scour and deposition of the riverbed around the piers, as well as the flow velocities, were compared under two conditions: with and without the quasi-stumps group. Three factors affecting the protective effect of the quasi-stumps group were studied: the horizontal distance L between the quasi-stump group and the pier, the height P of the quasi-stumps group, and the ratio S of the area of a single leaf on the quasi-stumps group to the cross-sectional area of a single pier. The effects of various factors on the siltation height at the centerline of the piers, the maximum siltation height, the siltation area, the change in quantity of the bedload sediment and suspended sediment, and the flow velocities around the piers were analyzed. The results are as follows:

(1) The results of the numerical simulations using FLOW 3D are in good agreement with the experimental results.
(2) The presence of the quasi-stumps group can effectively lower the flow velocities around the piers, promote the deposition of suspended sediment, and reduce the sediment scoured from the riverbed. It does not only help safeguard the riverbed around the piers against scouring but also forms siltation, which is beneficial to the stability of the bridge piers.
(3) As the distance L increases, the siltation height at the centerline of the piers group decreases gradually. However, the siltation area and the quantity of the suspended sediment deposition on the entire riverbed gradually increase.
(4) With the increase in the height P and ratio S, the siltation height at the centerline of the piers group, the maximum siltation height, the siltation area, and the quantity of the suspended sediment deposition gradually increase, while the flow velocities around the piers gradually decrease.
(5) The quantity of sediment scoured from the riverbed is less influenced by the L, P, and S.
(6) In this study, the combination of the quasi-stumps group with the best protective effect is P = H, S = 0.148, and L = D.

Author Contributions: Data curation, J.W. and W.T.; Supervision, Y.C.; Investigation, J.W. and W.T.; Software, Q.Z. and W.T.; Methodology, Y.Z.; Visualization, Y.C.; and Writing—review and editing, Y.Z. All authors have read and agreed to the published version of the manuscript.

Funding: This research received no external funding.

Data Availability Statement: The data that support the findings in this study are available from the corresponding author upon reasonable request.

Conflicts of Interest: The authors declare no conflict of interest.

Appendix A

The meanings of the symbols used in this paper are as follows:

Symbol	Meaning
D	Pier diameter
V	Average velocity upstream of the pier
H	Water depth
d_m	Maximum scouring depth
d_a	Average scouring depth
d_h	Maximum deposition height
A_h	Deposition area
A_D	Cross-sectional area of a single pier
S	Ratio of the area of a single leaf to the cross-sectional area of a single pier
P	Height of the stumps group in the water
L	Horizontal distance between the downstream edge of the stumps group and the upstream edge of Pier NO. 1
RANS	Abbreviation of the Reynolds-averaged Navier–Stokes equation model
RNG	Abbreviation of renormalization group
k	Turbulent kinetic energy of the turbulence model
ε	Turbulent dissipation rate of the k-ε turbulence model
ω	Turbulent dissipation rate of the k-ω turbulence model
X, Y, Z	Coordinate values in the x, y, and z directions of the coordinate axes are shown in Figure 5a

References

1. Smith, D.W. Why do bridges fail? *Civ. Eng.* **1977**, *47*, 58–62.
2. Lagasse, P.F.; Clopper, P.E.; Zevenbergen, L.W.; Girard, L.G. *NCHRP Report 593: Countermeasures to Protect Bridge Piers from Scour*; Transportation Research Board: Washington, DC, USA, 2007.
3. Arneson, L.A.; Zevenbergen, L.W.; Lagasse, P.F.; Clopper, P.E. *Evaluating Scour at Bridges*, 5th ed.; Bridge Superstructures; U.S. Department of Transportation: Washington, DC, USA, 2012.
4. Chiew, Y.M. Scour protection at bridge piers. *J. Hydraul. Eng. ASCE* **1992**, *118*, 1260–1269. [CrossRef]
5. Wang, C.; Yu, X.; Liang, F.Y. A review of bridge scour: Mechanism, estimation, monitoring and countermeasures. *Nat. Hazards* **2017**, *87*, 1881–1906. [CrossRef]
6. Jalal, H.K.; Hassan, W.H. Effect of Bridge Pier Shape on Depth of Scour. *IOP Conf. Ser. Mater. Sci. Eng.* **2020**, *671*, 12001. [CrossRef]
7. Jahangirzadeh, A.; Basser, H.; Akib, S.; Karami, H.; Naji, S.; Shamshirband, S. Experimental and Numerical Investigation of the Effect of Different Shapes of Collars on the Reduction of Scour around a Single Bridge Pier. *PLoS ONE* **2014**, *9*, e98592. [CrossRef] [PubMed]
8. Nazari-Sharabian, M.; Nazari-Sharabian, A.; Karakouzian, M.; Karami, M. Sacrificial Piles as Scour Countermeasures in River Bridges a Numerical Study using FLOW-3D. *Civ. Eng. J.* **2020**, *6*, 1091–1103. [CrossRef]
9. Dey, S.; Sumer, B.M.; Fredsoe, J. Control of scour at vertical circular piles under waves and current. *J. Hydraul. Eng.* **2006**, *132*, 270–279. [CrossRef]
10. Kumar, V.; Raju, K.G.R.; Vittal, N. Reduction of local scour around bridge piers using slots and collars. *J. Hydraul. Eng. ASCE* **1999**, *125*, 1302–1305. [CrossRef]
11. Mou, X.; Wang, D.; Ji, H.; Li, C.; Qiao, C. Resistance capability and hydraulic characteristics of ring-wing plates against local scour of round-ended piers. *South North Water Transf. Water Sci. Technol.* **2017**, *15*, 146–155. [CrossRef]
12. Zhang, Y.S.; Wang, J.F.; Zhou, Q.; Li, H.S.; Tang, W. Investigation of the reduction of sediment deposition and river flow resistance around dimpled surface piers. *Environ. Sci. Pollut. Res.* **2023**, *20*, 52784–52803. [CrossRef]
13. Chiew, Y.M. Mechanics of riprap failure at bridge piers. *J. Hydraul. Eng. ASCE* **1995**, *121*, 635–643. [CrossRef]
14. Parola, A.C.; Mahavadi, S.K.; Brown, B.M.; ElKhoury, A. Effects of rectangular foundation geometry on local pier scour. *J. Hydraul. Eng. ASCE* **1996**, *122*, 35–40. [CrossRef]
15. Li, Z.; Tang, H.; Dai, W. Case study of protective effects of ripraps or tetrahedron frame group with different densities on local scour around a pier. *J. Sediment Res.* **2011**, *6*, 75–80. [CrossRef]
16. Jones, J.S.; Kilgore, R.T.; Mistichelli, M.P. Effects of footing location on bridge pier scour. *J. Hydraul. Eng. ASCE* **1992**, *118*, 280–290. [CrossRef]
17. Shilong, F.; Hongwu, T.; Yilin, Z.; Kaixi, C.A.I. Experimental study on effect of local scour at piers and protection by tetrahedron frame. *Adv. Water Sci.* **2006**, *17*, 354–358. [CrossRef]
18. Tabarestani, M.K.; Zarrati, A.R. Design of Stable Riprap around Aligned and Skewed Rectangular Bridge Piers. *J. Hydraul. Eng.* **2013**, *139*, 911–916. [CrossRef]
19. Melville, B.W. Live-bed scour at bridge piers. *J. Hydraul. Eng. ASCE* **1984**, *110*, 1234–1247. [CrossRef]
20. Abt, S.R.; Clary, W.P.; Thornton, C.I. Sediment deposition and entrapment in vegetated streambeds. *J. Irrig. Drain. Eng. ASCE* **1994**, *120*, 1098–1111. [CrossRef]

21. Sun, Z.L.; Zheng, J.Y.; Zhu, L.L.; Chong, L.; Liu, J.; Luo, J.Y. Influence of submerged vegetation on flow structure and sediment deposition. *J. Zhejiang Univ. Eng. Sci.* **2021**, *55*, 71–80. [CrossRef]
22. Nabaei, S.F.; Afzalimehr, H.; Sui, J.Y.; Kumar, B.; Nabaei, S.H. Investigation of the Effect of Vegetation on Flow Structures and Turbulence Anisotropy around Semi-Elliptical Abutment. *Water* **2021**, *13*, 3108. [CrossRef]
23. Amir, M.; Hashmi, H.N.; Baloch, M.; Ehsan, M.A.; Muhammad, U.; Ali, Z. Experimental investigation of channel bank vegetation on scouring characteristics around a wing wall abutment. *Mehran Univ. Res. J. Eng. Technol.* **2018**, *23*, 16–22.
24. Wang, H.; Tang, H.; Zhao, H.; Zhao, X.; Lu, S. Incipient motion of sediment in presence of submerged flexible vegetation. *Water Sci. Eng.* **2015**, *8*, 63–67. [CrossRef]
25. Hongwu, T.; Shengqi, L.U.; Jianchuan, L. Settling velocity of coarse sediment particles in still water with rigid vegetation. *J. Hydraul. Eng.* **2007**, *38*, 1214–1220. [CrossRef]
26. Zhao, H.Q.; Yan, J.; Yuan, S.Y.; Liu, J.F.; Zheng, J.Y. Effects of Submerged Vegetation Density on Turbulent Flow Characteristics in an Open Channel. *Water* **2019**, *11*, 2154. [CrossRef]
27. Xiong, W.; Tang, P.B.; Kong, B.; Cai, C.S. Computational Simulation of Live-Bed Bridge Scour Considering Suspended Sediment Loads. *J. Comput. Civ. Eng.* **2017**, *31*, 12. [CrossRef]
28. Yu, S.L.; Dai, H.C.; Zhai, Y.W.; Liu, M.Y.; Huai, W.X. A Comparative Study on 2D CFD Simulation of Flow Structure in an Open Channel with an Emerged Vegetation Patch Based on Different RANS Turbulence Models. *Water* **2022**, *14*, 2873. [CrossRef]
29. Chen, M.; Lou, S.; Liu, S.; Radnaeva Larisa, D.; Nikitina, E.; Chalov Sergey, R. Numerical Simulation of Wave Propagation and Sediment Suspension Affected by Submerged Rigid Vegetation. *J. Tongji Univ. Nat. Sci.* **2022**, *50*, 861–870. [CrossRef]
30. Asadollahi, M.; Vaghefi, M.; Akbari, M. Effect of the position of perpendicular pier groups in a sharp bend on flow and scour patterns: Numerical simulation. *J. Braz. Soc. Mech. Sci. Eng.* **2020**, *42*, 422. [CrossRef]
31. Bozkus, Z.; Ozalp, M.C.; Dincer, A.E. Effect of Pier Inclination Angle on Local Scour Depth Around Bridge Pier Groups. *Arab. J. Sci. Eng.* **2018**, *43*, 5413–5421. [CrossRef]
32. Heidarpour, M.; Afzalimehr, H.; Izadinia, E. Reduction of local scour around bridge pier groups using collars. *Int. J. Sediment Res.* **2010**, *25*, 411–422. [CrossRef]
33. Saghravani, S.F.; Azhari, A. Simulation of clear water local scour around a group of bridge piers using an Eulerian 3D, two-phase model. *Prog. Comput. Fluid Dyn.* **2012**, *12*, 333–341. [CrossRef]
34. Voskoboinick, A.; Voskoboinick, V.; Turick, V.; Voskoboinyk, O.; Cherny, D.; Tereshchenko, L. Interaction of Group of Bridge Piers on Scour. In *Advances in Computer Science for Engineering and Education III. Advances in Intelligent Systems and Computing (AISC 1247)*; Springer: Berlin/Heidelberg, Germany, 2021; pp. 3–17. [CrossRef]
35. Jahangirzadeh, A.; Akib, S. Experimental study for determination of collar dimensions around bridge pier. *Balt. J. Road Bridge Eng.* **2015**, *10*, 89–96. [CrossRef]
36. Obied, N.A.; Khassaf, S.I. Experimental Study for Protection of Piers Against Local Scour Using Slots. *Int. J. Eng.* **2019**, *32*, 217–222. [CrossRef]
37. Debnath, K.; Chaudhuri, S. Effect of suspended sediment concentration on local scour around cylinder for clay-sand mixed sediment beds. *Eng. Geol.* **2011**, *117*, 236–245. [CrossRef]
38. Chiew, Y.M.; Melville, B.W. Local scour around bridge piers. *J. Hydraul. Res.* **1987**, *25*, 15–26. [CrossRef]
39. Mastbergen, D.R.; van den Berg, J.H. Breaching in fine sands and the generation of sustained turbidity currents in submarine canyons. *Sedimentology* **2003**, *50*, 625–637. [CrossRef]
40. Soulsby, R. Chapter 9: Bedload transport. In *Dynamics of Marine Sands*; Ice Publishing: Leeds, UK, 1997; pp. 155–170.
41. Meyer-Peter, E.; Müller, R. Formulas for bed-load transport. In Proceedings of the 2nd Meeting of the International Association for Hydraulic Structures Research, Delft, The Netherlands, 7 June 1948; pp. 39–64.
42. Daneshfaraz, R.; Ghaderi, A.; Sattariyan, M.; Alinejad, B.; Asl, M.M.; Di Francesco, S. Investigation of Local Scouring around Hydrodynamic and Circular Pile Groups under the Influence of River Material Harvesting Pits. *Water* **2021**, *13*, 2192. [CrossRef]
43. Kim, H.S.; Roh, M.; Nabi, M. Computational Modeling of Flow and Scour around Two Cylinders in Staggered Array. *Water* **2017**, *9*, 654. [CrossRef]
44. Wang, F.; Zhang, B.; Qi, J. 3D numerical investigation of bridge pier-scour development using a dynamic-mesh updating technique. *South North Water Transf. Water Sci. Technol.* **2017**, *15*, 132–137. [CrossRef]
45. Wang, S.H.; Yang, S.Y.; He, Z.G.; Li, L.; Xia, Y.Z. Effect of Inclination Angles on the Local Scour around a Submerged Cylinder. *Water* **2020**, *12*, 2687. [CrossRef]
46. Li, J.Z.; Kong, X.; Yang, Y.L.; Deng, L.; Xiong, W. CFD investigations of tsunami-induced scour around bridge piers. *Ocean Eng.* **2022**, *244*, 16. [CrossRef]
47. Li, A.; Zhang, G.; Chandara, M.; Zhou, S. Influence of sediment transport rate formula on numerical simulation of local scour around a bridge pier. *J. Sediment Res.* **2022**, *47*, 15–22. [CrossRef]
48. Sheppard, D.M.; Odeh, M.; Glasser, T. Large scale clear-water local pier scour experiments. *J. Hydraul. Eng.* **2004**, *130*, 957–963. [CrossRef]

water

MDPI

Article

Numerical–Experimental Study of Scour in the Discharge of a Channel: Case of the Carrizal River Hydraulic Control Structure, Tabasco, Mexico

Christian Caballero [1], Alejandro Mendoza [1,*], Moisés Berezowsky [1] and Abel Jiménez [1,2]

1 Instituto de Ingeniería, Universidad Nacional Autónoma de México, Coyoacán 04510, Mexico; ccaballeroc@iingen.unam.mx (C.C.); mberezowskyv@iingen.unam.mx (M.B.); ajimenezc@iingen.unam.mx (A.J.)

2 Facultad de Ingeniería, Universidad Nacional Autónoma de México, Coyoacán 04510, Mexico

* Correspondence: amendozar@iingen.unam.mx; Tel.: +52-5556233600

Abstract: Scour processes occur, among other causes, by the interaction of flow and sediment transport around structures built within a river. It is important to identify the location and quantify the magnitude of scour to avoid under or overestimation since the former puts the structure at risk and the latter demands an excessive economic cost. The data obtained from a physical model with a scale of 1:60 (without distortion) was used to assess the accuracy of a 2D free-surface numerical model (2DH) in predicting maximum scour. The physical model corresponds to the hydraulic-regulating structure located at the entrance of the Carrizal River, tributary to the Mezcalapa River bifurcation, located in Tabasco, Mexico. The scour is produced by the effect of a jet generated by a discharge channel. The maximum experimental scour was compared to the results of a numerical model and the estimation of four empirical equations: Breusers, Farhoudi and Smith, Negm, and Dietz. Setting an acceptance threshold of 75% accuracy, only the Breusers method provided values close to the measured values, and the numerical model failed to reproduce the location and magnitude of maximum scour, yet when the 2DH model was calibrated for secondary flow effects in the hydraulics. This indicates that the application of 2D models for the estimation of scour caused by jets may not provide good results when secondary currents are developed.

Keywords: jet scour; physical model; 2D numerical model; discharge channel

1. Introduction

Certain conditions in rivers generate convective accelerations in the flow, which produce a change in the sediment transport capacity; this leads to an imbalance between the average sediment transport in the stream and the transport capacity that is altered by the convective acceleration. In such conditions, a new equilibrium in the bed must be attained through bed adaptation generated by scour [1]. The phenomenon of scour in river beds is a complex problem of fluvial hydraulics, and it occurs not only in the neighborhood of structures and waterworks but also in river bends or meanders [2]. The estimation of the magnitude of scour is important from an engineering perspective because an underestimation can lead to structural failures of hydraulic works, and an overestimation has an adverse effect on costs [2,3]. In many cases, the derivation of equations for the calculation of bed-load transport was performed from data obtained in laboratory straight channels [4], and most of them are limited in the representation of more complex phenomena, such as secondary flows, transversal bed slopes, separation zones, turbulence effects, etc. Alternatives, for example, ref. [1] indicate that when there are no specific formulas for the conditions being analyzed, it is worthwhile to use physical modeling in scaled-down geometries, where similarity laws are applied; however, it is complex to try to reproduce certain characteristics inherent to the physics of the flow and the properties of the sediments simultaneously.

Scour is classified as local and general [1]. Local scour is caused by the interaction between flow and hydraulic structures, for example, in the vicinity of abutments, bridge piers, or downstream channels discharging into a river. In the latter case, scour occurs due to a redistribution of the velocity concentrated in a reduced area, where the turbulence characteristics are more complex. On the other hand, general scour is slower than local scour, and it consists of the general degradation of the bed of the channel, and there are different causes. In some cases, it occurs due to a reduction in the channel cross-section, which increases the velocity; it also appears in meanders and confluences due to flow redistribution by helical flow and velocity gradients. The magnitude of scour downstream of the structure depends on its geometrical characteristics, as well as the hydrodynamic conditions and the physical characteristics of the bed material in the zone [1]. The most relevant parameter to estimate is the maximum scour; however, ref. [5] highlights the importance of considering the temporal evolution of scour; they identify a phase, which they call the development phase, and another, which they call the developed phase [6].

In the last century, several empirical formulas have been proposed to estimate the scour equilibrium depth; most of these formulas have been derived based on dimensional analysis with a heuristic approach. The formulas may contain multiple exponents that have been determined from experimental data fitting [5]. The methods are categorized according to the sediment transport conditions; on the one hand, when there is no upstream transport, and the pool depth caused by scour reaches asymptotically in time, it is called clear water conditions. On the other hand, when there is upstream sediment transport, it is called active bed conditions; in this case, the pool depth is reached more rapidly and oscillates around a mean value [1]. A key factor used to calculate the local scour is the ratio U/UT_C, where U is the depth-averaged velocity immediately upstream of the structure and UT_C is the critical velocity for incipient bed-load transport, both in (m/s). Currently, the estimation of maximum scour through formulas still has uncertainty when compared with the actual scour, e.g., Ref. [7] presents a series of comparisons between measured data and estimations with formulas proposed in the literature for the case of scour by a jet, in such work a large scatter of real values versus observations is observed. Ref. [8] also proposed a formula to calculate the scour produced by circular piles for clear water conditions, with an error of less than 25%.

Computational Fluid Dynamics (CFD) models of local scour are gaining popularity in the engineering field and are a promising tool [9]. A numerical scour model requires three main components: a turbulent flow solver, a sediment transport solver, and a morphodynamical model for bed evolution [10]. Large Eddy Simulation (LES) models can calculate flow structures better than Reynolds Average Navier–Stokes (RANS) models; however, scour calculation with LES models is very expensive from a computational perspective to calculate a complete scour process. Therefore, RANS models requiring less computational capacity are more widely used, and it is currently accepted that RANS-type models will be the choice in the near future [10]. In this direction, for example, the work of [11] used a physical and a numerical model to determine the scour due to the discharge of a jump. However, it is still necessary to identify the capabilities of the RANS models and characterize their accuracy for estimating scour. In the direction of the use of 2D morpho-hydraulic models for scour computation, there is the work of [12] that determines the water depth to bendway weir height ratio, where 2D scour modeling produces reasonable results. Ref. [13] proposes a modification to a 2D mobile-bed model to improve the results of simulations of scour-around structures.

In the context of the use of a 2D free-surface morpho-hydraulic numerical model (NM-2DH) to calculate the maximum scour, the research questions are: (1) What is the capacity of a 2DH morpho-hydraulic model to reproduce hydrodynamic processes, local scour, and bed evolution, for the laboratory conditions, where secondary flow is present? (2) What is the accuracy in calculating the maximum scour with empirical equations vs. experimental measurements? (3) What is the accuracy of the numerical model for estimating the temporal dynamics of bed evolution at the point of maximum local scour?

(4) What are the differences of depth-averaged flow field computed with a 3D model with respect to the computation of the 2DH free surface model?

2. Methods

To answer the formulated questions, two techniques were utilized, physical and numerical modeling. The physical modeling is utilized to obtain the reference values of bed evolution (scour) and flow field (water depth, velocities). The numerical modeling solves the governing equations of flow, sediment transport, and bed evolution. A 2DH numerical model was used to solve the flow and scour, and a three-dimensional numerical model was used to characterize the flow structures.

2.1. Physical Model

The physical model is based on the Macayo Hydraulic Control Structure, which is located at the entrance of the Carrizal River in Tabasco, Mexico (Figure 1a–e). It has the purpose of regulating the flow distribution at the Mezcalapa River bifurcation. The physical model was built in the Engineering Institute, UNAM (Figure 1f–i).

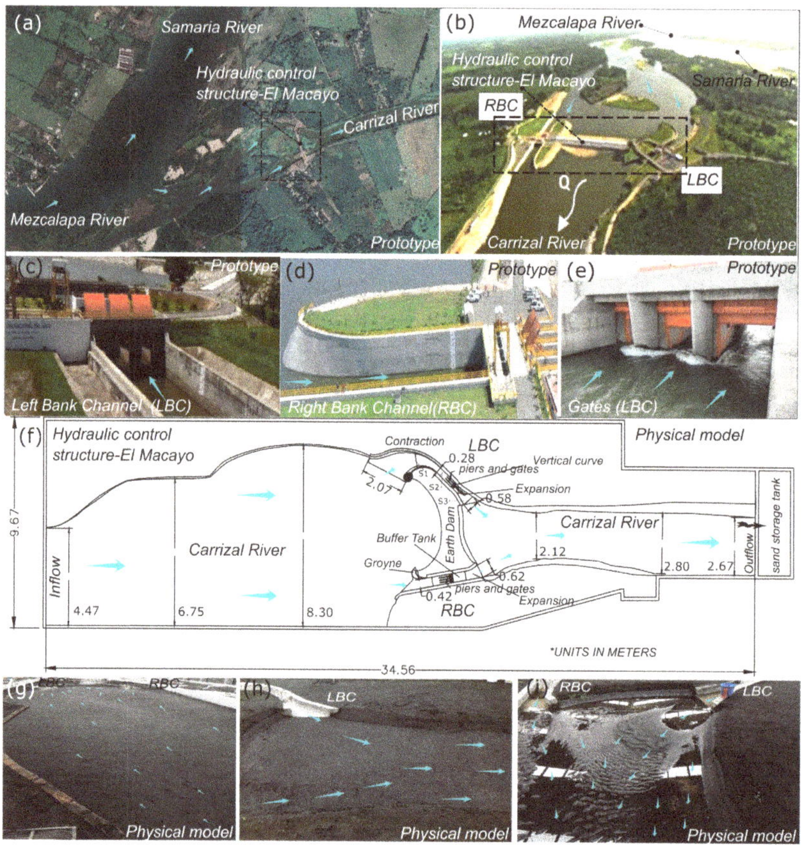

Figure 1. Prototype and experimental Physical Model (PM) of the Macayo Hydraulic Control Structure (MHCS). (**a**) MHCS location, (**b**) aerial view of the MHCS, (**c**) Left Bank Channel (LBC) and gates, (**d**) Right Bank Channel (RBC), (**e**) LBC gates seen upstream to downstream, (**f**) geometry of the experimental PM, (**g**) general view of the MHCS, (**h**) view of the discharge zone of the MHCS, (**i**) view of bed configuration at the end of the experiment. (**b**–**e**) Taken from [14] with authorization of the author.

The structure has a central dam and two discharge channels on both banks, the Left Bank Channel (LBC) and the Right Bank Channel (RBC). In each channel, radial gates are operated to regulate the flow captured by the Carrizal River. The physical model was built at a scale of 1:60. The design discharge in the real structure is 850 m^3/s; based on the Froude similarity principle, the discharge in the physical model corresponds to 30.48 L/s. The channels were made of concrete. Upstream and downstream, there is a layer of 20 cm of uniform fine sand, which has a median particle size of $D_{50} = 0.2664$ mm (from the particle size distribution curve, shown in Figure 2) and a density of 2723.75 kg/m^3. The standard method [15], was used to determine the particle size distribution analysis of the sediment used in the experiments. The uniformity of the fine sand particles was determined from the classification curve after calculating the geometric standard deviation (σ_g) with equation $\sigma_g = (d_{84}/d_{16})^{0.5}$ [16]. The diameters are $d_{16} = 14.44 \times 10^{-2}$ mm and $d_{84} = 42.88 \times 10^{-2}$ mm, and $\sigma_g = 1.72$. Therefore, the sand presents non-uniformity of the particles.

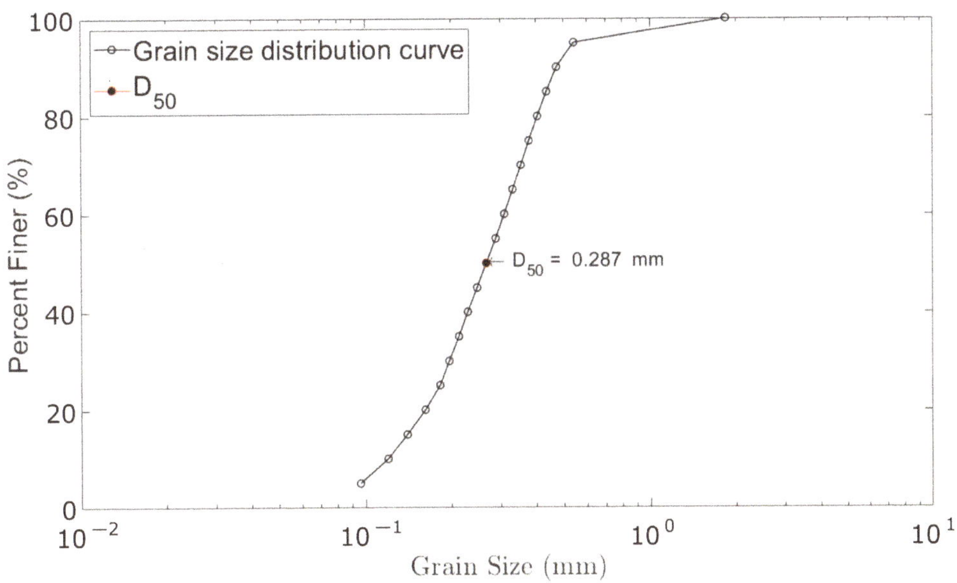

Figure 2. Bed sediment grain size distribution utilized in the physical model.

Flow Field Measurements

Measurements of the three velocity components were obtained with a Vectrino-II Doppler Acoustic Velocimetry Profiler. The volume of sampling is a cylindrical region located 50 mm below the central emitter, with a diameter of 6 mm and a standard height of 1–4 mm. The equipment measures three velocity components with sampling frequencies up to 100 Hz, and it can measure a complete 3 cm profile, over a range of 40 mm to 70 mm, from the central pulse emitter, with a resolution in the profile of up to 1 mm within this range. The measurements made here were used to calibrate the 2DH numerical model.

Three measurement sections were established in the LBC. The first one was located at the entrance of the channel, the second one just before the piles of the gates, and the third one at the exit, where the expansion of the channel ends. Figure 3a shows the location of the point of maximum scour. The velocity profile measured in Section 1 (S1) was measured 12 times, 5 times in Section 2 (S2), and 10 times in Section 3 (S3). In Figure 3a, the location of the measurement points in the three sections is also indicated. The Vectrino-II was placed on a stand along each of the three measurement sections, perpendicular to the main flow direction, and was moved vertically to obtain two profiles from the bed and determine the velocity field. Figure 3b shows the measurement zones (red dashed line box), as well as the water depth (blue line) for each Sections 1–3 (S1–S3). The water depth was 0.109 m, 0.098 m, and 0.103 m for S1, S2, and S3, respectively. Sampling at each point was performed for 180 s with a measurement frequency of 100 Hz. The velocities measured at 40% of the water depth from the bed were used to calibrate the 2DH numerical model. A detailed description of the operation of the measurement equipment can be found in [17].

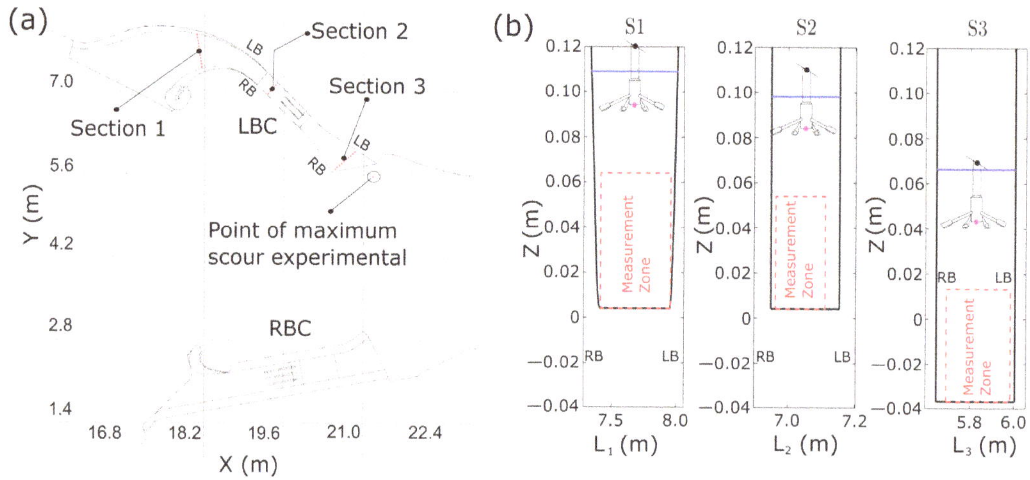

Figure 3. (**a**) Sections for velocity measurement in the LBC (S1, S2, S3) and maximum scour zone (SM). (**b**) Velocity measurement zone (highlighted in red) within the sections and position of the water surface.

In addition, velocity and bed elevation measurements were taken in the area where the maximum scour was determined; it was developed downstream of the outlet of the LBC. For this purpose, a previous experimental test was carried out to identify the zone of maximum scour. At this site, the Vectrino-II was placed at an elevation of 7 cm above the bed to measure the velocity profile with the initial bed (horizontal) before scour was developed. The objective was to obtain the velocity distribution in the first 3 cm above the original bed, with a vertical resolution of 1 mm, with a total of 30 measurement points. The equipment was oriented in the main flow direction and was configured for a measurement of 18 records (1R to 18R) with a frequency of 10 Hz for velocity and 2.5 Hz for measuring the position of the bed, with a duration of 15 min each, to obtain a continuous record during the time of the test (4.5 h).

2.2. Numerical Model

The numerical modeling was carried out with the Saint-Venant equations for the characterization of the flow, solved with the Telemac-2D module of the Telemac–Mascaret modeling system [18]. The Saint-Venant equations are expressed by Equations (1)–(3).

$$\frac{\partial h}{\partial t} + \boldsymbol{U} \cdot \nabla(h) + h\nabla \cdot (\boldsymbol{U}) = S_h, \quad (1)$$

$$\frac{\partial u}{\partial t} + \boldsymbol{U} \cdot \nabla(u) = \frac{1}{h} C_f u |\boldsymbol{U}| - g\frac{\partial \eta}{\partial x} + \frac{1}{h}\nabla \cdot (h\nu_t \nabla u), \quad (2)$$

$$\frac{\partial v}{\partial t} + \boldsymbol{U} \cdot \nabla(v) = \frac{1}{h} C_f v |\boldsymbol{U}| - g\frac{\partial \eta}{\partial y} + \frac{1}{h}\nabla \cdot (h\nu_t \nabla v), \quad (3)$$

where h is water depth, in m; S_h is the source or sink of fluid, in m/s; $\boldsymbol{U} = u\hat{\imath} + v\hat{\jmath}$ is the vector velocity with horizontal components u and v, in x and y directions, respectively, in m/s; η is the water free-surface elevation, in m; C_f is the coefficient of friction; t is time, in s; ν_t is the turbulent viscosity coefficient, in m^2/s.

The sediment transport and bed morphology were solved with the Sisyphe module of Telemac–Mascaret, using the mass conservation equation for sediment, or Exner equation [19], coupled with the Telemac-2D model [20]. The Exner equation is expressed by Equation (4)

$$1 - \lambda\frac{\partial Z_b}{\partial t} + \nabla \cdot \boldsymbol{Q}_b + (E - D)_{z=a} = 0, \quad (4)$$

where $\boldsymbol{Q}_b = Q_{bx}\boldsymbol{i} + Q_{by}\boldsymbol{j}$ is the vector of the unit volumetric transport rate without voids, in m^2/s; λ is the porosity of the bed material, dimensionless ($\lambda = 0.4$); Z_b is the bed elevation, in m; E and D are the entrainment and deposition rates at elevation $z = a$, the interface between the bed load and the suspended load mechanisms. The bed load is calculated as a function of sediment load capacity, where considerations of flow direction and bed slope are considered in the computation. In the case of coupling with Telemac2D, it considers the effect of secondary currents in the bed evolution. Sisyphe has been validated and calibrated [21,22]. Regarding bed load, the Meyer-Peter and Müller equation was utilized as a first approach, and the bed load was calibrated from the measured scour from the physical model experiments through the morphological factor explained in the next section.

The suspended transport concentration is calculated using the advection-diffusion equation.

$$\frac{\partial C}{\partial t} + U\frac{\partial C}{\partial x} + V\frac{\partial C}{\partial y} = \left[\frac{\partial}{\partial x}\left(\varepsilon_s\frac{\partial C}{\partial x}\right) + \frac{\partial}{\partial y}\left(\varepsilon_s\frac{\partial C}{\partial y}\right)\right] + \frac{(E - D)_{z=a}}{h}, \quad (5)$$

where C is the volumetric concentration of suspended sediment averaged vertically, ε_s is the diffusion coefficient for sediment, U and V are the depth-averaged flow velocities in the directions x and y, respectively, and h the water depth. The deposition (D) is determined by the reference concentration ($C_{z=a}$) located at an elevation a, the interface of the bed load and the suspended load depends on the fall velocity of the sediment and upstream conditions of suspended sediment given by Equation (5). While entrainment (E) is determined by the equilibrium concentration (C_{eq}) near the bed determined by an empirical formula, the Zyserman and Fredsoe equation is used [20]. Therefore, the net rate of sediment flux is determined based on the concept of equilibrium concentration, where the rate of sediment deposition is equal to the rate of sediment of the bed's entrainment.

The numerical analysis with the NM-2DH model was carried out through two processes, the first with a larger, general mesh of the domain, made up of 431,701 elements (Figure 4a), and the second with a reduced mesh, made up of 54,542 elements (Figure 4b). In the first analysis, the boundary conditions for the hydraulic calculation were established: (1) flow at the inlet and (2) the water level at the outlet of the downstream physical model, and the flow rates flowing through each of the channels were determined; then, the hydrodynamics calculation calibration process was developed (roughness coefficient and water levels), which is discussed later; for a hydrodynamics modeling computation time of 15 min, a processing time of 8.30 h was required, with a $\Delta t = 1 \times 10^{-3}$ s. The hydrodynamics modeling time was 8.30 h on a cluster with 60 AMD® processing cores, CentOS distribution, version 6.9, and 64-bit architecture. Subsequently, in order to reduce the calculation process times of the numerical analysis of scour and its subsequent 3D analysis, the reduced mesh was used, and the boundary conditions were established for each of the channels and a downstream level (data obtained from the first analysis), as well as the conditions of sediment transport in equilibrium. An important factor was to avoid erosion in the concrete channels, which was achieved with the NOEROD subroutine of Sisyphe. The computational meshes have the geometry and bed elevations associated with the physical model. An inlet boundary condition was established as a flow rate $Q = 30.48$ L/s ($Q_{RBC} = 0.54Q$; $Q_{LBC} = 0.46Q$, calculated by numerical integration, and the water levels were verified) and a water level at the outlet corresponding to $h = 10.25$ cm, and it was calculated with a Manning's roughness coefficient of 1.8×10^{-2} s/m$^{1/3}$ for the concrete channels (RBC and LBC) with less error relative, and in the rest of the model with a sand bed an n of 2.8×10^{-2} s/m$^{1/3}$, although a detailed discussion for determining roughness in erodible beds can be found, for example, in [19]. The processing time with the reduced mesh was reduced to 5.4 h.

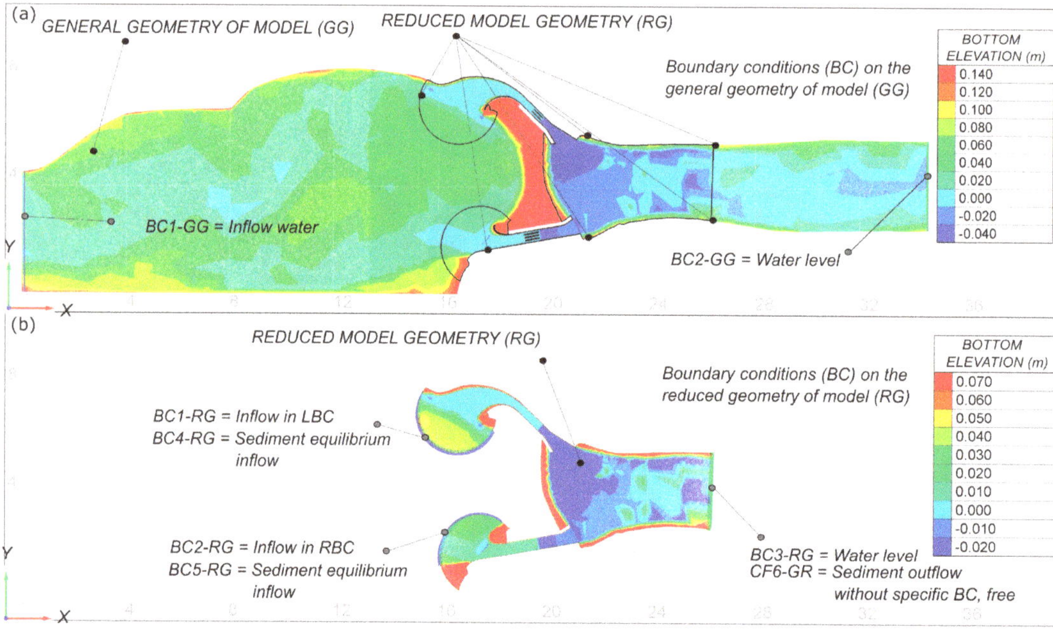

Figure 4. Computational meshes used for numerical simulations and boundary conditions, in (**a**) general mesh with triangular elements ranging from 5 cm to 0.5 cm, (**b**) reduced mesh with triangular elements ranging from 4 cm to 0.5 cm.

The 3D analysis was performed based on the results of the NM-2DH analysis, where the hydrodynamics were calibrated (first process), and the maximum scour was calculated (second process). An equivalent hydraulic roughness height of $k_s = 3 \times 10^{-3}$ m was determined for the concrete channels. The computation time to simulate 89 s of flow was five days, with a time step of $\Delta t = 5 \times 10^{-5}$ s using the same 60 processing cores and a three-dimensional mesh with eight planes in the vertical.

3. Results

3.1. Velocity Field Measurement

Figure 5 shows the distributions of the measured and computed velocity vectors in sections S1, S2, and S3 (the measured velocity was taken at the elevation $Z_m = 0.4$ h). The calculated velocity vectors were obtained with the Telemac-2D model. The cross-section averaged streamwise velocities were 0.236 m/s, 0.548 m/s, and 0.219 m/s for S1, S2, and S3. The velocity magnitude U increased by a factor of 2.3 for S2, and decreased slightly for S3, by a factor of 0.93, both with respect to the velocity measured in S1. It is important to highlight that a separation zone was developed near the right bank at the outlet of the channel, near S3 (gray oval, Figure 5), and downstream of the development of a recirculation zone with a very clear vortex, which is discussed below. Figure 5 also shows the point where the maximum scour was developed in the physical model, which is discussed in the next sections.

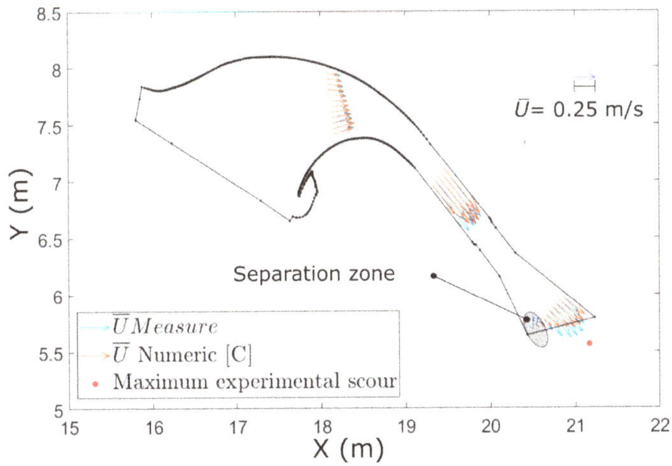

Figure 5. Velocity field (U) in m/s, measured at the LBC (blue) and calculated (orange) at S1, S2, and S3, and SM point. $\overline{U}$ is the time and depth-averaged velocity vector.

3.2. Numerical Model Calibration

A first computation of the flow with the NM-2DH was performed with the general mesh shown in Figure 4a; the boundary conditions were used for the flow rate at the inlet and the water level at the outlet. The development of secondary flows within the left bank channel was observed in the LBC, caused by its curvature, so the correction of this phenomenon in two-dimensional flow models was considered. Telemac2D uses the approach proposed by [23]. The velocity components, once the model was calibrated by adjusting the Manning and secondary flow production coefficients, are shown in Figure 5.

Figure 6a–c compare the measured versus calculated 2D velocity in three cross-sections (S1, S2, and S3). Three types of results were obtained: a model in which the Secondary Flow Correction (SFC) was not considered and two models with SFC using different secondary flow production coefficients. Figure 6a–c show the components $\overline{u}$ and $\overline{v}$ calculated with the numerical model 2DH, without considering SFC, and components calculated considering SFC, with a production coefficient of $\alpha = 14$ (recommended value, [18]), and $\alpha = 3.5$, in blue, green, and red colors, respectively. Experimental measurements are shown in black circles. Such calculations are identified as scenarios A, B, and C in Figure 5. A detailed analysis allowed us to identify whether the numerical results were improved by using any of the scenarios considering the SFC. For this purpose, the relative error (e_r) of the measured time and depth (0.4 h) averaged velocity with respect to each of the components ($\overline{u}, \overline{v}$) is calculated as $e_{rx} = (\overline{u}_m - \overline{u}_c/\overline{u}_m)$ and $e_{ry} = (\overline{v}_m - \overline{v}_c/\overline{v}_m)$, where the subscripts m and c stand for measured and calculated, respectively. Afterward, the average relative error in the cross-sections was obtained (e_{rxm}, e_{rym}). Finally, Figure 6d shows the generalized error (e_{rxmg}, e_{rymg}), which accounts for the behavior in the three sections S1, S2, and S3 of the 2DH model.

The overall mean relative errors (e_{rxmg}, e_{rymg}) in sections S1, S2, and S3 for scenarios A, B, and C are shown in Table 1. In general, the e_r is 88%, 24.7%, and 8.5% for component $\overline{u}$, while the $\overline{v}$ is 62.7%, 40.4%, and 29.4 % for the same scenarios. With respect to the velocity distributions, it is shown that the components for S1 have the lowest errors; in S2, there are slight differences with respect to measurements, while in S3, the distributions already have important differences. For example, the velocities for scenario A have a more uniform distribution across the width of the channel compared to scenario B, which fits better to the experimental measurements, and scenario C, has a better fit near the left margin for the component $\overline{u}$, and improves the approximation of $\overline{v}$.

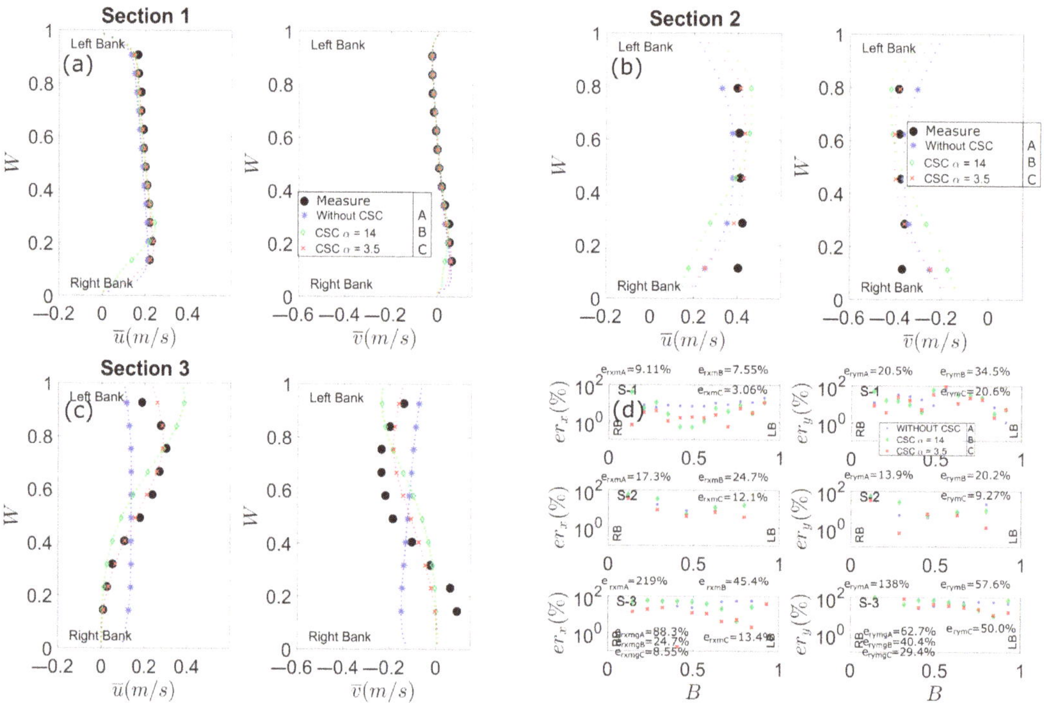

Figure 6. Measured and calculated velocity distributions, (**a**) in S1, (**b**) in S2, (**c**) in S3, and (**d**) relative errors.

Table 1. Mean and overall relative errors of numerical scenarios A, B, and C.

Section Scenario	S1 $\overline{u}$ (%)	S1 $\overline{v}$ (%)	S2 $\overline{u}$ (%)	S2 $\overline{v}$ (%)	S3 $\overline{u}$ (%)	S3 $\overline{v}$ (%)	General $\overline{u}$ (%)	General $\overline{v}$ (%)
A (without CFS)	9.11	20.5	17.3	13.9	219	138.0	88.3	62.76
B ($\alpha = 14$)	7.55	30.4	24.7	20.2	45.4	57.6	24.7	40.4
C ($\alpha = 3.5$)	3.06	20.6	12.1	9.27	13.4	50.0	8.55	29.4

The 3D computation of the flow field with the flatbed configuration is shown in Figure 7. In Figure 7a, the recirculation zones are identified with the help of streamlines. Also, the separation zone within the LBC is identified, which is an indicator that the flow is concentrated towards the left bank. Figure 7b shows the magnitude of the 3D velocity field in cross-sections S1, S2, and S3, where the separation zone and velocity reduction on the left of S3 are distinguished, as well as the concentration of the maximum velocity core at the outlet of the LBC. Figure 7c shows the streamlines with the vorticity magnitude, defined as $\boldsymbol{\Omega} = \nabla \times \boldsymbol{U}$, where $\boldsymbol{\Omega}$ is the vorticity vector (1/s) and $\boldsymbol{U}$ the velocity vector (m/s). The numerical results make it possible to identify zones where vortices and flow recirculation were developed downstream of the control structure; the lines of larger vorticity are in the narrowest zone of the LBC and in the transition zone of the material change (concrete–sand) where the maximum scour was developed.

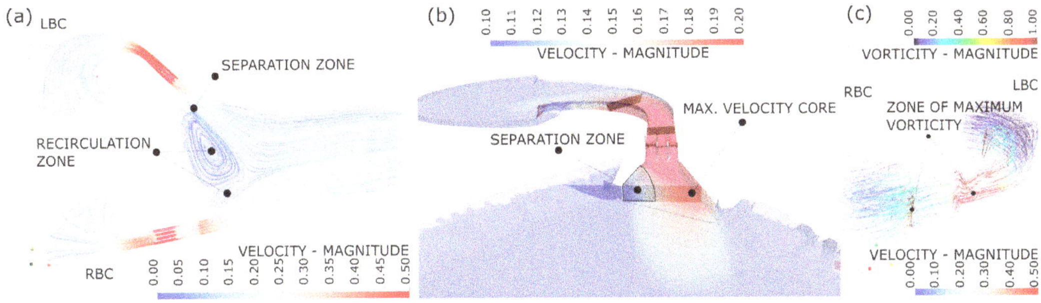

Figure 7. 3D numerical results with flatbed (no scour considered), (**a**) velocity streamlines (m/s), (**b**) 3D velocity magnitude field and separation zone, (**c**) streamlines mapped with vorticity (1/s).

The 3D flow field was also computed considering the bed configuration of the physical model developed after 4.5 h of bed evolution, where the maximum scours, the deposit formation, and the ripples were considered. Figure 8a shows the depth-averaged velocity computed by the NM-2DH model. In Figure 8c, the magnitude of the depth-averaged velocity computed from the numerical 3D velocity field is presented. Two differences are observed, on the one hand, the velocity field is more diffusive at the downstream outlet of the channels in the 2D results, and on the other hand, the flow at the outlet of the LBC computed by the 2D model leans towards the left bank. Figure 8b shows the depth-averaged turbulence kinetic energy (k) computed by the 2D model as $k = 0.5\left(\overline{u'^2} + \overline{v'^2} + \overline{w'^2}\right)$, where $\overline{u'^2}$, $\overline{v'^2}$, and $\overline{w'^2}$ are the time-averaged square of the fluctuations of the velocity caused by the turbulence. Figure 8d shows the depth-averaged horizontal value k computed from the 3D model results. Again, the 2DH model results are more diffusive. Figure 8e shows the Reynolds number (Re), and Figure 8f shows the general vorticity field.

Figure 9a shows the computed 3D velocity field vectors; the velocity core is concentrated near the surface, and the velocity is lower near the bed. A deviation is observed between the orientation of the scour hole and the direction of the flow. Figure 9b shows the vectors of the velocity field mapped with k located near the bed; the vectors indicate at least two interaction zones, one inside the scour hole in the inward direction and the other on the dune in the outward and downstream direction. These two zones have high turbulence kinetic energy, k. On the other hand, the vertical-oriented vorticity indicated at least two flow directions in the LBC outlet jet that interact with each other. Figure 8d shows the bed configuration analyzed, where the scour pool and the U-shaped deposition zone can be seen.

The turbulence kinetic energy (k) within the zone of maximum erosion is shown in Figure 10 at different horizontal planes. Figure 10a shows the complete volume ($-0.15 \leq Z \leq 0.17$); in Figure 10b, the volume ranges from $-0.15 \leq Z \leq 0.10$; in Figure 10c, the volume ranges from $-0.15 \leq Z \leq 0.05$; and in Figure 10d, the volume ranges from $-0.15 \leq Z \leq 0.025$. It can be seen how k is developed in the expansion inside the LBC and moves downstream; k increases between the recirculation zone (blue color) and the outflow jet. It also increases as it decreases in depth and increases in the dune zone.

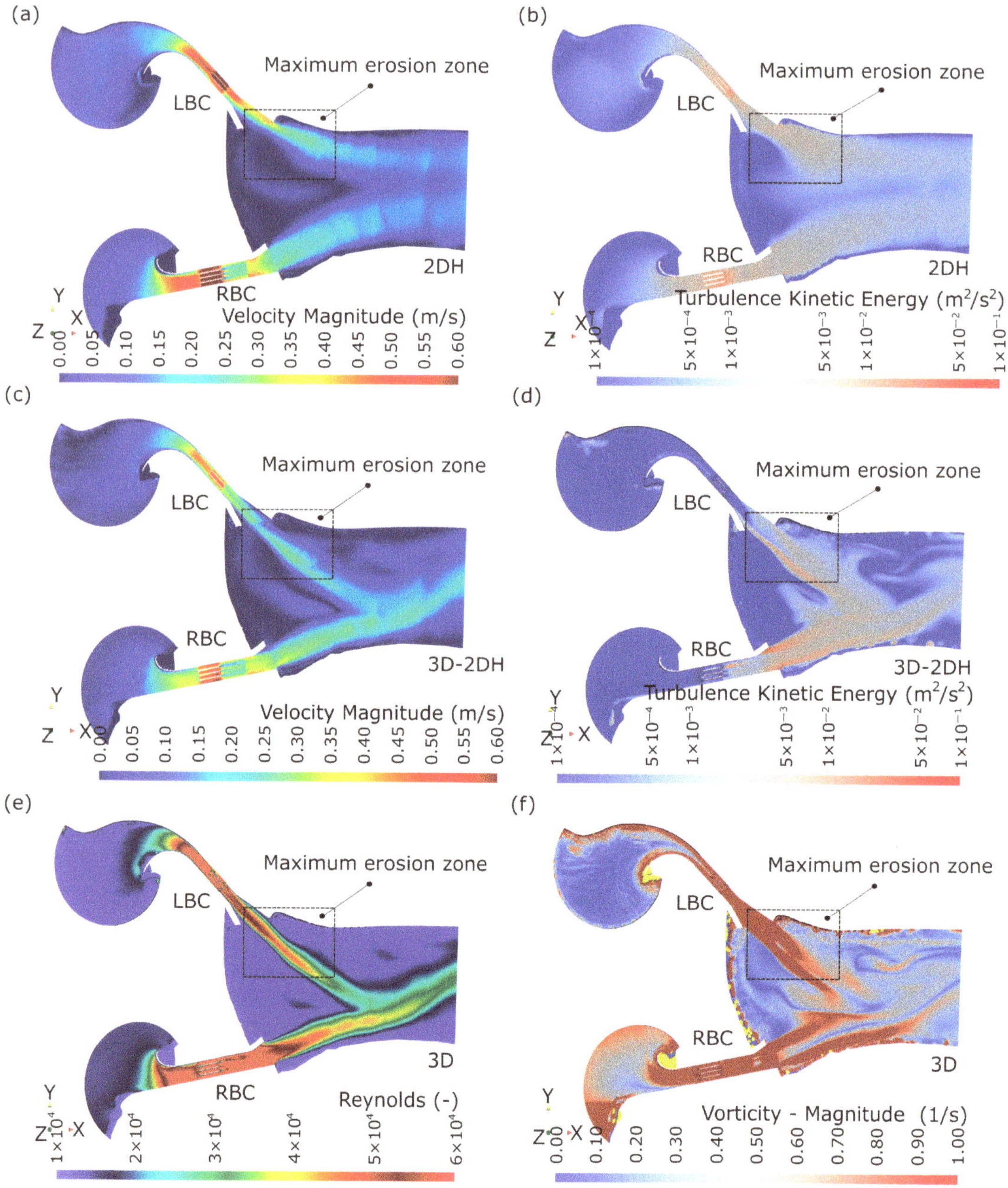

Figure 8. 2D and depth-averaged 3D numerical results with the developed bed configuration after 4.5 h, (**a**) 2DH velocity field (m/s), (**b**) depth-averaged turbulence kinetic energy computed with NM-2DH, k (m^2/s^2), (**c**) depth-averaged 3D velocity field (m/s), (**d**) depth-averaged turbulence kinetic energy computed from 3D model (m^2/s^2), (**e**) Reynolds number 3D, (**f**) Vorticity field from 3D model (1/s).

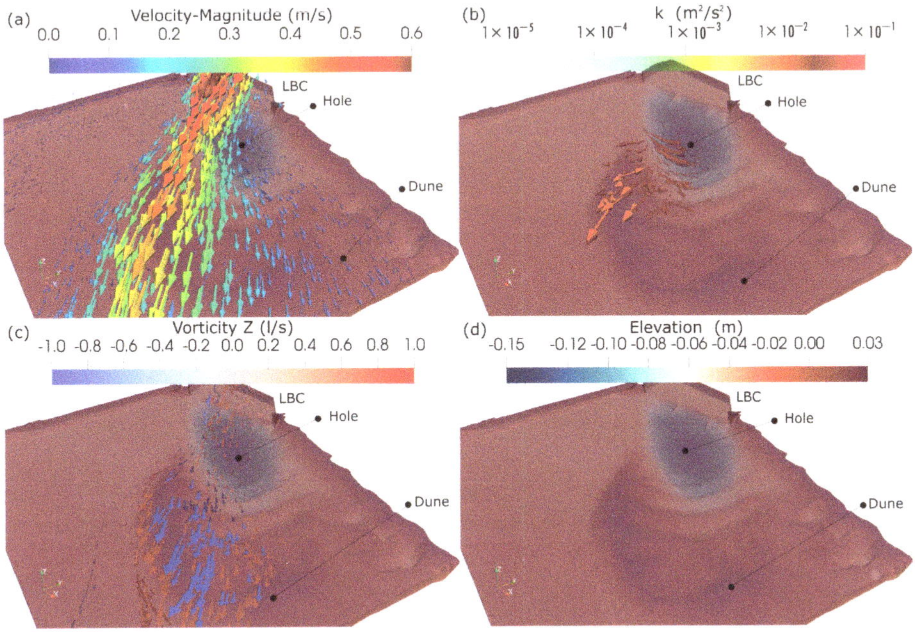

Figure 9. 3D numerical results with the bed configuration developed after 4.5 h, (**a**) 3D velocity field (m/s), (**b**) 3D velocity field mapped with k magnitude (m^2/s^2), (**c**) 3D vorticity field, Z-direction (1/s), and (**d**) elevation and bed configuration in maximum scour zone (m).

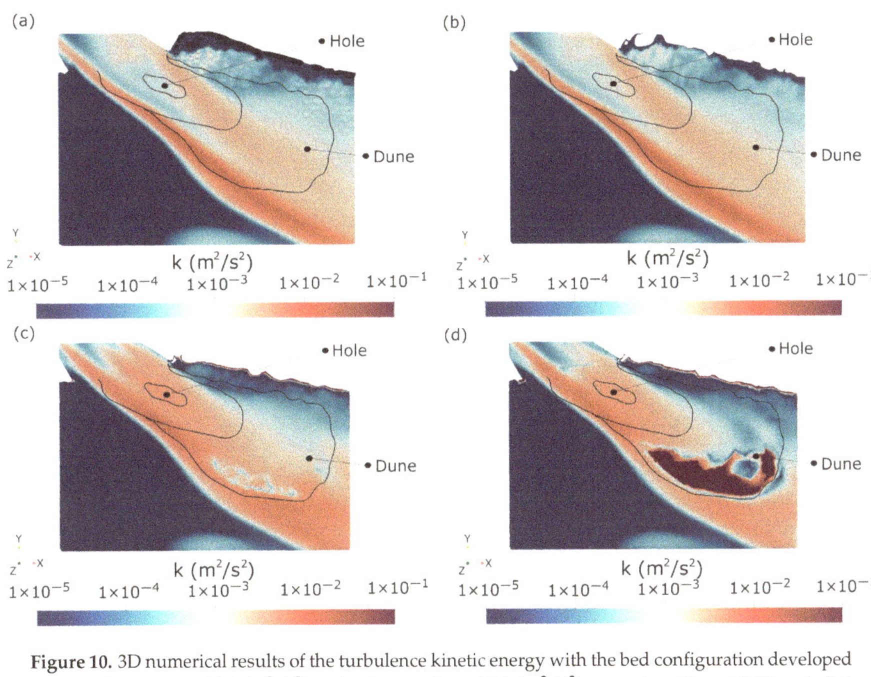

Figure 10. 3D numerical results of the turbulence kinetic energy with the bed configuration developed after 4.5 h, XY view, (**a**) k (m^2/s^2) at the free surface, (**b**) k (m^2/s^2) at an elevation of 0.10 m, (**c**,**d**) k (m^2/s^2) at an elevation of 0.05 m, and (**d**) k (m^2/s^2) at an elevation of 0.025 m elevation.

3.3. Scour

The initial configuration of the bed downstream of the control structure was horizontal, and after the 4.5 h that the experiment lasted, a new configuration developed due to the scour process downstream of the structure. The final bed configuration was measured with a Leica TS07 total station, with an angular resolution of 0.1″ of the screen and precision for horizontal and vertical angles (absolute, continuous, and diametral of 1″, 2″, and 3″), without a prism for bed measurement and with a prism for the control points. On the other hand, Figure 11 shows the average velocity recorded at five different elevations above the point of maximum scour. The data were divided into 18 stages (R1 to R18) of 15 min, where the velocity was averaged, with the same duration each (indicated in Figure 11, with vertical lines in gray).

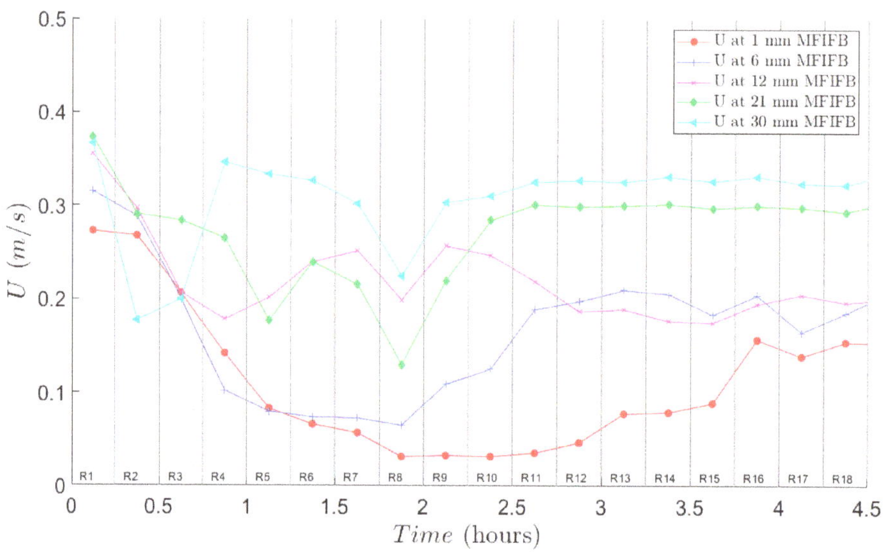

Figure 11. Velocity magnitude (m/s) records at different elevations above the point of maximum scour, measured from the initial flatbed (MFIFB) reference.

Figure 12 shows the turbulence kinetic energy (k) at five points at different elevations above the point of maximum scour in the same 18 stages of 15 min utilized for averaging velocity.

Figure 13 shows the evolution of the maximum scour downstream of the structure generated by the jet at the outlet of the channel. The point is located 59 cm downstream of the center of S3 (indicated in Figure 5). Sampling at this point was conducted for 4.5 h, with a measurement frequency of 10 Hz for velocity and 2.5 Hz for bed evolution (scour), and placed at an elevation to measure the first 3 cm above the original bed. The signal was filtered with correlations greater than 70% and SNR greater than 15 dB. Figure 13, in blue, shows the behavior of the bed evolution during the experiment, while the line point, in green, indicates the behavior of the scour calculated numerically with Telemac–Mascaret 2DH coupled with the Sysiphe module; the comparison is made at the point of maximum scour identified in the physical model. With the numerical model, the erosion was determined to be 8.4 cm, while the experimental value was 9.4 cm. After calibrating the bed-load transport with the morphological factor, there was a 1 cm error, which is equivalent to a 10.6% error with respect to the erosion measured in the physical model. However, the maximum erosion calculated numerically is in the vicinity of the hydraulic structure, near the left bank, in the area just below the outlet of the concrete structure where the sand begins; it was 4.6 times deeper than the maximum experimental erosion. Since no measurements of sediment transport were performed, a Morphological Factor (MF) of 5 was used to calibrate the scour process of the physical model; this factor has been analyzed by [24,25]. The computational time of modeling hydrodynamics and sediment transport of 1 h (considering the physical model) demanded a processing time of 94.6 h with a $\Delta t = 0.0005$ s using the same processors and cluster characteristics.

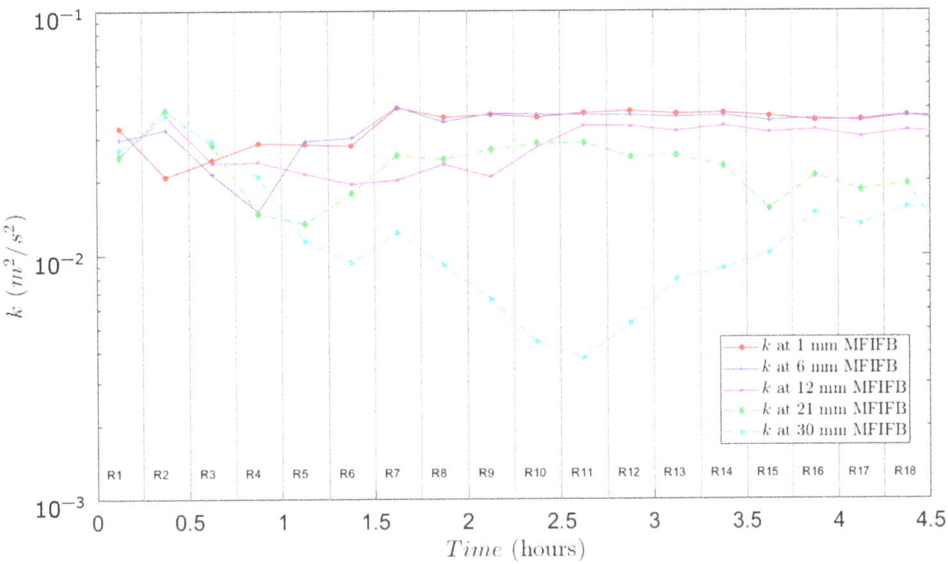

Figure 12. Records of turbulence kinetic energy (m^2/s^2) at different elevations above the point of maximum scour, measured from the initial flatbed (MFIFB) reference.

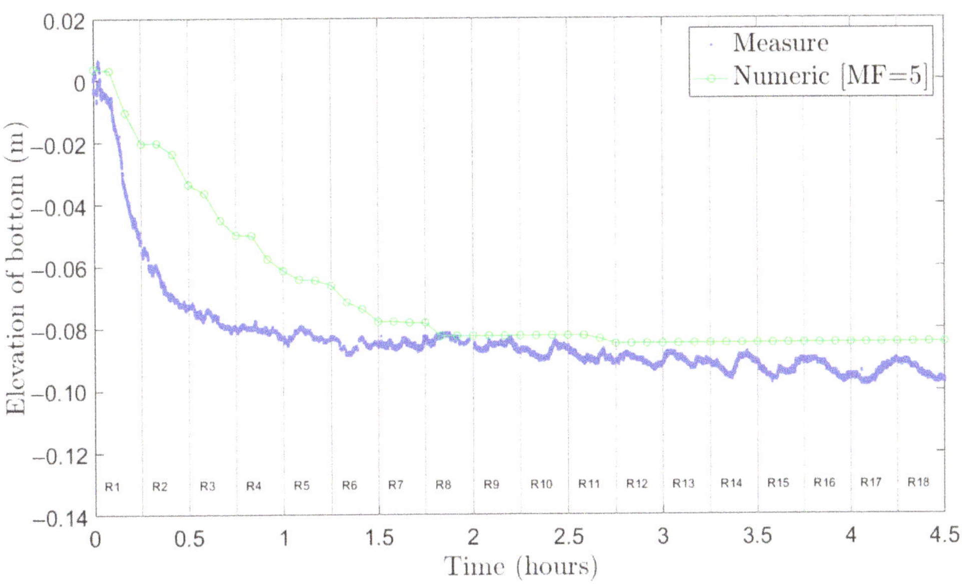

Figure 13. Temporal evolution of the bed (m) measured vs. calculated with the morpho-hydraulic model at the point of maximum experimental scour.

4. Discussion

4.1. Scour Computed by the Numerical Model

Figure 14a shows the final bed configuration calculated with Telemac–Mascaret 2DH, and Figure 14b shows the final configuration of the experiment. In Figure 14a, the maximum scour was developed just downstream of the outlet of the channel, and the sediment was deposited on the sides where the maximum velocity cores were developed. The erosion extends up to the border of the left bank. In the final part, a U-shape deposition pattern was developed. On the other hand, with the experimental results, which are shown in Figures 14b and 15, a less elongated erosion zone was followed by a dune-shaped deposit; ripples were formed on the deposit and along the channel (Figure 15a,b). The dune-shaped deposit is highlighted on the edge in red in Figure 15a. The maximum erosion zone was not located immediately downstream of the exit of the channels, as the numerical model predicted, but slightly further downstream. Also, the numerical model does not reproduce the ripples (Figure 14a).

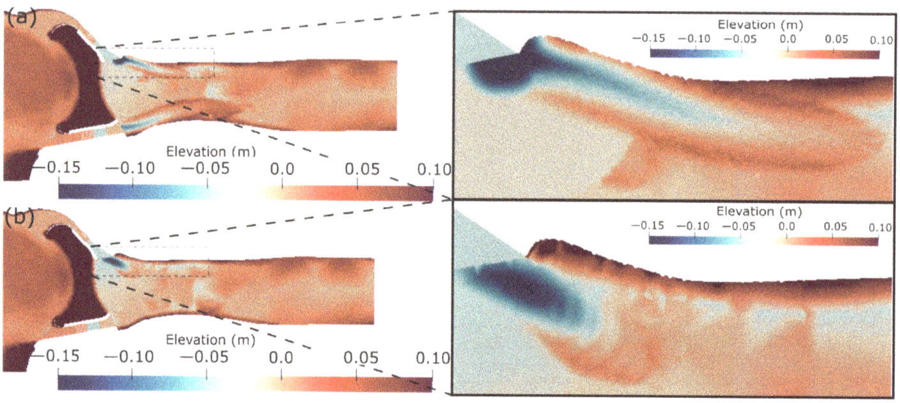

Figure 14. Final bed configuration (m) downstream of the structure (after 4.5 h), in (**a**) calculated with Telemac–Mascaret, and in (**b**) experimental measurement.

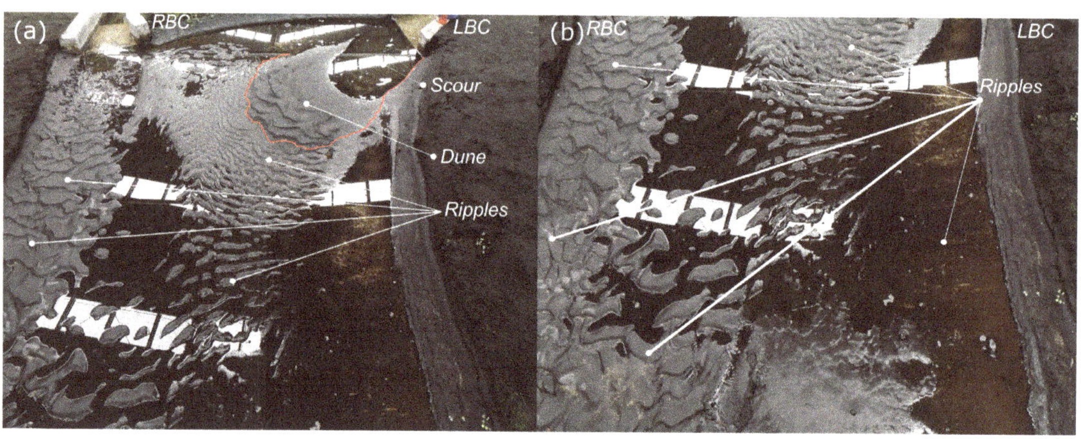

Figure 15. Final experimental bottom configuration (m) downstream of the structure (after 4.5 h), in (**a**) maximum scour and developed bottom forms (ripples), and in (**b**) developed bottom forms (ripples).

The difference in elevation between the Final Bed Configuration (FBC) of the numerical and experimental data was calculated as $Rz = (FBC_{exp} - FBC_{num})$. The results are shown in Figure 16a and zoomed in the scour zone in Figure 16b. The error relative to the measurement points (white dots in Figure 16a,b) is calculated as $Er = Rz/FBC_{exp}$. The errors for the 302 points considered are presented in Figure 16c. The average relative error is −3.6%.

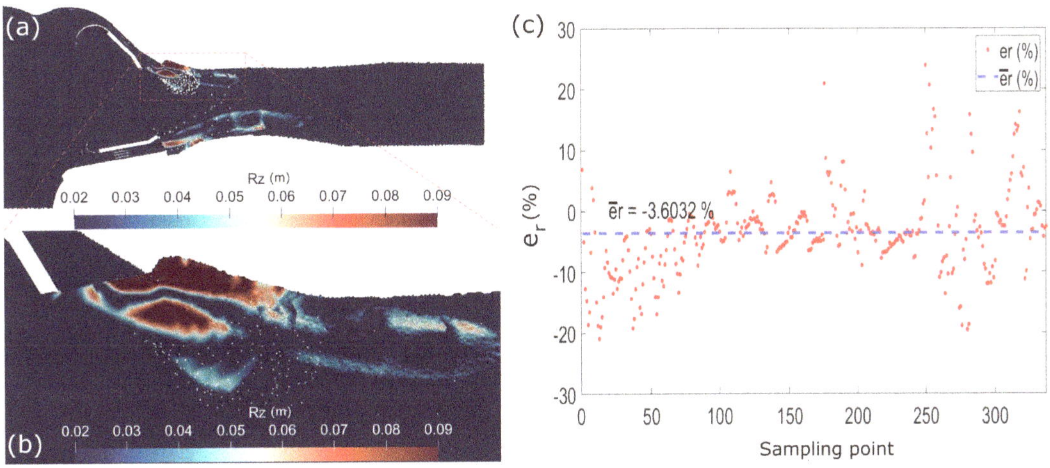

Figure 16. Elevation difference (Rz, m), in (**a**) general view, in (**b**) view near the CMI, and in (**c**) relative error (-).

4.2. Comparison with Empirical Formulas and Numerical Model Accuracy

To contrast the result of the numerical model, an analysis of the maximum downstream scour for low-head structures computed with formulas was performed. Ref. [26] analyzed several studies of local scour, the most important of which were those carried out initially by [27]. In their experiments, they determined a time scale between the physical model and the prototype for the development of local scour and proposed the expression given by Equation (6) for the calculation of the temporal evolution associated with the maximum scour depth, Ys_{max}, [27]:

$$\frac{Ys_{max}}{h_0} = \left(\frac{t}{t_0}\right)^{\beta}, \tag{6}$$

where t_0 is the time scale in hours to reach maximum scour Ys_{max}, h_0 is the depth at the end of the bed of the protected channel, t is the time in hours, and β = 0.38 is the adjustment coefficient. Ref. [26] determined the time scale expression given by Equation (7):

$$t_0 = 330\left(\frac{\rho_s - \rho}{\rho}\right)^{1.7} h_0^2(\alpha V - V_{cri})^{-4.3}, \tag{7}$$

where ρ_s is the density of the sediment, ρ is the density of the water, V_{cri} is the critical velocity determined with the critical shear stress (τ_c), α a factor that depends on the velocity distribution and whose value can be estimated by the expression given by Equation (8):

$$\alpha = 1 + 3\left(\frac{\overline{v}'}{V}\right), \tag{8}$$

where $\overline{v}$ is the relative average turbulent intensity and V the mean flow velocity near the bed. By applying this method to the data of the experiment, the maximum equilibrium scour is determined to be 7.64 cm (note that with active bed conditions, the bed configuration fluctuates, and the equilibrium refers to a temporal average). Paper [28] proposed a formulation analogous to that of [27], given by Equation (9):

$$\frac{Ys_{max}}{d_0} = \left(\frac{t}{t_0}\right)^{0.19}, \tag{9}$$

where d_0 is a characteristic length; here, the downstream balanced level was used. However, the time scale is determined by graphs, and for the example analyzed, a scale of $t_0 = 212$ h and, therefore, a maximum equilibrium scour of 5.87 cm were determined. Paper [29] conducted 210 laboratory experiments in a straight channel with dimensions of 0.3 m wide, 0.25 m deep, and 3.5 m long, with a sand layer of 7.5 cm thickness and a D_{50} of 1.77 mm. They used a floodgate that regulates the level upstream, at the entrance of the channel, and the development of a hydraulic jump; after a length L, the sandy bed is located, where the scour was developed. In addition, they performed a dimensional analysis to calculate the maximum scour downstream and obtained Equation (10):

$$\frac{ds_{max}}{G} = 1.13F_G - 28.9\left(\frac{D_{50}}{G}\right) + 0.26\left(F_G\left(\frac{B-b}{b}\right)\right) - 3.59\left(\frac{G}{H_u}\right) + 2.1, \tag{10}$$

where G is the gate opening, F_G is the Froude number downstream of the gate, H_u is the water head upstream of the gate, and b is the width of the upstream inlet channel (before expansion) and the downstream width (at expansion). B is the downstream width (at the expansion in the case of the experiment). Here, the gate opening G was substituted with the conjugate head (0.10 m) and H_u with the level located just upstream of the piles (0.12 m), thus determining a maximum equilibrium scour of 4.5 cm. Ref. [30] presented an equation for calculating the theoretical maximum scour, reported in [26], which is given by Equation (11):

$$\frac{y_{s\,max(equili)}}{y_0} = \frac{U_{max} - U_c}{U_c}, \tag{11}$$

with the ratio water depth to sediment grain diameter d/D of (0.0876 m/2.66 $\times$ 10^{-4} m) = 328.83 and a critical velocity (U_c) of 0.23 m/s (Equation (8.96), from [31]); therefore, from Equation (11), it was determined that the equilibrium maximum scour was 6.48 cm.

Figure 17 shows the temporal evolution of the scour measured in the physical model (blue dots), calculated numerically with Telemac–Mascaret 2DH coupled with the Sisyphe module (green line with unfilled circles), and the methods indicated above; results are synthesized in Table 2. In Figure 17, it is observed that the scour measured in the physical model developed temporal fluctuations. During the experiment, ripples were formed along the bed (see Figures 14 and 15b), which are related to 3D effects. The migration of the ripples produced the temporal fluctuations of the bed and the zone of maximum scour. On the other hand, a 2D flow model coupled with a bed evolution model (Telemac2D-Sisyphe) is not able to reproduce a ripple formation (see Figure 14a), and a consequence is the lack of fluctuations in the temporal evolution of scour (Figure 17).

4.3. Comparison of Computed 2D vs. 3D Flow Structures

The scour computed by the NM-2D leaned towards the left bank, while the experimental results are more centered with respect to the exit of the LBC (compare Figure 14a,b). This behavior is correlated to the 2D flow field; the higher velocities at the exit of the LBC also lean towards the left bank (Figure 8a). In contrast, the depth-averaged 3D flow field computed the maximum velocities at the exit of the LBC with a more consistent pattern with respect to the experimental maximum scour zone (Figure 8c). It is important to highlight that both computations were carried out using the experimental bed developed after 4.5 h. Another difference is the turbulence kinetic energy, k; the NM-2D is more diffusive (compare Figure 8b,d).

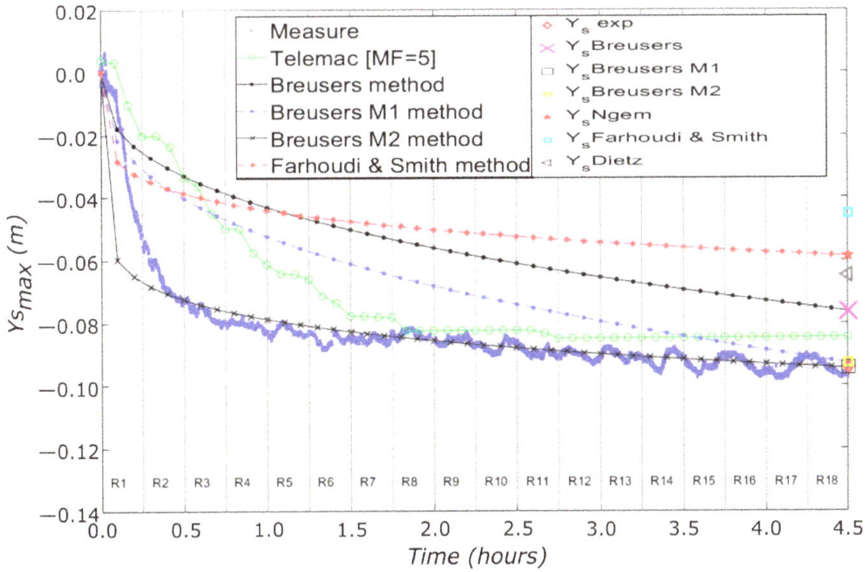

Figure 17. Temporal evolution of the bed (m) measured vs. calculated by the morpho-hydraulic model and the formulas at the point of maximum experimental scour.

Table 2. Experimental maximum scour vs. maximum scour calculated by four theoretical methods and errors. The values calculated with the 2D model are located at the point of maximum experimental scour (MES) and maximum modeled scour (MMS).

Method	Ys_{max} (cm)	Error $\left[1 - \left(\frac{Ys_{num}}{Ys_{exp}}\right)\right]$
Experimental	9.4	-
Telemac2D-Sishype (MES)	8.4	0.106
Telemac2D-Sishype (MMS)	43.2a	−3.6
Breusers M1 (1967)	7.64	0.187
Breusers M1 ($\overline{v'}$) (1967)	9.27	0.014
Breusers M2 (β) (1967)	9.44	0.004
Farhoudi and Smith (1982)	5.87	0.376
Negm (2002)	4.5	0.521
Dietz (1969)	6.48	0.311

5. Conclusions

The phenomenon of scouring downstream of low-head hydraulic structures is a prevalent problem for the safety of hydraulic works. Although there are numerous investigations on local scour around submerged structures within the flow, such as bridge piers or abutments, it is difficult to have direct and accurate solutions to estimate the maximum local scour that occurs downstream of hydraulic structures with particular geometrical characteristics, such as the El Macayo structure, where the discharge is influenced by an upstream curved channel. In such circumstances, the literature recommends the application of physical modeling. However, an area to explore is the applicability of numerical modeling to characterize scour processes quantitatively. In this study, the local maximum scour generated by a jet downstream of a discharge structure was calculated with a physical model of a low-head hydraulic structure. It was analyzed with a 2D depth-averaged model, and the flow field was characterized by a 3D RANS model. The accuracy of the 2D model coupled with the sediment transport and bed evolution module (Sisyphe) to reproduce the scour and the velocity field in the left channel in three sections was assessed.

The 2D model utilized secondary flow correction and was calibrated by means of the production coefficient [23]. After the calibration, the errors were diminished by more than an order of magnitude for the streamwise velocity component and by a factor of 2 for the normal component, compared to the results that did not use secondary flow correction. This is important to consider in engineering applications when using a 2DH model in cases where secondary-flow currents are expected to be formed. With respect to the scour, the 2D model coupled with Sisyphe calculated a maximum scour of 8.4 cm, where the experimental value of 9.4 cm was measured. Therefore, the 2D model reproduced 89% of the measured maximum scour. In addition, the bed configuration (the scour zone and its surroundings) was assessed after 4.5 h to compare the numerical and experimental results; an average relative error of −3.6% was determined, which indicates that the 2D model computes a higher global erosion compared to the measurements from the experiment. However, there was no good agreement on the final bed configuration calculated numerically with respect to the experimental development. Here, the largest errors were located in the zone of maximum scour and within the zone of the U-shaped deposition developed downstream of the pool generated by the scour. The numerical computation determined that the maximum scour was located in a different place with respect to the experimental results; the NM-2DH computed the maximum scour directly in the outlet of the left channel structure and was of the order of 4.6 times deeper than the reported experimental maximum scour. Therefore, it is a case of overestimation. In practical applications, it is recommended to perform sediment transport measurements to improve the approximation of the phenomenon by the sediment-load equations and to replicate its magnitude.

For contrasting purposes, the maximum scour was estimated with empirical equations, and it was compared with the experimental results. It was found that the Breusers method has an accuracy of 81.3%; however, using a modification (M1) of the local mean turbulent intensity ($\overline{v'}_b$) and close to the bed (1 mm elevation above the initial flatbed), the estimation was improved considerably, reaching 98.6% accuracy, and by performing a second modification (M2) of the β exponent of the equation, the estimation was improved to a 99.5% accuracy. However, it requires knowledge of the velocity and turbulence above this point, i.e., measurements. The application of the model (Equation (6)) and the modification M1 did not reproduce the magnitude of the maximum scour in time, but the modification M2 described it better. The application of the Farhoudi and Smith method is complicated since it requires the use of graphs and does not have an equation for the calculation of the time scale necessary for the development of maximum scour. With this equation, an accuracy of 62% was determined. On the other hand, the Negm method had an accuracy of 48%, while for the Dietz model, the accuracy was 69%. By establishing an acceptance condition of 75% accuracy for the calculation of the maximum scour (as it was established by [8]), only the Breusers method was acceptable. The numerical modeling was not acceptable when comparing the maximum erosion magnitude and location. The only way to accept it is by comparing the magnitude of erosion at the experimental point of maximum scour; however, this does not have practical application since it requires knowledge a priori of where the real maximum scour will occur. Something to highlight is that all the methods, except the 2DH numerical modeling, underestimated the maximum scour measured in the physical model, which would put the structure at risk; except for the M2 modification in the Breusers model, which developed a higher scour in the first 25 min of the experiment, but the fit after this time was improved considerably with respect to all the models implemented, which did not describe the scour process over time. Currently, there is no clear and accurate description of the fluid–sediment interaction in the contact or near-bed zone, so this field of research is still under development.

There are differences between the 2D flow field and the depth-averaged 3D results. The 2D model computes a velocity at the exit of the LBC that is leaned toward the left bank, while the 3D model computes a jet that is less leaned toward the left bank. It highlights the importance of secondary flow correction for 2D models. Even here, the 2DH model was calibrated for secondary flow correction with the help of the measurements in the bend of the LBC by the manipulation of the production coefficient. The 3D effects involved in the secondary flow, separation zone, and horizontal recirculation zones were not well represented by the 2D model in the zone of the scour pool or downstream of the channel. It indicates that 2DH models may not be able to compute the magnitude and location of scour when secondary currents are present in the flow, even if secondary correction sub-models are utilized. Additionally, the RANS model does not compute near-bed coherent turbulent structures; however, bursts and vortices as observed in the physical model. For this reason, 2D models for scour produced by jets require further research for engineering applications.

Author Contributions: Conceptualization and experimental work, C.C. and A.J.; formal analysis, C.C.; writing—original draft preparation, C.C.; writing—review and editing, A.M. and M.B.; funding acquisition, A.M. All authors have read and agreed to the published version of the manuscript.

Funding: The results presented here were supported by the UNAM-PAPIIT Grant IA102623.

Data Availability Statement: The data of the experiments are available upon reasonable request.

Acknowledgments: To the Dirección General de Cómputo y de Tecnologías de Información y Comunicación (DGTIC) of the Universidad Nacional Autónoma de México (UNAM), which provided the supercomputing resources for the development of the numerical calculations, assigned under the project LANDCAD-UNAM-DGTIC-389. To the Departamento de Sistemas Linux y Súper Cómputo of the Secretaría de Telecomunicaciones e Informática and the Unidad de Servicios de Información del Instituto de Ingeniería—UNAM. We also thank the Consejo Nacional de Ciencia y Tecnología (CONACYT) for the doctoral scholarship awarded to the first author.

Conflicts of Interest: The authors declare no conflict of interest.

References

1. Hoffmans, G.J.C.M.; Verheij, H.J. *Scour Manual*; Routledge: London, UK, 1997. [CrossRef]
2. Garde, R.J.; Ranga Raju, K.G. *Mechanics of Sediment Transportation and Alluvial Stream Problems/R.J. Garde, K.G. Ranga Raju*, 3rd ed.; New Age International, Ed.; Daryaganj: New Delhi, India, 2015.
3. Sousa, A.M.; Ribeiro, T.P. Local Scour at Complex Bridge Piers–Experimental Validation of Current Prediction Methods. *ISH J. Hydraul. Eng.* **2021**, *27*, 286–293. [CrossRef]
4. Vanoni, V.A.; Brooks, N.H. *Laboratory Studies of the Roughness and Suspended Load of Alluvial Streams*; California Institute of Technology: Pasadena, CA, USA, 1957; p. E-68.
5. Bombardelli, F.A.; Palermo, M.; Pagliara, S. Temporal Evolution of Jet Induced Scour Depth in Cohesionless Granular Beds and the Phenomenological Theory of Turbulence. *Phys. Fluids* **2018**, *30*, 085109. [CrossRef]
6. Palermo, M.; Bombardelli, F.A.; Pagliara, S.; Kuroiwa, J. Time-Dependent Scour Processes on Granular Beds at Large Scale. *Environ. Fluid Mech.* **2021**, *21*, 791–816. [CrossRef]
7. Zhao, P.; Yu, G.; Zhang, M. Local Scour on Noncohesive Beds by a Submerged Horizontal Circular Wall Jet. *J. Hydraul. Eng.* **2019**, *145*, 06019012. [CrossRef]
8. Franzetti, S.; Radice, A.; Rebai, D.; Ballio, F. Clear Water Scour at Circular Piers: A New Formula Fitting Laboratory Data with Less Than 25% Deviation. *J. Hydraul. Eng.* **2022**, *148*, 04022021. [CrossRef]
9. Ettema, R.; Constantinescu, G.; Melville, B.W. Flow-Field Complexity and Design Estimation of Pier-Scour Depth: Sixty Years since Laursen and Toch. *J. Hydraul. Eng.* **2017**, *143*, 03117006. [CrossRef]
10. Lai, Y.G.; Liu, X.; Bombardelli, F.A.; Song, Y. Three-Dimensional Numerical Modeling of Local Scour: A State-of-the-Art Review and Perspective. *J. Hydraul. Eng.* **2022**, *148*, 03122002. [CrossRef]
11. Castillo, L.G.; Castro, M.; Carrillo, J.M.; Hermosa, D.; Hidalgo, X.; Ortega, P. Experimental and Numerical Study of Scour Downstream Toachi Dam. In *Sustainable Hydraulics in the Era of Global Change—Proceedings of the 4th European Congress of the International Association of Hydroenvironment Engineering and Research, IAHR 2016, Liege, Belgium, 27–29 July 2016*; CRC Press: Boca Raton, FL, USA, 2016.
12. Garfield, M.; Ettema, R. Effect of Clearwater Scour on the Flow Field at a Single Bendway Weir: Two-Dimensional Numerical Modeling Supported by Flume Data. *J. Irrig. Drain. Eng.* **2021**, *147*, 04021055. [CrossRef]
13. Liao, C.T.; Yeh, K.C.; Lan, Y.C.; Jhong, R.K.; Jia, Y. Improving the 2d Numerical Simulations on Local Scour Hole around Spur Dikes. *Water* **2021**, *13*, 1462. [CrossRef]

14. Jiménez, A. *Calibración y Verificación de Las Leyes de Descarga de Una Estructura de Control*; Instituto de Ingeniería, Universidad Nacional Autónoma de México: Mexico City, Mexico, 2017.
15. *ASTM C778-13*; Standard Specification for Standard Sand. ASTM international: West Conshohocken, PA, USA, 2006.
16. Hamdan, R.K.; Al-Adili, A.; Mohammed, T.A. Physical Modeling of the Scour Volume Upstream of a Slit Weir Using Uniform and Non-Uniform Mobile Beds. *Water* **2022**, *14*, 3273. [CrossRef]
17. Nortek. The Comprehensive Manual for Velocimeters. Available online: https://www.nortekgroup.com/assets/software/N3015-030-Comprehensive-Manual-Velocimeters_1118.pdf (accessed on 28 May 2023).
18. EDF-R&Da. Telemac-2D. User Manual. Version V8p2; 2020. Available online: http://www.opentelemac.org/index.php/manuals/viewcategory/8 (accessed on 28 May 2023).
19. García, H.M. *Sedimentation Engineering*; Garcia, M., Ed.; MOP 110; American Society of Civil Engineers: Reston, VA, USA, 2008. [CrossRef]
20. EDF-R&Db. Sisyphe. User Manual. Version V8p2. 2020. Available online: http://www.opentelemac.org/index.php/manuals/viewcategory/8 (accessed on 28 May 2023).
21. Huybrechts, N.; Villaret, C.; Hervouet, J.M. Comparison between 2d and 3d Modelling of Sediment Transport: Application to the Dune Evolution. In Proceedings of the River Flow 2010: Proceedings of the International Conference on Fluvial, Braunschweig, Germany, 8–10 September 2010; Bundesanstalt für Wasserbau: Karlsruhe, Germany, 2010; pp. 887–893.
22. Cao, Z.; Carling, P.A. Mathematical Modelling of Alluvial Rivers: Reality and Myth. Part 1: General Review. In *Proceedings of the Institution of Civil Engineers—Water and Maritime Engineering*; ICE Virtual Library: London, UK, 2002; Volume 154. [CrossRef]
23. Finnie, J.; Donnell, B.; Letter, J.; Bernard, R.S. Secondary Flow Correction for Depth-Averaged Flow Calculations. *J. Eng. Mech.* **1996**, *125*, 848–863. [CrossRef]
24. Knaapen, M.A.F.; Joustra, R. Morphological Acceleration Factor: Usability, Accuracy and Run Time Reductions. In Proceedings of the XIXth Telemac-Mascaret User Conference, Oxford, UK, 18–19 October 2012.
25. Morgan, J.A.; Kumar, N.; Horner-Devine, A.R.; Ahrendt, S.; Istanbullouglu, E.; Bandaragoda, C. The Use of a Morphological Acceleration Factor in the Simulation of Large-Scale Fluvial Morphodynamics. *Geomorphology* **2020**, *356*, 107088. [CrossRef]
26. Breusers, H.N.C.; Raudkivi, A.J. Scouring. In *Dictionary Geotechnical Engineering/Wörterbuch GeoTechnik*; Springer: Berlin/Heidelberg, Germany, 1991. [CrossRef]
27. Breusers, H.C. Time Scale of Two Dimensional Local Scour. In Proceedings of the 12th Congress IAHR, Fort Collins, CO, USA, 11–14 September 1967; Volume 3.
28. Farhoudi, J.; Smith, K.V.H. Time Scale for Scour Downstream of Hydraulic Jump. *J. Hydraul. Div.* **1982**, *108*, 1147–1162. [CrossRef]
29. Negm, A.; Saleh, O.; Abdel-Aal, G.; Sauida, M. Investigating Scour Characteristics Downstream of Abruptly Enlarged Stilling Basins. In *River Flow Proceedings of the International Conference on Fluvial Hydraulic*; Didier Bousmar, Y.Z., Ed.; Balkema: Ghent, Belgium, 2002; pp. 1–1283.
30. Dietz, J.W.W. Kolkbildung in Feinen Oder Leichten Sohlmaterialien bei Strömendem Abfluß. Bachelor's Thesis, Versuchsanstalt für Wasserbau und Kulturtechnik, Theodor-Rehbock-Flußbaulaboratorium, Universität Fridericiana Karlsruhe, Karlsruhe, Berenz, 1969.
31. García Flores, M.; Maza Álvarez, J.A. *Inicio de Movimiento y Acorazamiento. Capítulo 8 del Manual de Ingeniería de Ríos*; Instituto de Ingeniería, Universidad Nacional Autónoma de México: Mexico City, Mexico, 1997.

water

MDPI

Article

Determination of the Canal Discharge Capacity Ratio and Roughness to Assess Its Maintenance Status: Application in Egypt

Mohamed Elsayed Gabr [1,*], Ehab M. Fattouh [2] and Mohamed K. Mostafa [3]

1 Civil Engineering Department, Higher Institute for Engineering and Technology, Ministry of Higher Education, New Damietta 34517, Egypt
2 Irrigation and Hydraulics Department, Faculty of Engineering, Ain Shams University, Cairo 11517, Egypt
3 Faculty of Engineering and Technology, Badr University in Cairo (BUC), Cairo 11829, Egypt
* Correspondence: mohamed.gabr@ndeti.edu.eg

Abstract: Waterlogging and soil salinity issues can be handled using surface or subsurface drainage networks, soil bed elevation, and soil and crop management patterns. A properly operating and maintained drainage system is important for both rural and urban inhabitants to protect lives and property from flooding and high groundwater levels, enhance health conditions, and safeguards water purity, soil salinity, and waterlogging. It also supports and increases crop yields and consequently rural incomes. This study assessed the maintenance condition of the main surface drains (Baloza and ELFarama) located in the Tina Plain (50,000 acres) and a portion of the Southeast El-Kantara regions (25,000 acres) in North Sinai, Egypt, based on the values of the Discharge Capacity Ratio (DCR) and Manning's roughness (n). Ten measurement locations at the drain cross-section were used in the investigation. For the ELFarama Drain, the average values of n and DCR were found to be 0.029 and 86.2%, and for the Baloza Drain, they were 0.032 and 78.6%, respectively. Compared to the design values, the actual Manning's roughness was higher, indicating that the drainage canals' capacities had been reduced and that their upkeep was inadequate. In both drains, sedimentation is present and they need to be maintained, according to the hydrographic surveying results for the actual cross-sections compared to the planned cross-sections. A methodology for the channel maintenance method is presented. For removing vegetation and dredging sediment, a long-boom mechanical hydraulic excavator with a bucket is suggested and to be conducted every two years. To the results of this study, the amount of weed infestation in vegetated channels is the main factor that affects Manning's roughness coefficient value. It is now easier to calculate the proportion of weeds that are submerged in vegetated channels using echo-sound sonar technology. The DCR is an affordable and simple methodology to assess the channel maintenance status for sustainable agriculture.

Keywords: water resources; canal roughness; sedimentation; vegetation management; Egypt

1. Introduction

Vegetation is a crucial factor from the perspective of catchment hydrology [1–3] as interception of raindrops, evapotranspiration, and infiltration are factors to take into account in surface and subsurface water balance. In terms of ecology and habitat, riparian vegetation is crucial and makes a considerable contribution to biodiversity. In fact, vegetation creates microenvironments that can house creatures and insects beneficial for farming and keeps pollutants and fertilizers from getting into streams. Due to how it affects the landscape, it also has a significant recreational impact [4]. Waterlogging and soil salinization issues are brought on by seepage from unlined canals, over-salinized groundwater inundation, and the use of ineffective irrigation techniques [5,6]. Since waterlogging and soil salinization issues are the root cause of anaerobic conditions and the growth of hydrophilic weeds, which result in nutrient shortages in the root zone, a sizable share of agricultural output declines in soggy soils worldwide [7,8]. However, it should be mentioned that a variety of programs and tactics have been employed to tackle these problems. For instance, the application of surface or subsurface drainage networks [9–11], soil bed elevation [12], and soil and crop management strategies [13,14].

In Egypt, the main causes of soil salinity issues are inadequate irrigation, shallow groundwater, and seawater intrusion [15,16]. Therefore, in the Delta, hard clay soil is the most prevalent soil-type [17,18].

More than 82% of Egypt's canals and drains are severely overrun by aquatic weeds of various types, making aquatic weeds the principal issue in open channels [19,20]. This percentage is expected to increase due to the recent change in water quality and the fall in actual water depths in the majority of Egyptian canals and drains. The three primary types of aquatic weeds are those that are submerged, floating, and ditch bank weeds. Open channels are most negatively impacted by aquatic weeds that are submerged. Worm infestation has two main effects: it increases flow resistance and reduces conveyance capacity [21].

Adequate drainage system operation and maintenance protects the resource base for food production, sustains and increases yields and rural incomes, protects irrigation investment, serves rural and urban residents as well as industrial activities, protects human lives and assets against flooding and high groundwater levels, improves health conditions, and protects the water quality [22,23]. Irrigation and drainage networks require periodic monitoring of size, water discharge, sedimentation, vegetation, water quality, and biodiversity to satisfy their high efficiency [24–26].

Vegetation management activities increase hydraulic efficiency and bank erosion protection. According to Wu et al. [27], the kind and distribution of vegetation in open channel cross-sections have an impact on velocity profiles due to plant drag, which significantly increases flow roughness. Five degrees of retardance for vegetated waterways were developed by Chow [28].

According to Han et al. [29], vegetation in open channels is a crucial design element. It has various impacts based on the type of vegetation and has an impact on the local water depth and the water velocity profile. Wan Yusof et al. [30] concluded that there is a significant association between Manning's roughness coefficient and drag coefficient in vegetated channels based on an experimental analysis employing natural vegetation to discover the relationship. Also, Pu et al. [31] concluded that the vegetation in open channels acts as a flow obstacle, which creates turbulence that alters the distribution of flow velocity and water depth as well as sedimentation in the canal cross-section.

Plant growth depends on many factors such as light, temperature, nutrient levels in the water column and streambed sediments, and hydrologic regime. Bed roughness for artificial and natural channels is affected by several factors such as the shape of a canal, silting and scouring, vegetation, canal irregularity, obstruction, seasonal change, bed load, and suspended material [32]. A numerical model was developed to evaluate the state of canal maintenance based on the canal carrying capacity ratio (DCR) [33]. The DCR is defined as the actual capacity of the selected canal over its designed capacity.

The Nile Delta of Egypt is a fertile cultivated area that depends on surface irrigation for cultivated crops; recently, the soil salinity rates were raised to 30% [34,35]. As reported, [14], in Egypt's Nile West Delta, soil salinization degrees according to Russian classification were at 71%—non-saline, 10.5%—slight saline, 9%—moderate saline, 3.8%—strongly saline, and 5.7%—very strongly saline in the investigated area. The El-Salam Canal Project for reclamation and cultivation of 260 thousand hectares in the North East of Egypt is considered to be the Delta extension [36,37]. It consisted of two phases; phase one lies in the Nile West Delta to the west of the Suez Canal for an area that served 220 thousand acres. On the other hand, phase two lies east of the Suez Canal for an area that served 400 thousand acres at the North Sinai Development Project (NSDP) [38]. This project faced waterlogging and salinity problems and solved them by constructing subsurface and surface drainage networks [37].

The objective of this work was to apply the DCR method to assess the channel maintenance status for sustainable agriculture. The Baloza and ELFarama surface drains in the NSDP, North Sinai Egypt, were used as a case study. To achieve this objective (i) the studied drains were described and the water levels and channel roughness were collected and investigated, (ii) the actual velocities during the maximum water level period were measured in the field, (iii) the actual Manning's roughness coefficient (n) was determined and compared with the literature, (iv) hydrographic surveying to measure the drains actual cross-sections was carried out, (v) surveying for vegetation was also carried out, and (vi) a methodology for the channel maintenance method was determined.

2. Materials and Methods

2.1. Study Area

Egypt's Nile West Delta is located between the latitudes of 30.33° and 31.166° N and longitudes of 31.5° and 32° E. The Nile West Delta elevations range from 1 to 50 m above the main sea level [39]. The soil's texture ranges from sand to clay, with clay content increasing nearer the center of the Nile Delta and declining toward the delta's boundaries and the desert [40]. Fluvio-marine flats and river terraces are the two dominant terrain types in the region east of the Nile Delta. Additionally, the primary geomorphic units of the East Nile Delta are as follows: clay flats in the north, gypseous flats in the northwest of Ismailia, and historic river terraces with sandy soils in the south-west and east of El-Sharkia and the governorate [41]. The Nile West Delta experiences arid and semiarid weather [42]. Additionally, it has meteorological characteristics such as significant daily temperature swings between the lowest and highest temperatures. The temperature varies greatly between summer and winter, with an average yearly temperature of roughly 22 °C [42–44]. Additionally, 33.3 mm of rainfall precipitates on average throughout the winter months [41].

The studied surface drains Baloza and ELFarama are located in the Tina Plain region at the borders of the Egypt's Nile West Delta as shown in Figure 1. El-Sheik Gaber Canal is the main feeder for the NSDP, the annual water resource for the NSDP is 4.5 km^3. The canal starting from Suez Canal extended to the east with a length of 86.5 km; the canal was lined with gabions to protect and stabilize inside slopes in the Tina Plain region (clay soils) for 24.5 km. The remaining length is lined with plain concrete in the Romanna, Rabaa, and Bir EL-Abd regions (sandy soil). The topography of the Tina Plain region is flat with mean elevations of 1.0 m above mean sea level, and the predominate soil type is clay.

The area served by the drains is 75,000 acres, which has been operated and maintained by the Sector for Irrigation, Water Resources, and Infrastructures at North Sinai (SIWRI) Ministry of Water Resources and Irrigation (MRWI) since 1998. Potatoes, wheat, beans, tomatoes, and sugar beet are the major crops grown besides poplar trees. About 55,000 acres are served by the Baloza Drain, which is 17.480 km long, 2.0 m wide at the entrance, and 10 m wide at the outlet, with corresponding water levels of 1.22 m and 1.97 m and has an inside slope of 1 to 3. There is a pump station at 3.500 km from the drain outlet that pumps the irrigation drainage water into a 3.0 km long carrier channel that empties into the Suez Canal. On the other hand, about 20,000 acres are served by the ELFarama Drain, which is 20 km long, 3.0 m wide at the entry and 7.0 m wide at the outflow, with corresponding water levels of 4.17 m and 4.52 m, and has an inside slope of 1 to 3. A pump station is located at 3.500 km from the drain outlet and pumps drainage water into a 3.0 km long carrier channel that empties into the Suez Canal. According to [45], the soil classification for the studied drains is shown in Table 1.

Figure 1. Location of the studied drains and the hydrographic survey cross-sections.

Table 1. Soil classification for the studied drains.

Drain	Km	Sample Location	Soil Classification			Soil Type
			Sand (%)	Silt (%)	Clay (%)	
ELFarama	0.25	Bed	4.02	88.75	10.23	Silty clay
	7.5	Right bank	59.59	32.18	8.23	Silty sand
	15.0	Right bank	16.67	75.10	8.23	Silty clay
	19.5	Bed	20.41	71.36	8.23	Silty clay
Baloza	0.3	Bed	98.50	1.5	-	Sandy
	7.3	Right bank	99.69	0.31	-	Sandy
	13.3	Right bank	99.27	0.73	-	Sandy
	17.3	Bed	98.29	1.71	-	Sandy

2.2. Research Methodology

Figure 2 shows a flow chart for the research methodology. Therefore, (1) the studied drains were described and the design (projected) water levels, bed width, bed slope, and channel roughness ($n_{projected}$) were collected, (2) the actual velocities during the maximum water level period were measured, (3) the actual Manning's roughness coefficient (n_{actual}) is determined, (4) the canal Discharge Capacity Ratio (DCR) is estimated, and (5) a methodology for the channel maintenance method was selected.

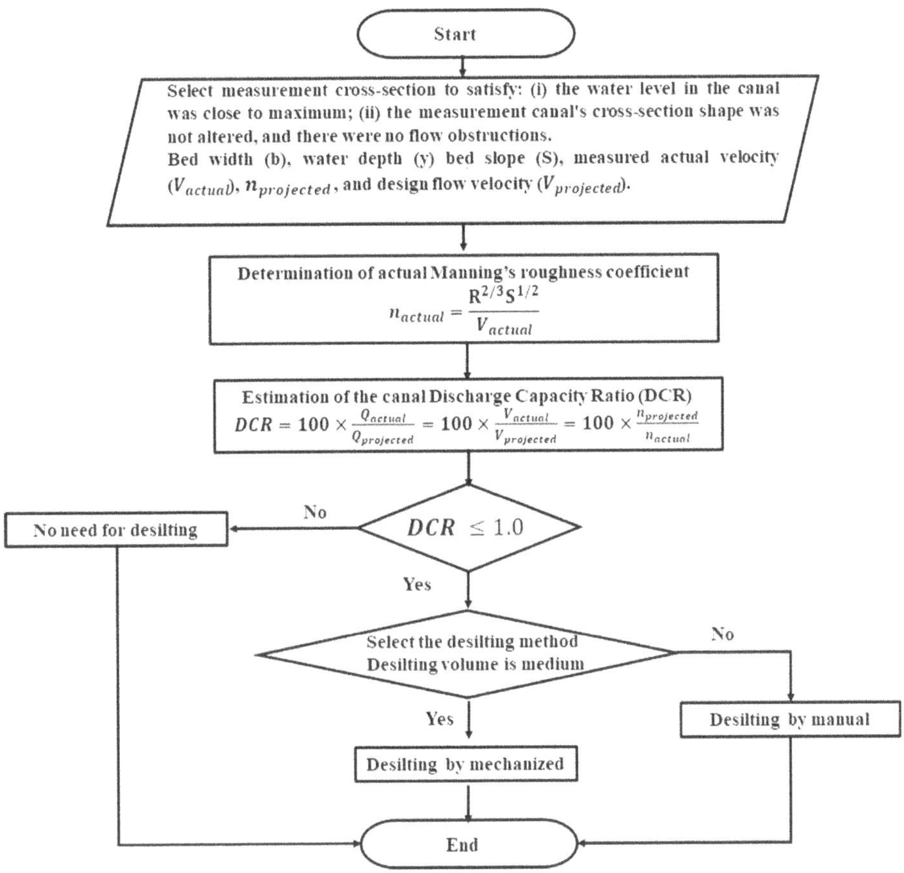

Figure 2. Flow chart for the research methodology.

2.3. Hydrographic Surveying for the Studied Drains

Measurement points were selected according to the following criteria: (1) the water level in the canal was close to maximum; (2) nothing happened to change the water levels while measurements were being taken; and (3) the cross-section geometry of the drainage canal being measured was not deformed and did not cause any obstruction to flow (no vegetation). The study was carried out according to these criteria at ten measurement points, five points on the ELFarama Drain, and five points on the Baloza Drain. A series of 18 cross-sections along the examined drains were chosen in accordance with the existing vegetation and hydraulic structures in order to examine the changes in the drains' cross-section with respect to the specified cross-sections (Figure 1). The selected reaches were measured three times in one year, at the beginning of February, August, and February in the next year, using an echo sounder and total station survey equipment to gauge the rates of sedimentation. Additional information about the hydrographic survey is available in [46]. The actual cross-sections were compared to the ones that were intended, and the annual rate of sand accumulation as a percentage of the intended cross-section area during the six months was calculated. The actual drain velocity is measured as follows: the canal cross-section at the measurement point was first divided into subsections (4 subsections were selected), and velocity values were measured for each subsection by the two-points method, using a propeller current meter calibrated by the hydraulic laboratory at the National Water Research Center, Egypt. Then, the average measured velocity was calculated according to the area of each subsection provided. This was checked by measuring flow depths while measurements were being performed.

2.4. Determination of Manning's Roughness Coefficient (n)

Manning's roughness (n) actual values were determined at the selected measurement points on the studied main drains using Manning's velocity Equation (1):

$$\mathrm{n} = \frac{\mathrm{R}^{2/3}\mathrm{S}^{1/2}}{\mathrm{V}_{actual}} \tag{1}$$

where R is the hydraulic radius (m), S is the channel bed slope (m/m), and V_{actual} is the measured velocity in (m/s). The value of V_{actual} was defined as the average velocity at the measurement points on these drains.

The value of S for each canal was taken from SIWRI records. In order to evaluate the physical condition of the drains included in the study, values of n obtained were compared with the design values of n in Table 2, which were selected from SIWRI and MRWI records.

Table 2. Projected values of Manning roughness for ELFarama and Baloza drains with no vegetation [45].

Manning's Roughness (n) ELFarama Drain	Manning's Roughness (n) Baloza Drain	Canal Condition
0.025	0.028	(Best exceptional (projected))
0.028	0.031	Good
0.031	0.034	Fair
0.035	0.037	Bad

2.5. Estimation of Discharge Capacity Ratio (DCR) for the Studied Drains

The ideal ratio of DCR would be one to one. Estimation of Discharge Capacity Ratio (DCR) for the studied drains is displayed in Equation (2):

$$DCR = 100 \times \frac{Q_{actual}}{Q_{projected}}\% = 100 \times \frac{V_{actual}}{V_{projected}}\% = 100 \times \frac{n_{projected}}{n_{actual}}\% \tag{2}$$

where DCR = Discharge Capacity Ratio (%); Q_{actual} = actual canal discharge capacity in m^3/s; $Q_{projected}$ = projected canal discharge capacity in m^3/s; V_{actual} = actual water velocity in m/s; $V_{projected}$ = projected water velocity in m/s; n_{actual} = actual Manning's roughness coefficient; and $n_{projected}$ = projected Manning's roughness coefficient.

In applying this indicator, flow measurements were carried out during the month of August in the summer season for the selected section, where conditions for designated water levels (maximum levels) for unchanged channel cross-sections with no vegetation and sediment deposition existed (field design cross-section). Further, assuming that for the field design cross-section, the canal slope values in Manning's velocity equation were the same as the designed values. With these conditions, the second part of the DCR equation was obtained from the first part. In the DCR equation, the value of n_{actual} indicates that values were obtained from a previous Manning's equation. The value of n designed was based on the value of $n_{projected}$ = 0.028 for the Baloza Drain and $n_{projected}$ = 0.025 for the ELFarama Drain as observed in the drainage system planning in SIWRI and MRWI

(Table 2). Values of Manning's roughness coefficient (n) and DCR along with the values of other hydraulic parameters, such as flow average velocity (V), wetted perimeter (P), cross-section (A), and hydraulic radius (R), were used in the calculations. These parameters were obtained at the ten measurement points determined for ELFarama and Baloza Drains. The impact of weed infestation on the hydraulic efficiency and other hydraulic parameters of vegetated channels was investigated through a number of experiments.

Mohamed and El-Samman [19] developed an equation to forecast the value of Manning's roughness coefficient in the vegetated channel based on measured parameters, such as weed infestation percentage and water surface slope, using a physical model and laboratory study. The Manning roughness coefficient in vegetated open channels was calculated using two straightforward formulas as follows [Equations (3) and (4)]:

$$\frac{n_v}{n_o} = 0.089W_{inf} + 1.037 \quad (3)$$

$$\frac{n_v}{n_o} = 0.067W_{inf} + 240.43S + 0.969 \quad (4)$$

where n_v is the Manning roughness coefficient of vegetated open channel, n_o is the Manning roughness coefficient of non-vegetated open channel, S is the water surface slope, and W_{inf} is the percentage of weed infestation in open channel determined using Equation (5):

$$W_{inf} = \frac{Weeds\ area\ of\ the\ channel\ cross-section}{Total\ area\ of\ the\ channel\ cross-section} \quad (5)$$

2.6. Monitoring the Aquatic Weeds

A survey to measure the type of vegetation and the area of infection for the studied drains was carried out using a total station survey device; additional details on the hydrographic survey are provided. However, employing conventional variables like topography, overstory, and soils has proven to be quite challenging when attempting to construct spatial projections of plant cover or density in open channels [46]. LiDAR (Light Detection and Ranging) is an active remote-sensing technology that could be utilized to produce estimates to solve this problem. LiDAR can estimate the terrain, understory vegetation, and three-dimensional forest structure, including the canopy structure. LiDAR is a surveying technique that calculates the amount of time it takes for a laser light pulse to return after reflecting off of the ground or other solid objects. An image of the forest in three dimensions is produced by these laser returns. In their efforts to manage the vegetation in open streams' forests effectively and sustainably, researchers, conservationists, and forest managers have benefited greatly from this skill. Numerous forest structural characteristics, such as canopy height and cover [47–49], basal area, and tree density [50], as well as understory structure, may be accurately and robustly estimated using LiDAR.

3. Results

3.1. Maintenance Status for the ELFarama and Baloza Drains

The hydrographic surveying results showed that the flow velocities in ELFarama Drain are low ranging from 0.38 to 0.41 m/s, while for the Baloza Drain flow velocities ranged from 0.3 to 0.41 m/s. Values of n for the ELFarama Drain ranged from 0.028 to 0.03 with an average of 0.029, while n for the Baloza Drain ranged from 0.031 to 0.033 with an average of 0.032. The DCR value for the ELFarama Drain was calculated as 83.3–88.5%, with an average of 86.2%. The DCR value for Baloza Drain was calculated as 85.5–89.6%, with an average of 87.6%. Table 3 summarizes the hydraulic parameters used in the calculation of velocity, water depth, n, and DCR at the ELFarama and Baloza Drains. Values of n for the ELFarama Drain, when compared with the design rate, are located "between fair and bad" ($0.033 < n < 0.037$) (Table 2). This also has an adverse effect on the discharge capacity ratio (DCR < 100%). In this way, designed water transmission capacities are reduced by an average of 13.8% (DCR = 86.2%). While the Baloza Drain results indicated a fair canal condition (n = 0.032). This also has an adverse effect on the discharge capacity ratio (DCR < 100%).

Table 3. Hydraulic parameters used in the calculation of n_{actual}, and DCR at ELFarama and Baloza Drains.

Drain Name	Measurement Section Location (km)	A (m^2)	P (m)	R (m)	y (m)	S (m/m)	V_{actual} (m/s)	n_{actual}	DCR (%)
ELFarama Drain	0.180	19.21	17.25	1.114	1.62	0.00012	0.41	0.029	87.1
	2.600	15.97	15.25	1.048	1.62	0.00012	0.4	0.028	88.5
	7.450	14.35	14.25	1.008	1.62	0.00013	0.4	0.029	87.2
	11.150	12.25	13.11	0.934	1.52	0.00015	0.39	0.030	83.3
	16.600	9.74	11.66	0.835	1.37	0.00016	0.38	0.030	84.7
	Average				0.4	0.00014	0.40	0.029	86.2
Baloza Drain	1.600	31.34	22.46	1.4	1.97	0.00011	0.42	0.031	89.6
	4.300	20.19	17.26	1.17	1.78	0.00012	0.38	0.032	87.5
	7.500	16.72	15.56	1.07	1.67	0.00012	0.35	0.033	85.5
	11.200	9.52	11.54	0.82	1.35	0.00014	0.32	0.032	86.4
	15.300	6.91	9.72	0.71	1.22	0.00014	0.3	0.031	89.2
	Average				0.35	0.00013	0.35	0.032	87.6

Note: A, cross-section area; P, wetted perimeter; V_{actual}, actual velocity (measured in the field); y, water depth; S, channel slope; n_{actual}, Manning's coefficient; and DCR, discharge capacity ratio.

The results obtained from the study showed that the general state of maintenance of the studied drains is not good concerning roughness and water conveyance capacity. It was determined that roughness is caused by factors such as increased silting and vegetation.

3.2. Drain Cross-Section Area Deformation

According to the hydrographic survey findings, for the ELFarama Drain, the critical sedimentation reach was between kilometer 0.250 and kilometer 2.250, where the sedimentation accumulation rate was 11.8% of the designed cross-section area, and between kilometer 14.950 and kilometer 19.50, where it ranged between 10.55 and 12.64% of the designed cross-section area. The estimated total annual sedimentation in the ELFarama Drain was 34,369 m^3. For the Baloza Drain, the critical sedimentation reach was between kilometer 0.300 and kilometer 3.300, where the sedimentation accumulation rate was 7.74% of the designed cross-section area. For the ELFarama Drain, the estimated total annual sedimentation was 29,153 m^3, with rates ranging from 9.64 to 7.51% of the designed cross-section area. The windblown fine sands from the northwest during the month of March are to blame for the rising sedimentation rates for the stated reaches [51,52].

3.3. Vegetation Infection

The Baloza Drain has a 28–30% weed infection ratio at the reach from kilometer 0.00 to kilometer 3.30, a 36% weed infection ratio at the reach from kilometer 3.30 to kilometer 5.50, a 27% weed infection ratio at the reach from kilometer 5.50 to kilometer 7.0, and a 22% weed infection ratio at the reach from kilometer 7.00 to kilometer 15.30. The extant vegetation included Phragmites australis, Cyperus articulates, and Ceratophyllum demersum. The existing vegetation in the ELFarama Drain between kilometer 0.00 and kilometer 2.60 consisted of Phragmites australis, Cyperus articulates, and Ceratophyllum demersum, with a weed infection ratio of 0.28–30%. There was a 35% weed infection ratio at the reach from kilometer 2.60 to kilometer 7.45, a 20% weed infection ratio at the reach from kilometer 7.45 to kilometer 11.15, and a 19% weed infection ratio at the reach from kilometer 11.15 to kilometer 16.6. The vegetation in Egypt's Damietta branch of the River is an expansion of the plant diversity in the studied drains [53,54].

4. Discussion

4.1. Comparison between the Literature and the Measured DCR for ELFarama and Baloza Drains

Applying Equations (3) and (4) developed by Mohamed and El-Samman [19] to forecast the value of Manning's roughness coefficient in the vegetated channel was based on the measured parameters, weed infestation percentage, and water surface slope. The results show an average Manning's roughness coefficient in the vegetated channel for the ELFarama and Baloza Drains of 0.27 and 0.03, respectively. Table 4 summarizes a comparison between the literature and the measured DCR for ELFarama and Baloza Drains.

Table 4. Comparison between the literature and the measured DCR for ELFarama and Baloza Drains.

Drain Name	Measurement Section Location (km)	W_{inf} (%)	S (m/m)	Projected (n_o)	DCR According [19]				DCR Current Study
					n_{v1} Equation (3)	n_{v2} Equation (4)	DCR1 (%)	DCR2 (%)	DCR (%)
ELFarama Drain	0.180	28	0.00012	0.025	0.027	0.027	94.2	92.2	87.1
	2.600	30	0.00012	0.025	0.027	0.027	94.0	92.1	88.5
	7.450	35	0.00013	0.025	0.027	0.027	93.6	91.6	87.2
	11.150	20	0.00015	0.025	0.026	0.027	94.8	92.0	83.3
	16.600	19	0.00016	0.025	0.026	0.027	94.9	91.9	84.7
	Average	26.4	0.00014	0.025	0.027	0.027	94.3	92.0	86.2
Baloza Drain	1.600	28	0.00011	0.028	0.030	0.030	94.2	92.4	89.6
	4.300	36	0.00012	0.028	0.030	0.031	93.5	91.7	87.5
	7.500	27	0.00012	0.028	0.030	0.030	94.2	92.3	85.5
	11.200	22	0.00014	0.028	0.030	0.030	94.6	92.1	86.4
	15.300	20	0.00014	0.028	0.030	0.030	94.8	92.2	89.2
	Average	26.6	0.00013	0.028	0.030	0.030	94.3	92.2	87.6

Note: W_{inf}, the percentage of weed infestation in open channel given by Equation (5); S, water surface slope; n_o, Manning roughness coefficient of non-vegetated open channel (projected); n_{v1}, Manning roughness coefficient of vegetated open channel calculated by Equation (3); n_{v2}, Manning roughness coefficient of vegetated open channel calculated by Equation (3); DCR1, discharge capacity ratio for n_{v1}; DCR2 discharge capacity ratio for n_{v2}; and DCR, discharge capacity ratio for current study.

The maintenance problem picture gives the idea that the responsibility and corresponding institutions should be specified explicitly before decisions on a finance strategy are made. These problems are complicated in Egypt by the division between the main irrigation system, outflow command, and drainage. Determining the farmers' participation, or alternatively, the extent of government engagement in funding as well as in duties for execution and control of maintenance at the outflow command level and in downstream drainage would be crucial subjects [55,56].

Additionally, the particular climate, soil, and hydrological conditions are important. For instance, the kind and rate of growth of the plants may be influenced by the soil and climate. In some regions, vegetation regeneration is particularly quick, which necessitates frequent upkeep.

There are many different types of vegetation in some ecosystems, while only one dominates in others. While desilting is hardly ever necessary in some locations where the irrigation water is clear, in others it is substantially silt-loaded and necessitates periodic desilting. There are some maintenance variations between peat, clay, and sandy environments in the Netherlands. Different alluvial, loamy, or clay soil may exist in the Egypt Delta [57,58].

Any maintenance policy needs a choice between two options. With or without reliable cost and benefit information, this decision may be taken based solely on political or national economic considerations. More specifically, the problem is impacted by a number of things. Examples include maintenance procedures, methodologies, and physical circumstances. The work's quality, and hence the frequency needed, are influenced by the methods utilized, but so are the prices [59].

Because an increase in roughness reduces the canal's capacity, it gives rise to operating problems with regard to sufficiency and flexibility. The fact that an important part of the irrigation area's crop pattern is cotton means that irrigation is concentrated in specific periods. At these times, the reduction in canal capacity due to roughness increases the problem of insufficiency in canal capacity [60]. Surface roughness has an impact on water transmission since it sits at the soil-atmosphere interface. Surface roughness can improve infiltration, reduce overland flow, and store water in puddles. It may also have an impact on the direction, depth, and speed of overland flow. Surface roughness can exhibit a variety of characteristics as a result of field activity and rainfall, and this causes its impact on water transport to change across time and space. This variability and its time-consuming measurement contribute to the complexity of determining roughness effects. However, a correlation between surface roughness parameters and water transfer can be seen [61].

4.2. Assessment of the Hydraulic Roughness and Vegetation Management

From a spatial and temporal perspective, riparian vegetation, especially that found along floodplains, exhibits quite varied properties. The hydraulic and hydrological models need to effectively account for these properties. Traditional ground-based monitoring is frequently impractical since it takes a lot of time and money [62], especially for huge areas and inaccessible areas. Remote sensing, which has advanced significantly over the past few decades and has seen an increase in use in the environmental field, presents new prospects. The use of remote sensing in fluvial investigations has been discussed in few papers [63,64], with a focus on mapping riparian vegetation and estimating biomechanical parameters [65]. From multispectral satellite data, Forzieri et al. [66] proposed a method to estimate vegetation height and flexural rigidity for herbaceous patterns as well as plant density, tree height, stem diameter, crown base height, and crown diameter of high-forest and coppice consociations for arboreal and shrub patterns (SPOT 5). The method is created in four consecutive steps: (1) five land cover classifications are derived from the classification of pixel surface reflectance: mixed arboreal, shrubby, herbaceous, bare soil, and water habitats; (2) data transformation based on principal component analysis of the original multispectral bands and use of only the first principal component since it explains a lot of variances; (3) identification of significant correlation structures between the main components and biomechanical properties; (4) use of the principal component analysis of the original multispectral bands to transform the data; and (5) determination, estimation, and validation of the simple tri-parametric power law relationship between the normalized principal component and the biomechanical characteristics. By comparing the vegetation hydrodynamic maps with simulated water stages, it is demonstrated that they can also accurately reflect the comparable Manning's roughness coefficient.

Although the vegetation's vertical structure cannot be determined from satellite photos, its spatial variability can be determined. Information on the three-dimensional structure of vegetation is available because of LiDAR technology. Terrain-based (TLS), airborne (ALS), and mobile (MLS) platforms all use Laser Scanning (LS). A realistic representation of the forest canopy and ground elevations is provided by the Aerial Laser Scanner (ALS), which creates a digital terrain model and a digital surface model. The tree heights are determined by the variation between the digital surface model and the digital terrain model. In order to calculate the total plant area of herbaceous vegetation and the vertical distribution of the total plant area of foliated woody vegetation at various levels of submersion, Jalonen et al. [67] used multistation TLS in both field and laboratory circumstances. Using airborne LiDAR data, Forzieri et al. [68] created a model to pinpoint specific tree locations, crown limits, and plant density. It requires a preliminary calibration stage based on a basic additive multiple attribute decision-making algorithm. One of the first to offer some instances of managing vegetation in both man-made and naturally occurring trapezoidal-shaped channels were Phillips and Tadayon [69]. The design discharge for the canal in full-grown vegetation conditions would flood nearby areas, according to simulation results from the Hydrologic Engineering Center's River Analysis System (HEC-RAS) program. Instead, simulations under post-vegetation maintenance settings that included only partial removal of the vegetation show that the intended discharge would stay inside the channel.

Verifying whether the design flow has the ability to lay over bushes was a recurring theme in a few of the examples. If this happened, the accompanying roughness component was regarded as insignificant and was not taken into account when calculating the Manning coefficient. Additionally, a minimum of 30 cm of freeboard above the planned sea surface height was taken into account. This freeboard's function is to reduce risk by adding a margin of safety. Phillips and Tadayon [69] have discussed how having the vegetation randomly scattered is more aesthetically pleasing, but maintaining the trees and bushes grouped together may create a better habitat environment for wildlife.

According to Luhar and Nepf [70], the most efficient mowing patterns for lowering hydraulic resistance are those that produce less interfacial area per channel length (for example, a single continuous cut on one side of the channel). With regard to a gravel–cobble river in California, Abu Aly et al. [71] examined how vegetation affected flow rates, depths, and the size of the flooded areas for flows between 0.2 and 20 times the bankfull discharge (BFQ). They examined a section of the water flow that was 28.3 km long with a mesh of 1–3 m using a two-dimensional finite volume model (SRH-2D) that solves the vertical-mediated Reynolds equations. They used LiDAR data to determine the height of the vegetation in each cell using the method devised by Katul et al. [72] described above to estimate the Manning coefficient. For a flow of four times the BFQ, they were able to accomplish an increase in the mean water depth of 7.4% and a drop in the mean velocity of 17.5% when compared to the case of no vegetation; these values grow to 25% and 30%, respectively, for a flow rate of 22 times the BFQ. The model also demonstrates how vegetation strongly channels the flow; in fact, flow is diverted away from densely vegetated places and the gap between the mid-channel and bank velocities widens. According to the model's findings, Benifei et al. [73] demonstrated that removing the high vegetation (trees) inside the flooding area with a return period of two years and preventing the establishment of bushes (5-year-old trees) results in the most efficient flood management approach.

4.3. Procedure for Channel Maintenance Method

Six steps are important for performing channel maintenance for the purpose of achieving sustainable agriculture.

Step 1: Describe the associated geometry and characteristics of the Egyptian canal

The geometry of the cross-section, the bed slope of the reach, and the type of vegetation, which define the roughness coefficient, are identified for all the major watercourses in Egypt using data from the national database for the Egyptian Ministry of Water Resources and Irrigation (MWRI). Although it is understood that local information, when accessible, may provide more trustworthy and accurate inputs, the national database is used to obtain these data.

Step 2: Describe maintenance scenarios (comprising sediment and vegetation management)

Procedures are established to assess the potential management scenarios and the variations they cause on the features of the watercourse utilizing the parameters of the watercourse and existing advice [23].

Step 3: Determine the conveyance capacity of each maintenance scenario

A hydraulic model such as HEC-RAS is used to estimate the conveyance capacity based on the cross-section, roughness, and slope of the watercourse as calculated in Step 1. Both roughness coefficients and cross-sectional area are primarily affected by the management of the sediment and vegetation, respectively.

Step 4: Compute the impact of maintenance

To calculate the effects of maintenance, the variance in the main channel's conveyance capacity (reported as a percentage) is used. To provide a quick and accurate way to gauge the effects of maintenance activities, this parameter is approximated for all watercourses.

Step 5: Attribute benefits to various lengths of watercourses

Applying a streamlined hazard attribution methodology to watercourses, river lengths of 50 m are given characteristics in the floodplain that could make them candidates for channel maintenance. It should be assumed that the closest watercourse will most likely cause flooding in every given impact cell since we do not know which length of a watercourse would cause flooding. There are presumptions made regarding the flooding extension.

Step 6: Detecting the ideal management option

The likelihood of the characteristics in a watercourse length having the highest impact on conveyance management and highest benefits connected with the watercourse is used to determine the best maintenance option. The MWRI data are used to estimate the price of vegetation maintenance work, which is measured in pounds per square meter. The benefits to costs ratios are used to select the preferred maintenance option for a given channel, and in the case of dredging, the option with the highest ratio is chosen. However, it is impossible to state the cost of the works at the national level due to the significant variations depending on the site's features. The disposal of materials off-site, which may have significant variances, must also be taken into account when estimating the cost of this type of job. In this study, we propose a maintenance schedule for the investigated drains that calls for the use of a hydraulic excavator with a long boom to remove vegetation and silt from the most exposed sections every two years (Table 5).

Table 5. Suggested maintenance program for the studied surfaces drains.

Drain	Maintenance Type					
	Sedimentation Removal			Vegetation Removal		
	Procedure	Machine	Annual Removal Rate	Procedure	Machine	Annual Removal Rate
ELFarama and Baloza	Mechanical	Long-boom mechanical hydraulic excavator with bucket for sediment dredging	Every two years	Mechanical	Long-boom mechanical hydraulic excavator with bucket for sediment dredging	Every two years

5. Conclusions

The key factor that determines the value of Manning's roughness coefficient is the percentage of weed infestation in vegetated channels. By using echo-sound sonar equipment, it is now simpler to determine the percentage of weeds that are submerged in vegetated channels. Average actual Manning's roughness values of 0.029 were found for the EL-Farama Drain, and 0.032 for the Baloza Drain, which are greater than the design n values of 0.025 and 0.028, respectively. Siltation and weed and algal growth all served to increase this roughness. The fact that the actual roughness was higher than the designed values meant that canal capacities were reduced. The study found that this reduction-averaged value is 13.8% (DCR = 86.2%) for the ELFarama Drain and 12.4% (DCR = 87.6%) for the Baloza Drain. In addition, the hydrographic surveying results showed high measured annual sedimentation volumes for the ELFarama Drain and Baloza surface Drains. Vegetation surveying results showed that the ELFarama and Baloza Drains have average weed infection ratios of 30%. It is suggested to utilize a long-boom mechanical hydraulic excavator with a bucket for clearing vegetation and dredging sediments. The DCR is a simple tool for evaluating the state of channel maintenance for sustainable agriculture.

Author Contributions: Conceptualization, M.E.G.; methodology, M.E.G.; software, M.E.G.; validation, M.E.G. and M.K.M.; formal analysis, E.M.F.; supervision, E.M.F.; writing—review and editing, M.E.G. and M.K.M.; investigation, M.E.G. and M.K.M.; resources, M.E.G. and M.K.M.; data curation, M.E.G.; writing—original draft preparation, M.E.G.; writing—review and editing M.E.G., E.M.F. and M.K.M. All authors have read and agreed to the published version of the manuscript.

Funding: This research received no external funding.

Data Availability Statement: Not applicable.

Acknowledgments: The authors would like to thank the Ministry of Water Resources and Irrigation (MWRI) in Egypt for their special logistical and technical support.

Conflicts of Interest: The authors declare no conflict of interest.

References

1. Ferro, V. Assessing flow resistance law in vegetated channels by dimensional analysis and self-similarity. *Flow Meas. Instrum.* **2019**, *69*, 101610. [CrossRef]
2. D'Ippolito, A.; Calomino, F.; Alfonsi, G.; Lauria, A. Flow resistance in open channel due to vegetation at reach scale: A Review. *Water* **2021**, *13*, 116. [CrossRef]
3. Hinsberger, R.; Biehler, A.; Yörük, A. Influence of water depth and slope on roughness experiments and roughness approach for Rain-on-Grid Modeling. *Water* **2022**, *14*, 4017. [CrossRef]
4. Penman, H.L. *Vegetation and Hydrology*; Commonwealth Agricultural Bureau: Harpenden, UK, 1963; p. 124.
5. Waller, P.; Yitayew, M. (Eds.) *Subsurface Drainage Design and Installation BT-Irrigation and Drainage Engineering*; Springer International Publishing: Cham, Germany, 2016; pp. 531–544.
6. Imran, M.A.; Xu, J.; Sultan, M.; Shamshiri, R.R.; Ahmed, N.; Javed, Q.; Asfahan, H.M.; Latif, Y.; Usman, M.; Ahmad, R. Free discharge of subsurface drainage effluent: An alternate design of surface drain system in Pakistan. *Sustainability* **2021**, *13*, 4080. [CrossRef]
7. Cannell, R.Q.; Belford, K.; Gales, K.; Dennis, C.W.; Prew, R.D. Effects of waterlogging at different stages of development on the growth and yield of winter wheat. *J. Sci. Food Agric.* **1980**, *31*, 117–132. [CrossRef]
8. Tejaswini, V.; Babu, G.R.; Kumar, H.V.H.; Prasad, B.V.S.; Rao, C.S. Status of Water Quality from Agriculture Drains in Guntur District, Andhra Pradesh, India. *Curr. J. Appl. Sci. Technol.* **2022**, *41*, 1–9. [CrossRef]
9. Velmurugan, A.; Swarnam, T.P.; Ambast, S.K.; Kumar, N. Managing waterlogging and soil salinity with a permanent raised bed and furrow system in coastal lowlands of humid tropics. *Agric. Water Manag.* **2016**, *168*, 56–67.
10. Dennis, W.; Manzoor, Q. Achieving sustainable irrigation requires effective management of salts, soil salinity, and shallow groundwater. *Agric. Water Manag.* **2015**, *157*, 31–38.
11. Liu, H.; Li, M.; Zheng, X.; Wang, Y.; Anwar, S. Surface salinization of soil under mulched. *Water* **2020**, *12*, 3031. [CrossRef]
12. Mohamed, E.S.; Belal, A. Saleh assessment of land degradation east of the Nile Delta, Egypt using remote sensing and GIS techniques. *Arab. J. Geosci.* **2013**, *6*, 2843–2853. [CrossRef]
13. Hammam, A.A.; Mohamed, E.S. Mapping soil salinity in the East Nile Delta using several methodological approaches of salinity assessment. *Egypt. J. Remote Sens. Space Sci.* **2020**, *23*, 125–131. [CrossRef]
14. Ministry of Water Resources and Irrigation (MWRI). *Egypt's Water Resources Plan for 2017–2037*; Planning Sector, Ministry of Water Resources and Irrigation (MWRI): Giza, Egypt, 2017. Available online: https://www.mwri.gov.eg (accessed on 10 June 2022).
15. Gabr, M.E. Study of reclaimed water reuse standards and prospects in irrigation in Egypt. *Port-Said Eng. Res. J.* **2020**, *24*, 65–75. [CrossRef]
16. François, M.; Ibrahim, G.; Doaa, E.E.; Edwin, R. The Nile delta's water and salt balances and implications for management. *Agric. Water Manag.* **2018**, *197*, 110–121.
17. FAO. *Irrigation in Africa in Figures: AQUASTAT Survey*; FAO: Rome, Italy, 2005.
18. Mohamed, N.N. Land degradation in the Nile Delta. In *The Nile Delta. The Handbook of Environmental Chemistry*; Negm, A., Ed.; Springer: Cham, Switzerland, 2016; Volume 55.
19. Mohamed, S.A.; El-Samman, T.A. Manning roughness coefficient in vegetated open channels. *Water Sci.* **2020**, *34*, 121–128. [CrossRef]
20. El Samman, T.A.; Abou El Ella, S.M. Aquatic weeds monitoring and associated problems in Egyptian channels. In Proceedings of the Thirteenth International Water Technology Conference, IWTC 13 2009, Hurghada, Egypt, 12–15 March 2009.
21. FAO. Aquatic Weeds and Associated Problems in Irrigation Systems. Available online: https://www.fao.org/3/T0401E/T0401E04.htm (accessed on 15 March 2023).
22. Zachary, M.E.; Emily, B.; Amy, S.C. Factors When Considering an Agricultural Drainage System. Produced by Communications and Marketing, College of Agriculture and Life Sciences. *Virginia Tech*. 2016. Available online: www.ext.vt.edu (accessed on 15 March 2023).
23. Anthony, M.; Kevin, B.; Bob, B. Sustainable Drainage Design & Evaluation Guide. Citywest Campus, Dublin. *Courtesy of Davy Hickey*. 2021. Available online: https://www.dublincity.ie/sites/default/files/2021-12/dcc-suds-de-guide-2021.pdf (accessed on 25 February 2023).
24. Abdel-Dayem, S. An integrated approach to land drainage. *Irrig. Drain.* **2006**, *55*, 299–309. [CrossRef]
25. Abdeldayem, O.M.; Eldaghar, O.; Mostafa, M.K.; Habashy, M.M.; Hassan, A.A.; Mahmoud, H.; Morsy, K.M.; Abdelrady, A.; Peters, R.W. Mitigation plan and water harvesting of flashflood in arid rural communities using modelling approach: A Case Study in Afouna Village, Egypt. *Water* **2020**, *12*, 2565. [CrossRef]

26. Essex County Council Environment (ECC). Sustainable Drainage Systems Design Guide. *Sustain. Highw.*; 2016. Available online: www.essex.gov.uk/flooding (accessed on 20 March 2023).
27. Wu, F.C.; Shen, H.W.; Chou, Y.J. Variation of roughness coefficients for unsubmerged and submerged vegetation. *J. Hydraul. Eng.* **1999**, *125*, 934–942. [CrossRef]
28. Han, L.; Zeng, Y.; Chen, L.; Huai, W. Lateral velocity distribution in open channels with partially flexible submerged vegetation. *J. Environ. Fluid Mech.* **2016**, *16*, 1267–1282. [CrossRef]
29. Wan Yusof, K.; Mujahid Muhammad, M.; Ul, R.; Mustafa, M.; Azazi Zakaria, N.; Gahani, A.A. Analysis of Manning's and drag coefficients for flexible submerged vegetation. *IOP Conf. Ser. Mater. Sci. Eng.* **2017**, *216*, 012046. [CrossRef]
30. Jaan, H.P.; Khalid, H.; Song-dong, S.; Yuefei, H. Shallow sediment transport flow computation using time-varying sediment adaptation length. *Int. J. Sediment Res.* **2014**, *29*, 171–183. [CrossRef]
31. Chow, V. *Open Channel Hydraulics*; McGraw-Hill: New York, NY, USA, 1973.
32. Ijir, T.A.; Burton, M.A. Performance assessment of the Wurno irrigation scheme, Nigeria. *ICID J.* **1988**, *47*, 31–46.
33. Gabr, M.; Rageh, O. Strategic planning model for the construction and remediation of irrigation networks: A case study for Egypt. *Delta Univ. Sci. J.* **2023**, *6*, 85–102. [CrossRef]
34. Mohamed, E.S.; Belal, A.; Shalaby, A. Impacts of soil sealing on potential agriculture in Egypt using remote sensing and GIS techniques. *Eurasian Soil Sci.* **2015**, *48*, 1159–1169. [CrossRef]
35. Gabr, M. Evaluation of irrigation water, drainage water, soil salinity, and groundwater for sustainable cultivation. *Irrigat Drain. Sys. Eng.* **2018**, *7*, 224.
36. Gabr, M. Drainage management problems evaluation: Case Study Baloza and EL-Farama Drains, North Sinai, Egypt. *J. Water Resour. Prot.* **2019**, *11*, 675–689. [CrossRef]
37. Gabr, M.E. Proposing a constructed wetland within the branch drains network to treat degraded drainage water in Tina Plain North Sinai, Egypt. *Arch. Agron. Soil Sci.* **2021**, *67*, 1479–1494. [CrossRef]
38. Gabr, M.E. Modelling net irrigation water requirements using FAO-CROPWAT 8.0 and CLIMWAT 2.0: A case study of Tina Plain and East South ElKantara regions, North Sinai, Egypt. *Arch. Agron. Soil Sci.* **2022**, *68*, 1322–1337. [CrossRef]
39. Gabr, M.E.; Ehab, M.F. Assessment of irrigation management practices using FAO-CROPWAT 8, case studies: Tina Plain and East South El-Kantara, Sinai, Egypt. *Ain Shams Eng. J.* **2021**, *12*, 1623–1636. [CrossRef]
40. Gabr, M.E.; El-Ghandour, H.A.; Elabd, M. Prospective of the utilization of rainfall in coastal regions in the context of climatic changes: Case study of Egypt. *Appl. Water Sci.* **2023**, *13*, 19. [CrossRef]
41. Gabr, M.E.S. Management of irrigation requirements using FAO-CROPWAT 8.0 model: A case study of Egypt. *Model. Earth Syst. Environ.* **2022**, *8*, 3127–3142. [CrossRef]
42. Gabr, M.E. Land reclamation projects in the Egyptian Western Desert: Management of 1.5 million acres of groundwater irrigation. *Water Int.* **2023**, *48*, 240–258. [CrossRef]
43. Gabr, M.E.; Al-Ansari, N.; Salem, A.; Awad, A. Proposing a Wetland-Based Economic Approach for Wastewater Treatment in Arid Regions as an Alternative Irrigation Water Source. *Hydrology* **2023**, *10*, 20. [CrossRef]
44. Gabr, M.E. Study of lowlands drainage problems, case study Kamal El-Den Hessen Reclaimed Area, North Sinai, Egypt. *J. Water Resour. Prot.* **2018**, *10*, 857–869. [CrossRef]
45. Channel Maintenance Research Institute, National Water Research Center, Ministry of Water Resources and Irrigation (MRWI), Assessing of the Irrigation and Drainage Network Maintenance at the North Sinai Project. In *Technical Report submitted to the Sector of Irrigation, Water Resources and Infrastructures in the North of Sinai*; Channel Maintenance Research Institute, MWRI: El-Kanater El-Khairia, Egypt, 2010. (In Arabic)
46. Eskelson, B.N.I.; Madsen, L.; Hagar, J.C.; Temesgen, H. Estimating riparian understory vegetation cover with beta regression and copula models. *For. Sci.* **2011**, *57*, 212–221.
47. Lim, K.; Treitz, P.; Baldwin, K.; Morrison, I.; Green, J. LiDAR remote sensing of biophysical properties of tolerant northern hardwood forests. *Can. J. Remote Sens.* **2003**, *29*, 658–678. [CrossRef]
48. Naesset, E. Practical large-scale forest stand inventory using a small-footprint airborne scanning laser. *Scand. J. For. Res.* **2004**, *19*, 164–179. [CrossRef]
49. Thomas, V.; Treitz, P.; McCaughey, J.H.; Morrison, I. Mapping stand-level forest biophysical variables for a mixedwood boreal forest using lidar: An examination of scanning density. *Can. J. For. Res.* **2006**, *36*, 34–47. [CrossRef]
50. Venier, L.A.; Swystun, T.; Mazerolle, M.J.; Kreutzweiser, D.P.; Wainio-Keizer, K.L.; McIlwrick, K.A.; Woods, M.E.; Wang, X. Modelling vegetation understory cover using LiDAR metrics. *PLoS ONE* **2019**, *14*, e0220096. [CrossRef]
51. Gad, A. Sand dune distribution and related impacts on agricultural resources of Sinai Peninsula, Egypt, using integrated remote sensing-GIS techniques. *Glob. Adv. Res. J. Agric. Sci.* **2016**, *5*, 42–50.
52. Gabr, M.E. Magnitude and characteristics of sand dunes encroachment towards El-Sheikh Gaber Canal, North Sinai, Egypt. In Proceedings of the Twenty-First International Water Technology Conference, IWTC 21, Ismailia, Egypt, 28–30 June 2018.
53. Alaa, M.M. Comparative view for the impact of five eco factors on species distribution and weed community structure in Isthmus of Suez and adjoining farmland east Nile Delta, Egypt. *Heliyon* **2019**, *5*, e02161.
54. Yasser, A.; Mahmoud, A.; Shaymaa, O. Plant diversity of the Damietta Branch, River Nile, Egypt: An ecological insight. *Mesop. Environ. J.* **2015**, *1*, 109–129.

55. Huan, W.; Jiejun, H.; Han, Z.; Chengbin, D.; Chuanglin, F. Analysis of sustainable utilization of water resources based on the improved water resources ecological footprint model: A case study of Hubei Province, China. *J. Environ. Manag.* **2020**, *262*, 110331.
56. Roca, M.; Rushworth, A.; Dennes, D. Supporting decision-making for channel conveyance maintenance. In Proceedings of the 3rd European Conference on Flood Risk Management, Lyon, France, 17–21 October 2016; pp. 2–7. [CrossRef]
57. Barnes, J. States of maintenance: Power, politics, and Egypt's irrigation infrastructure. *Environ. Plan. D Soc. Space* **2017**, *35*, 146–164. [CrossRef]
58. Salah, M.; Zayed, M. An experimental study of the influence of open channel hydraulic parameters and soil gradation on jet pump sand removal's efficiency. *Br. J. Appl. Sci. Technol.* **2016**, *14*, 1–10. [CrossRef]
59. Mohamed, S.A. Improving the carrying capacity of irrigation canals: Al-Tawfiky diversion. *Water Sci.* **2021**, *35*, 89–99. [CrossRef]
60. Osman, I.S.; Schultz, B.; Osman, A.; Suryadi, F.X. Effects of different operation scenarios on sedimentation in irrigation canals of the Gezira Scheme, Sudan. *Irrig. Drain.* **2017**, *66*, 82–89. [CrossRef]
61. Darboux, F. Surface Roughness, Effect on Water Transfer. In *Encyclopedia of Agrophysics*; Encyclopedia of Earth Sciences Series; Gliński, J., Horabik, J., Lipiec, J., Eds.; Springer: Dordrecht, The Netherlands, 2011. [CrossRef]
62. Dudley, S.J.; Bonham, C.D.; Abt, S.R.; Fischenich, J.G. Comparison of methods for measuring woody riparian vegetation density. *J. Arid Environ.* **1998**, *38*, 77–86. [CrossRef]
63. Hohenthal, J.; Alho, P.; Hyyppa, J.; Hyyppa, H. 2011—Laser scanning applications in fluvial studies. *Prog. Phys. Geog.* **2011**, *35*, 782–809. [CrossRef]
64. Tomsett, C.; Leyland, J. Remote sensing of river corridors: A review of current trends and future directions. *River Res. Appl.* **2019**, *35*, 779–803. [CrossRef]
65. Huylenbroeck, L.; Laslier, M.; Dufour, S.; Georges, B.; Lejeune, P.; Michez, A. Using remote sensing to characterize riparian vegetation: A review of available tools and perspectives for managers. *J. Environ. Manag.* **2019**, *267*, 110652. [CrossRef]
66. Forzieri, G.; Castelli, F.; Preti, F. Advances in remote sensing of hydraulic roughness. *Int. J. Remote Sens.* **2012**, *33*, 630–654. [CrossRef]
67. Jalonen, J.; Järvelä, J.; Virtanen, J.P.; Vaaja, M.; Kurkela, M.; Hyyppa, H. Determining characteristic vegetation areas by terrestrial laser scanning for floodplain flow modeling. *Water* **2015**, *7*, 420–437. [CrossRef]
68. Forzieri, G.; Degetto, M.; Righetti, M.; Castelli, F.; Preti, F. Satellite multispectral data for improved floodplain roughness modelling. *J. Hydrol.* **2011**, *407*, 41–57. [CrossRef]
69. Phillips, J.V.; Tadayon, S. *Selection of Manning's Roughness Coefficient for Natural and Constructed Vegetated and Non-Vegetated Channels, and Vegetation Maintenance Plan Guidelines for Vegetated Channels in Central Arizona*; Scientific Investigations Report 2006-5108; U.S. Geological Survey: Reston, VA, USA, 2007; p. 41.
70. Luhar, M.; Nepf, H.M. From the blade scale to the reach scale: A characterization of aquatic vegetative drag. *Adv. Water Resour.* **2013**, *51*, 305–316. [CrossRef]
71. Abu-Aly, T.R.; Pasternack, G.B.; Wyrick, J.R.; Barker, R.; Massa, D.; Johnson, T. Effect of LiDAR-derived, spatially distributed vegetation roughness on two-dimensional hydraulics in a gravel-cobble river at flows of 0.2 to 20 times bank full. *Geomorphology* **2014**, *206*, 468–482. [CrossRef]
72. Katul, G.; Wiberg, P.; Albertson, J.; Hornberger, G. A mixing layer theory for flow resistance in shallow streams. *Water Resour. Res.* **2002**, *38*, 1250. [CrossRef]
73. Benifei, R.; Solari, L.; Vargas-Luna, A.; Geerling, G.; Van Oorschot, M. Effects of vegetation on flooding: The study case of the Magra river. In Proceedings of the 36th IAHRWorld Congress, The Hague, The Netherlands, 28 June–3 July 2015.

water

MDPI

Article

Multifrequency Downstream Hydraulic Geometry of Alluvial Mountain Rivers Located on the Qinghai–Tibet Plateau

Chao Qin [1,2,3], Baosheng Wu [1,2,3], Yuan Xue [1,2,3,*], Xudong Fu [1,2,3], Guangqian Wang [1,2,3] and Ge Wang [1,2,3]

1 State Key Laboratory of Hydroscience and Engineering, Tsinghua University, Beijing 100084, China; glqinchao@nwsuaf.edu.cn (C.Q.); xdfu@tsinghua.edu.cn (X.F.); dhhwgq@mail.tsinghua.edu.cn (G.W.); wangg19@mails.tsinghua.edu.cn (G.W.)
2 Key Laboratory of Hydrosphere Sciences of the Ministry of Water Resources, Tsinghua University, Beijing 100084, China
3 Department of Hydraulic Engineering, Tsinghua University, Beijing 100084, China
* Correspondence: xueyuan_thu@163.com

Abstract: Downstream hydraulic geometry (DHG) for multiple discharge frequencies remains poorly investigated. This paper seeks to clarify the DHG relations of different discharge frequencies and proposes the definition, mathematical expression, and geomorphological interpretation of multifrequency DHG (MFDHG). It also verifies the existence of DHG and MFDHG in the six major exoreic rivers located in the Qinghai–Tibet Plateau. MFDHG can be depicted with (1) log-linear plots between DHG coefficients and exponents for multiple discharge frequencies and (2) independent DHG curves intersecting near congruent discharge, width, depth, or velocity. The results show that rivers in the study area exhibit strong DHG relations. The variations in the DHG coefficients and exponents usually exhibit opposite trends with increasing discharge frequency. The MFDHG of a river reach is generally stronger than that of a river basin. Congruent hydraulics, as indices of geometric variability and hydraulic self-similarity, reflect consistent changes in hydraulic variables downstream. MFDHG is a novel geomorphic phenomenon that bridges spatiotemporal dimensions in HG systems and provides a basis for establishing an overall HG relationship.

Keywords: multifrequency downstream hydraulic geometry MFDHG; discharge frequency; river network; Qinghai–Tibet Plateau

1. Introduction

Empirical hydraulic geometry (HG) describes the quantitative relationship between river morphology (cross sections or longitudinal profiles) and basin characteristics. The relationships between discharge (Q), river width (W), average flow depth (H), and flow velocity (V) of an individual cross section are defined as At-a-station HG (AHG), while similar trends between Q and W, H, and V among cross sections in the downstream direction are termed Downstream HG (DHG). Specifically, strong DHG relations have mostly been verified along river reaches. For different river systems, data from various watersheds with bankfull discharge were also combined to reveal good DHG relations between channel geometry and discharge [1].

An empirical engineering procedure for the design of unlined canals was developed by Lindley [2] early in the 20th century. This engineering procedure, from which HG was directly derived, was summarized by Leopold and Maddock [3] in the form of a power law in alluvial channels:

$$W = aQ^b \quad (1)$$

$$H = cQ^f \quad (2)$$

$$V = kQ^m \quad (3)$$

where a, c, and k are the HG coefficients, and b, f, and m are the HG exponents of rivers that conform to the regime theory.

Previous studies of HGs have primarily focused on their existence under various geomorphic conditions [4–7] and their theoretical bases [8–11], denotations and connotations [8,12,13], simulations [13,14], and applications. Although HG relations were treated as purely empirical, they were widely used for streamflow measurements and then for routing in hydrologic models [15,16], geomorphological assessment [10,11], river engineering and stream restoration [17–19], discharge estimation [12,20], and river carbon emissions [21,22].

Establishing an overall HG relationship is a challenging topic that has attracted much attention. The aims of this endeavor are (1) to build dimensionless analytical HG expressions, (2) to include as large an area as possible covering diverse boundary conditions in the spatial dimension, and (3) to simultaneously simulate variations between channel geometry and discharge along the length of the river and the water level in the spatiotemporal dimension. Consistent channel-bounding (bed and bank) material contributes to the constant variation between channel morphology and hydraulic variables; this has been emphasized through regime theory in many previous studies [1]. Dimensionless DHGs across morphologically similar sand beds [11] or gravel riverbeds [10] have been proposed to construct an overall HG relationship. Recent studies have injected fresh perspectives into classic HG studies. For instance, all anabranches of one cross section of a river with multiple channels can be plotted to yield an interchannel HG [23]. In the downstream direction, based on the idea that "cross sections of a given stream system are interrelated" [24], basin HG has been proposed to define the average values of hydraulic variables for a given streamflow and drainage area in a hydrologically homogeneous basin [8,25]. Furthermore, a specific cross section can be linked to the outlet of the river network by the same runoff event [26]. Local variations in cross sectional form are possible sources of scatter in DHG [18]. Therefore, relationships between the AHG of channels in a downstream direction are expected. At-Many-stations HG (AMHG) refers to the paired coefficients and exponents of AHG from many cross sections of a given river reach that are functionally related to one another and exhibit a log-linear relationship [27]. AMHG encompasses variations in both water levels (multiple discharge frequencies) and cross sections (multiple locations) [12]. It extends the spatial dimension from a single AHG cross section to strong relations across a whole river network. All these efforts pave the way for creating a possible overall HG relationship. More facets of HG than previously thought can be uncovered by closely linking DHG and AHG [28].

Previous DHG studies have usually considered bankfull discharge to be the formative discharge (channel-forming discharge); this highlights the dominant role of bankfull flow in shaping channel morphology, as it is the transition from shaping channels to shaping floodplains based on discharge and sediment load. However, a single discharge frequency cannot quantify the variations in hydraulic variables of individual cross sections. DHG exponents are significantly influenced by discharge frequencies [29]. From the limited data presented in Table 1, we find that when discharge frequencies varied from 2–50%, the differences in DHG width coefficients, and exponents changed sharply in the ranges of 13–69% and 32–375%, respectively. However, no universal trend in the variations in coefficients and exponents with changing discharges has been identified to date [30] (Table 1). Based on these facts and fresh perspectives on the classic HG topic, we will examine whether the change in DHG exponents with discharge frequency is a universal rule and whether the DHG coefficients regularly change with discharge frequency. AMHG depicts HG variations in both the spatial and temporal dimensions across river networks on the basin scale [7]. Similarly, we can hypothesize that a series of DHGs with multiple discharge frequencies will include as large an area as possible with similar boundary conditions on the basin scale.

Table 1. Values, maximum differences, and variation trends of DHG exponents at different discharge frequencies from the literature.

Country	References	Discharge Frequency/%	*a*	*b*	*f*	*m*	Maximum Difference/%				Variation Trend		
							a	*b*	*f*	*m*	*b*	*f*	*m*
US	[31]	50 15 2		0.34 0.38 0.45	0.45 0.42 0.43	0.32 0.32 0.17		32.4	7.1	88.2	increase	no trend	no trend
United Kingdom	[29]	50 15 2		0.46 0.54 0.61	0.16 0.23 0.31	0.38 0.23 0.08		32.6	93.8	375.0	increase	increase	decrease
Canada	[32]	83 50 20 10	5.6 4.8 3.8 3.3	0.51 0.53 0.54 0.54			68.5	5.4			increase		
Puerto Rico	[33]	70 50 30		0.46 0.44 0.46	0.27 0.30 0.32	0.27 0.35 0.25		4.5	18.5	40.0	no trend	increase	no trend

Thus, the key findings of previous DHG studies and the remaining gaps are summarized as follows: (1) discharge frequency significantly influences DHG coefficients and exponents, but research on its variation characteristics is lacking, and whether DHG coefficients and exponents are strongly correlated is unclear; and (2) to extract the common characteristics of existing HG relations, a combination of AHG and DHG features with spatiotemporal dimensions is necessary to contribute to the construction of an overall HG relationship. Due to the limitations of previous studies, this article aims to (1) explore relationships between DHG coefficients and exponents under multiple discharge frequencies; (2) identify any apparent trend in DHG coefficients and exponents based on changes in discharge frequencies; and (3) propose a framework system of multifrequency DHG (MFDHG) that may be the foundation of an overall HG relationship.

In Section 3, the definition and mathematical expression for MFDHG is proposed, along with its geomorphological interpretation along a river reach and for different reaches located in the same river basin based upon the hypothesis of the DHG series with multiple discharge frequencies. Then, we verify DHGs based on bankfull discharge and present MFDHGs in nine scenarios in river systems originating from the QTP in Sections 4.1 and 4.2. The typical methods for estimating bankfull discharge and fitting DHGs are presented in detail in Sections 2.2 and 2.3.

2. Data and Methods

2.1. Data and Study Area

2.1.1. Study Area

The study area is located in the eastern and southern portions of the Qinghai–Tibet Plateau (QTP) and includes three national river basins (the upper Yellow River (YR), the Yalong River (YLR), and the upper Jinsha River (JSR), all of which flow to the Pacific Ocean) and three international river basins (the Lantsang River (LCR), which flows to the Pacific Ocean, and the Nu River (NR) and the Yarlung Zangbo River (YLZBR), which flow to the Indian Ocean) (Figure 1). The total area of the study area is 130.787 $\times$ 10^4 km^2. To maintain the integrity of the basins and expand the database, portions of the connecting regions between the QTP and the Loess Plateau, as well as the Yunnan–Kweichow Plateau, are included (Figure 1).

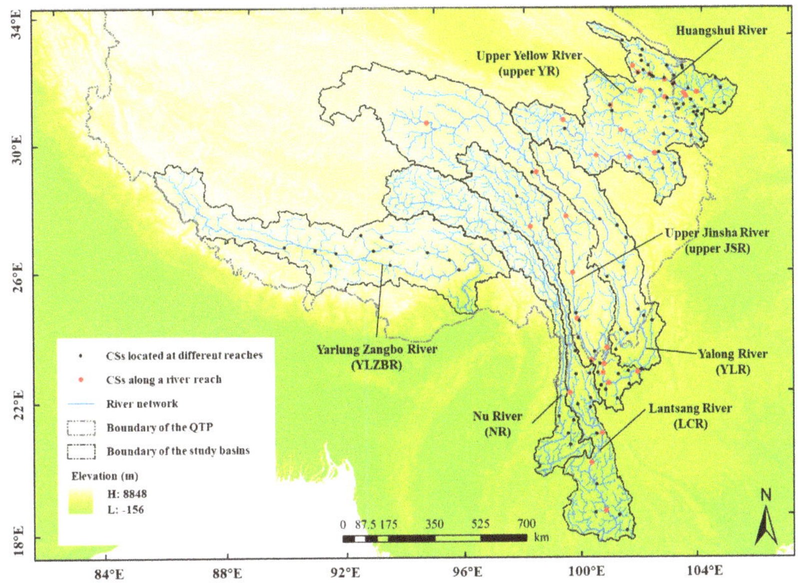

Figure 1. Location of the study area and cross sections (CSs). Red dots represent cross sections along a river reach (the 1st–4th scenarios of Table 2). Black dots represent cross sections located at different reaches of the same river basin (the 5th–9th scenarios of Table 2).

Table 2. DHG coefficients, exponents, and strength R^2 of the six rivers and their tributaries originating from the QTP under bankfull conditions. DHG exponents of the theoretical solutions and the global rivers are presented for comparison with the data generated from the QTP. The $a \times c \times k$ represents the product of DHG coefficients, while the $b + f + m$ represents the sum of DHG exponents; R^2 indicates the determination coefficient between discharges and river width/water depth/flow velocities under bankfull conditions.

Sources	Scenarios	Width-DHG			Depth-DHG			Velocity-DHG			Relations	
		a	b	R^2	c	f	R^2	k	m	R^2	$a \times c \times k$	$b + f + m$
Along a river reach	Main stream of the YR	35.62	0.22	0.54	0.12	0.48	0.86	0.24	0.30	0.65	1.00	1.00
	Main stream of the JSR	45.01	0.15	0.02	0.13	0.51	0.39	0.18	0.34	0.29	1.01	1.00
	Main stream of the LCR	0.11	0.88	0.76	0.13	0.50	0.57	77.66	-0.38	0.88	0.98	1.00
	Main stream of the Huangshui River	4.57	0.46	0.57	0.25	0.33	0.59	0.88	0.21	0.74	1.01	1.00
Different reaches located in the same river basin	Tributaries of the JSR	7.38	0.31	0.65	0.23	0.43	0.58	0.60	0.26	0.55	1.00	1.00
	Tributaries of the YLR	2.18	0.54	0.46	0.19	0.44	0.29	2.41	0.02	0.01	1.01	1.00
	Tributaries of the LCR	2.36	0.55	0.85	0.40	0.34	0.79	1.05	0.11	0.12	1.00	1.00
	Main stream and tributaries of the NR	2.01	0.55	0.99	0.22	0.42	0.94	2.27	0.03	0.12	1.01	1.00
	Main stream and tributaries of the YLZBR	7.58	0.43	0.50	0.12	0.44	0.70	1.11	0.14	0.45	1.01	1.00
Theoretical solutions	Minimum variance theory [34]		0.50			0.38			0.13			1.01
	Momentum diffusion [35]		0.50			0.42			0.08			1.00
	Minimum stream power [36]		0.47			0.42			0.11			1.00
	Threshold theory [37]		0.46			0.46			0.08			1.00
Global	72 streams around the world-range [4]		0.03–0.89			0.09–0.70			−0.51–0.75			
	72 streams around the world-model class [4]		0.4–0.5			0.3–0.4			0.1–0.2			

2.1.2. Data Collection

In situ-measured discharges and corresponding river widths, average flow depths, flow velocities, and cross sectional morphologies were all acquired from the Annual Hydrological Reports of P.R. China (1967–2019). As indicated by Qin et al. [7], the dataset covers a wide range of stream patterns, which have high representation of the rivers in this area. Sixty cross sections were located within the area of the QTP, while another 69 cross sections were located outside the southeastern and northeastern margins of the QTP. The HydroSHEDS River Network (https://www.hydrosheds.org/page/hydrorivers, accessed on 31 May 2023) was used as a basic dataset to locate the in situ-measured cross sections.

2.1.3. Data Screening

The total number of cross sections with long-term measurements was 209, and 129 cross sections were selected for study. The criteria for selecting the cross sections were as follows: (1) having a consecutive hydrological record exceeding 10 years; (2) experiencing relatively low anthropogenic influence (e.g., no hydropower station and artificial diversion 10 km upstream and downstream of the measured cross section; located outside the backwater zone of a dam) during the study time period to minimize the external disturbance; (3) located at a distance from the regions that might be affected by extreme events such as landslides, debris flows, and glacial outbursts; (4) for cross sections that experienced notable bed elevation changes greater than 1 m between years, we took into account data with evident deposition or scouring and removed them from the following analysis; and (5) acting as a natural riverway with perennial drainage [7].

While alluvial reaches and bedrock-constrained reaches alternated with transient reaches in the rivers we studied, the selected cross sections were predominantly located in alluvial reaches, which were deemed more favorable for establishing hydrological stations. These reaches, which were in the quasi-equilibrium state, were characterized by a stable morphology and exhibited relatively minor fluctuations in erosion and sediment deposition both within and between years. Therefore, the relationships between water level and discharge were found to be curves rather than loops. The bed and bank materials of these cross sections are similar and mainly consist of gravel and sand. Table S1 shows a qualitative description of the bed and bank materials of the 129 studied cross sections.

2.2. *Estimation of Discharges Based on Different Frequencies and Bankfull Conditions*

2.2.1. Estimation of Discharges Based on Different Frequencies/Recurrence Intervals

Annual maximum peak discharges (over no less than 20 years) of each cross section were used to quantify discharge–frequency relations via Pearson III curves. For each cross section, the corresponding discharges from a 95.2% discharge frequency (equal to a recurrence interval of 1.05) to the minimal discharge frequency (equal to the recurrence interval of bankfull discharge) of a studied cross section group were estimated. Using the Huangshui River (Figure 1 and Table 2) as an example, there were five consecutive cross sections in this river reach (the number of cross sections per river reach in this study ranged from 1 to 11). The recurrence intervals of these cross sections (cross section IDs 20–24 in Table S1) under bankfull conditions were 2.28, 2.58, 2.34, 2.16, and 2.34, respectively. The value of the minimal recurrence interval under bankfull conditions for all cross sections of the Huangshui River cross section group was 2.16. Therefore, discharges of the recurrence intervals of 1.05, 1.10, 1.15, 1.20, 1.30, ... , 2.0, 2.1, 2.16 for the five cross sections were estimated.

2.2.2. Estimation of Bankfull Discharge

There are two primary ways to define the bankfull discharge: one is based on the discharge frequency or flow duration; the other is based on the river cross section morphology [1].

To illustrate the estimation based on cross section morphology, the morphologies of 129 cross sections were first depicted to detect turning points (Figure 2). The potential bankfull positions (turning points) of each cross section were recorded if an evident break or discontinuity existed (Figure 2). In detail, bed elevation was in situ measured for every 1 m to the left bank. The indicator α is defined as the elevation change rate: $\alpha_i = (|h_{i+1} - h_i|)/h_i$, where $i = 0, 1, 2, \ldots, n$ m is the distance to the left bank, and h_i is the bed elevation at the ith distance to the left bank (m). If $\alpha_i > 20\%$, then the first point from the left bank at the ith distance was defined as a potential bankfull position. If

$\alpha_i \leq 20\%$, then the point at the ith distance to the left bank was recognized as an integral part of the river bed or river bank. Then, the discharge frequencies of the potential bankfull positions were checked to determine whether they met general requirements. Potential bankfull positions with a recurrence interval of more than 8 years were removed [38]. The determined bankfull turning point and corresponding bed elevation (equals to water level) were recorded. Discharge corresponding to bankfull water level was then determined based on the in situ-measured records or water level–discharge curve.

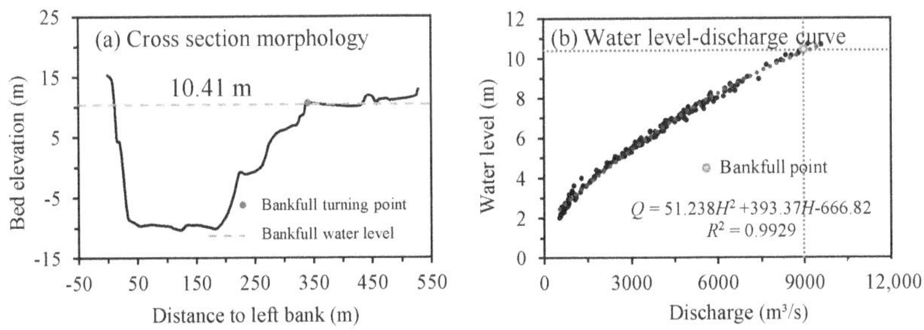

Figure 2. Typical cross section morphology (**a**), Nuxia cross section located at the main stream of the YLZBR), and corresponding water level–discharge rating curve (**b**).

Estimation for cross sections that have no evident flood plain (represented by the above-mentioned turning point) should account for the following factors: stream order, contributing area, upstream and downstream relationships, and main stream and tributary relationships. Previous studies have indicated that bankfull floods are often associated with a nearly constant discharge frequency (often in the range of 1–2 years) [39,40]. Therefore, the main steps in the estimation procedure are as follows: (1) use the mean of known bankfull discharge frequencies in the same river reach to impute the bankfull discharge frequencies of cross sections with unknown values; (2) if there is no known bankfull discharge frequency in the same reach, use the bankfull discharge frequency of the same stream order as an imputed value; (3) check the bankfull discharge of the frequencies determined by the above two steps using the Pearson III discharge–frequency curve; (4) the estimation is reliable if the bankfull discharge increases from upstream to downstream, and if this is not the case, check whether the discharge increases downstream and the cross section is influenced by evident anthropogenic disturbances, such as water divisions or reservoirs.

2.3. Calculation of DHGs under Different Discharge Frequencies and Bankfull Conditions

In situ-measured hydrological data with recurrence intervals equal to and less than the bankfull discharge were used to calculate DHGs. Bankfull discharge is the turning point at which the dynamic action of water and sediment transitions from shaping river channels to shaping floodplains. It dominates the shaping of river morphology as a reference discharge. Due to the sudden widening of a cross section above the bankfull water level, the morphology of the flood plain is evidently different from that of the main channel. Therefore, this study mainly focused on DHGs below bankfull water levels.

In situ-measured Q, W, H, and V under bankfull conditions and of the same discharge frequency were first sorted within the same group. Then, we fitted the DHG relations under bankfull conditions and different discharge frequencies. Two cases were considered (Figure 1 and Table 2): the same river reach and different reaches located in the same river basin. For the reaches with $\geq$5 cross sections, the width-, depth- and velocity-DHG of different discharge frequencies were fitted along the reach (Figure 1 and Table 2). For reaches with <5 cross sections, cross sections located in both the main stream and tributaries of a river basin were used to fit the DHG relations (different reaches located in the same river basin in Figure 1 and Table 2).

3. MFDHG Relations

3.1. From DHG to MFDHG

DHG quantifies the relationship between a reference discharge (usually represented by channel-forming discharge or bankfull discharge) and the corresponding hydraulic variables of multiple cross sections. For a particular cross section, one reference discharge corresponds to only one river width, average flow depth, and flow velocity. This leads us to ask how the hydraulic variables of multiple cross sections would change with multiple water levels (discharge frequencies) across the watershed. The AMHG considers both the spatial connections of individual cross sections and multiple discharge frequencies. Another question is whether an AMHG-like relationship exists when DHG is the basis.

Based on the above hypotheses, we explored the relations among DHGs at multiple discharge frequencies to fully capture the variations in W, H, V, and Q across the river network. The intent was to promote the scientific understanding of HGs by identifying a previously unnoticed correlative relationship between a river's DHG coefficients and their corresponding exponents. This was achieved by plotting a–b, c–f, and/or k–m DHG pairs for multiple temporally distributed discharge frequencies, for example, from thousands of in situ measurements of Q, W, H, and V collected from 1967–2019 at 129 cross sections located at national hydrological stations. Because these correlations were obtained by simply aggregating DHG parameter pairs from many discharge frequencies, they are here referred to as MFDHGs.

3.2. Two Presentations of MFDHG

Similar to AMHG, the MFDHG correlation is driven by two facts: one is the mathematical construct with DHG exponents in both the regressor and the regressand, and the other is geomorphological coevolution among discharge, cross sectional shape, and hydraulic variables [41,42]. This paper defines the MFDHG from two aspects: mathematical expression, and geomorphological significance. It is noteworthy to state that only DHGs with $R^2 > 0.5$ were considered for the construction of MFDHGs, thereby ensuring the reliability and accuracy of the analysis.

(1) MFDHG expressed as log-linear relations between DHG coefficients and exponents

Inspired by the idea of the AMHG, which relates AHG coefficients and exponents using a log-linear relation [12], we employed various functions, including linear, exponential, quadratic, and log-linear models, to fit the DHG coefficients and exponents. Our analysis revealed that the log-linear function demonstrated the most promising results, thus enabling the definition of the MFDHG:

$$b = \alpha_1 \ln(a) + \beta_1 \tag{4}$$

$$f = \alpha_2 \ln(c) + \beta_2 \tag{5}$$

$$m = \alpha_3 \ln(k) + \beta_3 \tag{6}$$

where α_1, α_2, and α_3 are slopes of the MFDHG, and β_1, β_2, and β_3 are intercepts of the MFDHG.

Exponents b, f, and m were shown to be functions of coefficients a, c, and k, which reduced the number of unknown parameters in the DHG system by half. This feature also suggested that the DHG is not temporally specific, as previously theorized [3], but rather is dependent on the DHG of other discharge frequencies in a given river reach or different reaches in the same river basin. This also showed that MFDHG is not contradictory to the findings of DHG research over the past few decades, as suggested by a previous empirical analysis.

(2) MFDHG expressed as congruent hydraulics

MFDHG is a function of the integral geomorphology of a river basin. We found that MFDHG appeared when individual DHG rating curves for each discharge frequency in a given river reach or at different reaches in the same river basin converged at the same values of $W/H/V$ and Q (Figure S1); this relation is presented as the congruent hydraulic pairs $Q_{cW} - W_c$, $Q_{cH} - H_c$, and $Q_{cV} - V_c$:

$$W_c = a_{p1,\ p2,\ \ldots,\ px} Q_{cW}^{b_{p1,\ p2,\ldots,\ px}} \tag{7}$$

$$H_c = c_{p1,\ p2,\ldots,\ px} Q_{cH}^{f_{p1,\ p2,\ldots,\ px}} \tag{8}$$

$$V_c = k_{p1,\ p2,\ldots,\ px} Q_{cV}^{m_{p1,\ p2,\ldots,\ px}} \tag{9}$$

where subscripts $p_1, p_2 \ldots\ p_x$ are temporally indexed discharge frequencies for all cross sections in a certain study area; subscript c is termed "congruent hydraulics", the empirically fit parameters that define the MFDHG; and Q_{cW}, Q_{cH}, Q_{cV}, W_c, H_c, and V_c are congruent hydraulics determined by the internal geomorphic characteristics of a river basin. When plotting all DHG curves of a river under different discharge frequencies on the same coordinate system, an interesting phenomenon is observed—all the curves intersect at a single point (Figure S1), resulting in a convergence within the log-linear hydraulic coordinate space. These points of intersection represent congruent hydraulic pairs, indicating a consistent relationship between hydraulic parameters across various flow conditions.

Similar to the AMHG, congruent hydraulics for the MFDHG can be estimated by the intercepts and slopes of the MFDHG. Taking width-MFDHG as an example (Equation (7)), if any two discharge frequencies (p_1, p_2) share W_c and Q_{cW}, then we can solve Equation (1) for W_c at each discharge frequency and equate the two expressions:

$$b_{p1}\log(Q_{cW}) + \log(a_{p1}) = b_{p2}\log(Q_{cW}) + \log(a_{p2}) \tag{10}$$

Moving Q_{cW} to the left side of the equation gives the slope (α_1) of the MFDHG by two cross sections:

$$-\frac{1}{\log(Q_{cW})} = \frac{b_{p2} - b_{p1}}{\log(a_{p2}) - \log(a_{p1})} = \alpha_1 \tag{11}$$

Finally, the intercept (β_1) of MFDHG can be defined empirically as

$$\beta_1 = \frac{\log(W_c)}{\log(Q_{cW})} \tag{12}$$

Ideally, based on Equations (11) and (12), congruent values of Q_{cW} and W_c for multiple discharge frequencies can be estimated. However, MFDHG slopes and intercepts for any two discharge frequencies are not equal in actual river systems. Congruent hydraulics estimated based on any two discharge frequencies are also not equal. Therefore, the mean intercepts and slopes of all possible combinations of MFDHG slopes and intercepts should be used to estimate congruent hydraulics (e.g., if a given river reach has 20 DHGs under 20 discharge frequencies, this results in 190 slope and intercept pairs corresponding to 190 discharge frequencies). Subsequently, the mean slope and intercept of these 190 pairs were used to estimate the congruent hydraulics of the river reach.

4. Results

4.1. DHGs under Bankfull Conditions

The existence of DHG under bankfull discharge is a prerequisite for the further study of MFDHG. Four river reaches were used to study DHGs along a river reach (Figure 1). The main streams of the Lantsang River, the upper Yellow River, and its tributary, the Huangshui River, exhibited relatively strong DHGs with $R^2 > 0.5$ (Table 2). The DHG of the Jinsha River was weak overall, which may be attributed to the influence of the braided Tuotuo River reach. A relatively large bed slope and W, as well as small Q and H values, weakened the consistent adjustments of W, H, and V to Q variations (Table 2).

Five river basins, including both the main stream and tributaries, were used to study the DHGs of different reaches located in the same basin (Figure 1 and Table 2). Generally, the DHG strengths of different reaches located in the same river basin were weaker than those along a single river reach. In particular, the DHG strengths of tributaries of the Yalong River were nearly the weakest among all scenarios. Data from cross sections located within 10 km upstream and downstream of hydropower stations and artificial diversions were removed through data screening processes. However, large-scale hydropower development and construction projects (22 cascade hydropower stations along the 1571 km main stream) still contributed to changes in the discharge–sediment relationship and further to the weak DHG relations. The shaping of river morphology was no longer dominated by natural discharge–sediment processes but was largely influenced by artificial runoff regulation. Adjustments of W, H, and V to Q were impaired, and the DHG relation was weakened.

The cross section morphology is shaped by discharge and sediment load and influenced by boundary conditions such as geology and geomorphology [43]. Variations in climate and underlying surfaces are smaller within one river basin than across basins, which results in a strong correlation between discharge and hydraulic variables in one basin. This provides a basis for studying the MFDHG for different reaches located in the same river basin and emphasizes the importance of similar boundary conditions under regime theory in HG studies.

Although within the range (−0.51–0.75) proposed by Park [4] for 72 streams around the world, the velocity exponent m of the Lantsang River was <0, which is rare (Table 2). In addition, the DHG exponents of our research showed relatively large differences compared to the four theoretical solutions presented in Table 2. This may be attributed to the fact that DHG relations may not be observed in steep mountain streams unless certain criteria, such as stream power/grain size threshold standards, are met [44]. We did not verify the standards proposed by Wohl and Wilcox [44], but the results confirmed the existence of DHGs in the study area, which supports the exploration of the relationships of DHGs under multiple discharge frequencies and the development of the MFDHG concept.

4.2. MFDHG under Different Scenarios

To verify the hypotheses proposed in Section 3, we tested the MFDHG relations in two sets of data obtained along a single river reach and from different river reaches located in the same river basin (Figure 1 and Table 2).

4.2.1. MFDHG along a River Reach

DHG coefficients and exponents under multiple discharge frequencies showed good log-linear correlations along the river reaches. The determination coefficients R^2 of MFDHG along the four river reaches were >0.82 (Figure 3 and Table 3). Depth-MFDHGs exhibited the strongest correlations, while width- and velocity-MFDHGs had lower strengths (Table 3). Analytically, depth was more responsive in adjusting to accommodate changes in discharge and stream power, while width may have been more prone to random variations in boundary conditions such as geologic constraints and human modifications [41]. Variations in the velocity exhibited relatively strong randomness and less consistency. The slopes of mountain rivers changed significantly along different reaches; as a result, the regularity of velocity change along the reach was not sensitive. This led to a nonsignificant adjustment in velocity with variable discharge and low exponents m and R^2 of MFDHG. Furthermore, the MFDHG slopes of the three major rivers (Yellow River, Jinsha River, Lantsang River) were smaller than that of the tributary Huangshui River (Figure 3).

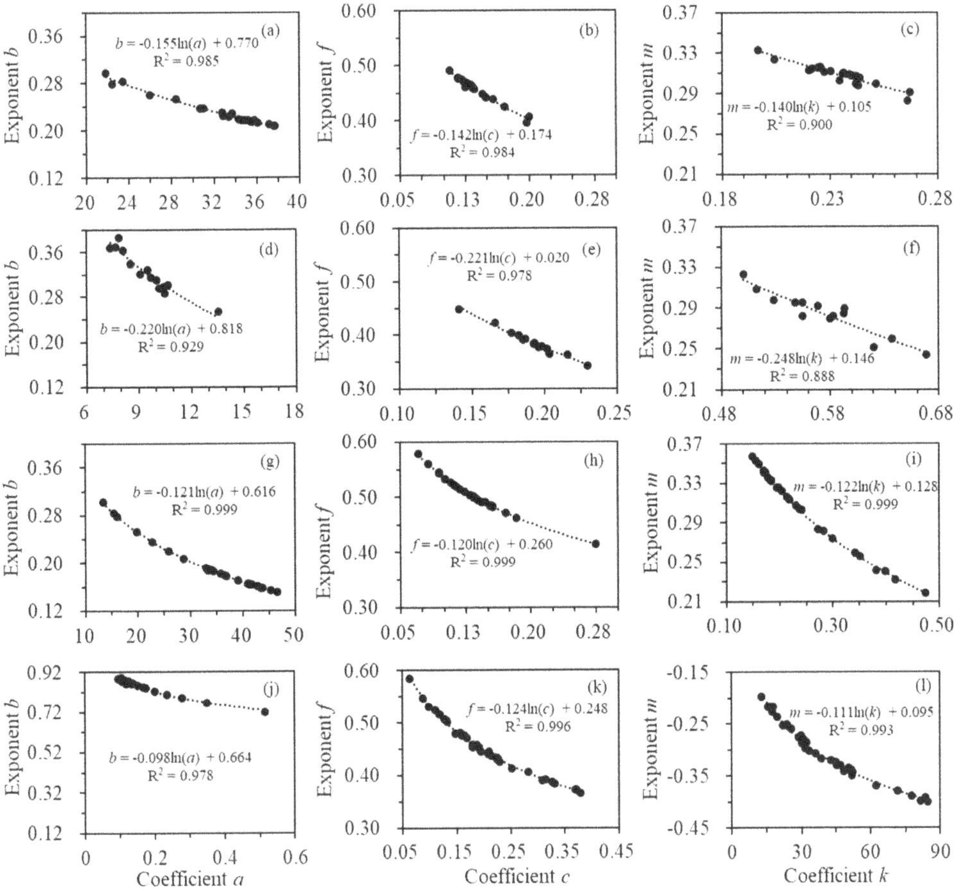

Figure 3. Relations between DHG coefficients and exponents at multiple discharge frequencies for the main streams of the upper YR (**a–c**), Huangshui River (**d–f**), upper JSR (**g–i**), and LCR (**j–l**). As an example, subfigure (**a**) encompasses data from cross sections 2 through 12 (Table S1). Width-DHG relationships for discharge frequencies of 5%, 10%, 15%, ..., 95% were first fitted. Then, a total of 20 pairs of DHG coefficients and exponents was determined and utilized to generate the MFDHG relation.

The strength R^2 of the MFDHG was determined by the degree of DHG curve convergence in double logarithmic coordinate systems [27]. R^2 can be interpreted as a geomorphic index indicating the variability of water surface morphologies for different discharge frequencies (taking width-DHG as an example) and the hydraulic self-similarity of a certain study area. Both the log-linear relations between DHG coefficients and exponents and the convergence of DHG rating curves reflected hydraulic self-similarity and consisted of variations in channel morphology, which enabled the extraction of DHG common features and the achievement of an overall HG. In addition, the strong MFDHG correlations also indicated the equilibrium state of the studied river reaches, which could be reflected by the similar boundary sediments of the studied cross sections (Table S1).

Table 3. Z-values of the Mann–Kendall test in detecting changes in DHG coefficients and exponents with discharge frequency and strengths R^2 of MFDHG along a river reach and different reaches in the same river basin. The number of asterisks represents the level of significance: * is 0.05, ** is 0.01, and *** is 0.001. $|Z| > 1.96$, $|Z| > 2.58$, and $|Z| > 3.30$ indicate that the trends are significant at the $\alpha = 0.05$, $\alpha = 0.01$, and $\alpha = 0.001$ levels, respectively. The positive and negative signs indicate increasing and decreasing trends in the time series, respectively. R^2 indicates the determination coefficient between DHG coefficients and exponents.

Sources	Scenarios	Mann–Kendall Test Z		R^2 of Width-MFDHG	Mann–Kendall Test Z		R^2 of Depth-MFDHG	Mann–Kendall Test Z		R^2 of Velocity-MFDHG	Mean R^2 of MFDHG
		a	*b*		*c*	*f*		*k*	*m*		
Along a river reach	Main stream of the YR	1.44 *	−1.93 **	0.985	−2.68 ***	2.68 ***	0.984	0.40	0.74	0.900	0.956
	Main stream of the JSR	5.01 ***	−5.01 ***	0.999	0.09	−0.17	0.999	−4.15 ***	4.11 ***	0.999	0.999
	Main stream of the LCR	−5.43 ***	3.49 ***	0.978	−1.92 **	2.16 **	0.996	5.14 ***	−4.70 ***	0.993	0.989
	Main stream of the Huangshui River	1.98 **	−2.08 **	0.929	−0.40	0.49	0.978	−1.78 **	2.47 ***	0.888	0.932
	Mean			0.973			0.989			0.945	0.969
Different reaches located in the same river basin	Tributaries of the JSR	−0.53	−3.01 ***	0.448	−1.32 *	2.27 **	0.928	1.74 **	−0.90	0.933	0.770
	Tributaries of the YLR	−3.66 ***	−0.06	0.511	−1.30 *	1.39 *	0.991	3.02 ***	−1.65 **	0.939	0.814
	Tributaries of the LCR	−7.56 ***	6.03 ***	0.932	0.51	1.40 *	0.761	5.60 ***	−3.74 ***	0.963	0.885
	Main stream and tributaries of the NR	−8.08 ***	7.66 ***	0.976	2.72 ***	−2.91 ***	0.972	3.89 ***	−2.66 ***	0.924	0.957
	Main stream and tributaries of the YLZBR	−2.45 ***	2.37 ***	0.993	1.75 **	−2.71 ***	0.944	0.14	1.02	0.833	0.923
	Mean			0.772			0.919			0.918	0.870
Mean				0.861			0.950			0.930	0.914

4.2.2. MFDHG for Different Reaches Located in the Same River Basin

The MFDHG strengths for different reaches located in the same river basin were high but were generally lower than those along the same reach (Figure 4 and Table 3). The mean R^2 values for MFDHG along a river reach and on different reaches in the same river basin in this study were 0.969 and 0.870 (Table 3), respectively, suggesting that rating curve convergence was widespread in our data. Similar to the observations from AMHG research [41], width-MFDHG strength with a mean R^2 of 0.861 arose mainly from the mathematical construct, had only weak geomorphological significance, and was weaker than those of depth- and velocity-MFDHG (with mean R^2 values of 0.950 and 0.930, respectively), which were dominated by geomorphological coevolution (Table 3).

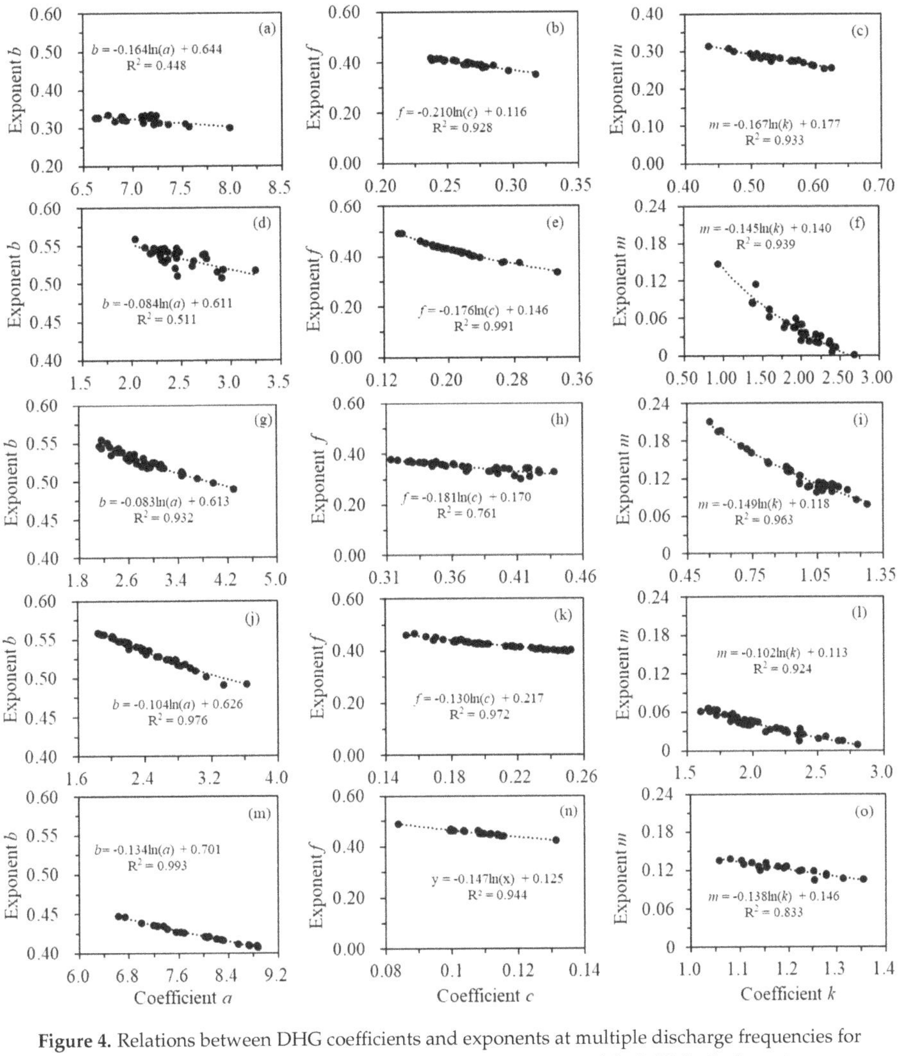

Figure 4. Relations between DHG coefficients and exponents at multiple discharge frequencies for tributaries of the JSR (**a–c**), tributaries of the YLR (**d–f**), tributaries of the LCR (**g–i**), the main stream and tributaries of the NR (**j–l**), and the main stream and tributaries of the YLZBR (**m–o**).

Correlations between DHG coefficients and exponents under multiple discharge frequencies reflected the spatial consistency of HG parameters in a certain study area. The spatial continuity of water and sediment was strong along a river reach, which contributed to a greater MFDHG strength. The spatial heterogeneity of the geology and geomorphology of different river reaches was generally high, and the spatial continuities of discharge and sediment were lower than those along the same river reach. This contributed to the decrease in MFDHG strength for the scenario of different river reaches. Similar conclusions were introduced by Qin et al. [7], who extended the AMHG concept from the reach scale to the watershed scale. These authors noted that cross sections across river reaches showed strong AMHG relations, although weaker than those along a reach, which also reflects the consistency in flowing water and sediment as well as local climate, landscape, soil, vegetation, etc.

4.3. Coordinated Variations in DHG Coefficients and Exponents

Correlations between DHG coefficients/exponents and discharge frequencies for nine scenarios were explored (Table 3). Coefficient a represents river widths for unit-width discharge along a reach. This reflects the ability of the unit-width flow to shape river morphology. Variations in river widths under unit-width discharge along a reach can be mainly attributed to variable boundary conditions. Exponent b represents the rate of increase in river width with increased discharge along the reach under a certain discharge frequency. This reflects the consistency of the shaping power of flowing water on river morphology.

Taking the main stream of the upper Yellow River as an example, exponent b decreased with increasing coefficient a, which could be expressed as a good log-linear relation with an R^2 of 0.985 (Table 3 and Figure 5a). With increased discharge frequency, coefficient a and exponent b showed an increase and a decrease, respectively, but the variations with discharge frequency were not strictly one-to-one (Figure 5a–c). Specifically, with increasing discharge recurrence intervals, coefficients and exponents for width- and depth-MFDHG exhibited statistically significant variations (passed the M–K test at the 0.01 confidence level, Figure 5b,c,e,f), while a trend was not evident for velocity-MFDHG (Figure 5h,i). For the other eight scenarios, DHG coefficients and exponents exhibited the opposite variation trends with increasing discharge recurrence intervals if both passed the M–K test at the 90% confidence level (Table 3). For a coefficient and exponent pair that passed the 99% confidence level, the MFDHG strength varied from 0.920 to 0.999 with a mean of 0.974; if one of the coefficients or exponents did not pass at the 90% confidence level, the MFDHG strength varied from 0.448 to 0.999 with a mean of 0.795 (Table 3). Therefore, consistent variations in DHG coefficients and exponents with discharge frequency, represented by the M–K test Z-value, were one of the determinants of MFDHG strength.

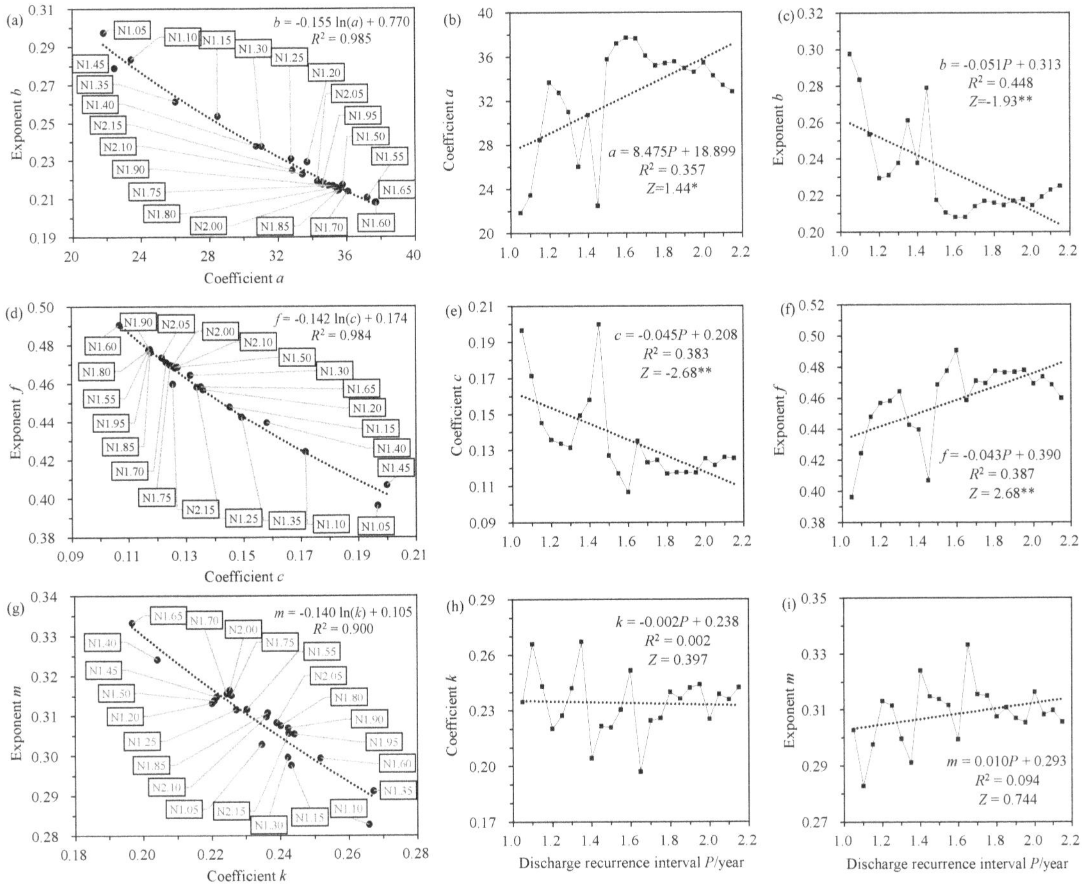

Figure 5. Relations between DHG coefficients and exponents (**a**,**d**,**g**) and their variations depending on discharge recurrence intervals (**b**,**c**,**e**,**f**,**h**,**i**) for the main stream of the upper YR. In subfigures (**a**,**d**,**g**), dashed lines represent the variation trends between DHG coefficients and exponents, and discharge recurrence intervals. N1.05 inside the black box of subfigure (**a**) represents the discharge recurrence interval is 1.05 years. In subfigures (**b**,**c**,**e**,**f**,**h**,**i**), solid polygons represent the values of DHG coefficients or exponents under different discharge recurrence intervals, dashed lines represent the variation trends between DHG coefficients and exponents, and discharge recurrence intervals, the number of asterisks represents the level of significance: * is 0.05, ** is 0.01.

5. Discussion

5.1. Explanations of Congruent Hydraulics

A river reach exhibits a perfect MFDHG (with $R^2 = 1$) when individual DHG rating curves converge exactly at a congruent point (Figure S1). Understanding the meaning and variation trends of congruent hydraulics contributes to revealing the physical basis of the MFDHG concept. Unlike AMHG research, which mainly focuses on short discharge-conserving reaches [27], the congruent hydraulics of MFDHG can be explained with the conceptual model in Figure 6 for relatively long reaches with tributaries. The flows used to generate the MFDHG in this study ranged from a 1.05-year recurrence interval to bankfull conditions (Sections 2.2 and 2.3), but in this section, the flow range is extended from extreme minimum to extreme maximum discharge to construct a conceptual model and fully illustrate the variations in hydraulic variables when P_c, Q_c, W_c, and H_c are requested:

(1) Congruent discharge frequency (P_c): for cross sections along a reach that share a P_c (e.g., 2.5 years recurrence interval discharge), Q and the corresponding W, H, and V increase with increasing drainage area and confluence of tributaries (Figure 6a). This scenario can be described with DHG.
(2) Congruent discharge ($Q_{cW}/Q_{cH}/Q_{cV}$): for cross sections along a reach that share a Q_c, the value should be equal to both the discharge of one extreme flood event of the uppermost cross section and the discharge of one extreme low water event of the lowermost cross section. The probability of Q_c occurrence increases first and then decreases for cross sections along the reach (Figure 6b).
(3) Congruent average flow depth (H_c): for cross sections along a reach that share an H_c, river widths increase with the gradual change in cross sectional morphology from narrow V-shaped to wide U-shaped along the reach (Figure 6c).
(4) Congruent river width (W_c): for cross sections along a reach that share a W_c, average flow depths decrease along the reach (Figure 6d).

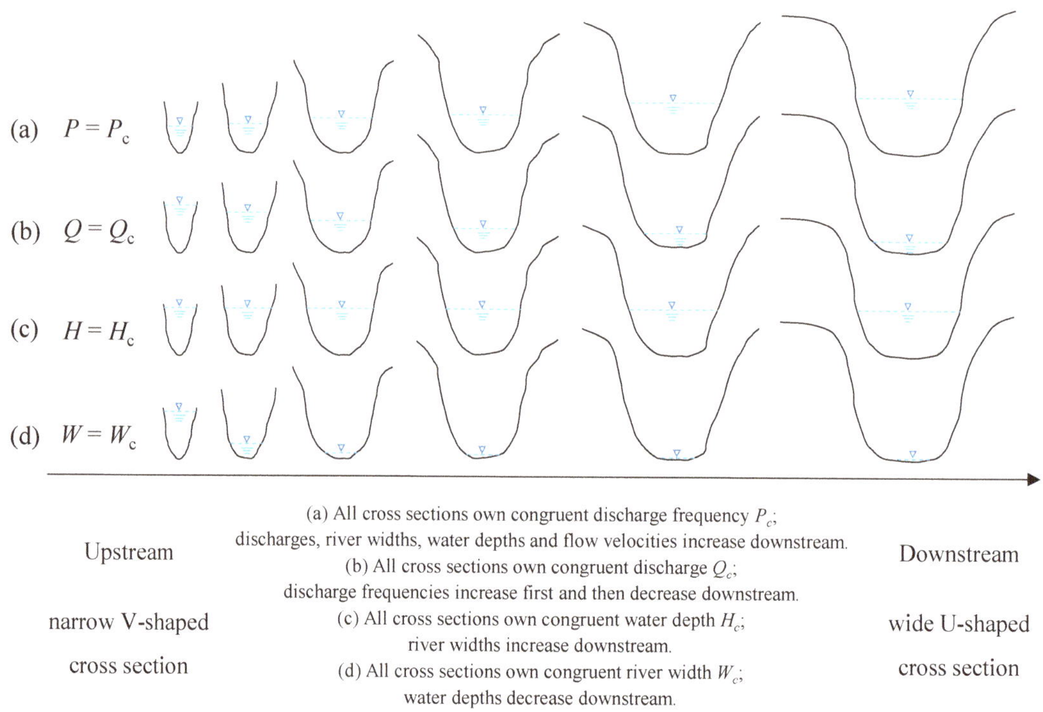

Figure 6. Conceptual model for congruent hydraulics of (**a**) DHG and (**b**–**d**) MFDHG.

Q_{cW}, Q_{cH}, and Q_{cV} represent three congruent discharges when the W, H, and V of different cross sections are equal. These discharges may not be equal to each other or lie far outside the range of observed values given the very large variations in W, H, and V downstream. Furthermore, the conceptual model can be extended for different river reaches located in the same river basin due to the dominant effects of flow and sediment processes on shaping river morphology. This dominant shaping effect can be maximized when river boundary sediments are similar and the river is in an equilibrium state, which usually occurs in lowland alluvial rivers or alluvial reaches of mountain rivers. The convergence of DHG curves indicates hydraulic self-similarity of river morphology in a certain study area induced by discharge and sediment load; this provides part of the theoretical basis for congruent hydraulic variables.

5.2. Significance of MFDHG for the HG System

Based on former studies of AHG, DHG, and AMHG [3,5,7,12,45], the concept of MFDHG, with respect to the sequence of temporal and spatial HG dimensions, has compensated for the shortcomings of a single DHG time dimension and the impossibility of describing consistent variations in hydraulic variables along a river reach under different discharge frequencies. MFDHG is practically useful because it is convincingly aligned with the actual character of the hydraulic geometry system. The relations among AHG, DHG, AMHG, and MFDHG in terms of temporal and spatial dimensions are outlined as follows (Figure 7):

(1) DHG quantifies the spatial distributions of channel morphology shaped by channel-forming discharge (bankfull discharge). It reflects variations in hydraulic variable-discharge relations downstream of a reach or at different reaches located in the same river basin. Multiscale variations in river morphology in the spatial dimension can be depicted by DHG, but the time dimension has only one scale.
(2) AHG quantifies the morphological characteristics of a single cross section under different discharge frequencies. It reflects variations in hydraulic variable-discharge relations for individual cross sections and depicts multiscale changes in cross sectional morphology in the temporal dimension, but the spatial dimension has only one scale.
(3) AMHG relates the AHG of different cross sections along a river reach and extends the one-dimensional AHG in space. A multidimensional spatiotemporal connection between the geometric parameters of the cross section and discharge is reflected.
(4) MFDHG extends the multiscale spatial attributes of DHG in the temporal dimension and contributes to achieving the same goal of relating river morphological parameters and discharges in both spatial and temporal dimensions.

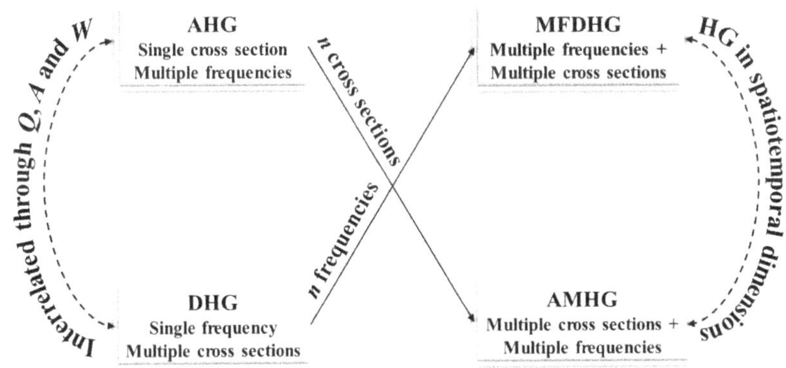

Figure 7. Mutual transformations among AHG, DHG, AMHG, and MFDHG. AHG and DHG can be linked through discharge Q, drainage area A, and river width W [26].

Since DHG relies on bankfull variables at a cross section and cross sectional shape (which determines AHG) determined by bank strength and sediment transport properties, DHG and AHG are implicitly linked [28]. For an ideal channel whose riverbanks and bed are composed of homogeneous noncohesive material, the channel morphology adjusts readily to changes in discharge, and the AHG and DHG do not differ [29]. A natural river may scale consistently with flow and sediments in the AHG and DHG if the river is in an equilibrium state and within the constraints of similar channel-bounding materials [46]. Previous studies have shown that strong AHG and AMHG relations exist in disconnected alluvial reaches of mountain rivers located in the QTP [6,7]. This paper focuses on these alluvial reaches and verifies the dominant role of the power conveyed by water and sediment in shaping river morphology. The alluvial characteristics of mountain rivers can be characterized by the existence of DHG and the coordinated variations in DHG coefficients and exponents of the disconnected alluvial reaches. Therefore, for a natural alluvial river or an alluvial reach located in a mountain river, common features can be extracted from both the AHG and DHG, which provides a basis for the study and interpretation of the AMHG and MFDHG.

5.3. Differences and Relations between AMHG and MFDHG

Based on the above analysis, the main differences between MFDHG and AMHG can be summarized as follows:

(1) Focus: AMHG examines the relationships between AHG coefficients and exponents, whereas MFDHG explores the relationships between DHG coefficients and exponents.
(2) Spatial and temporal extension: AMHG extends the one-dimensional AHG concept to spatial scales by considering different cross sections within a river. In contrast, MFDHG extends the one-dimensional DHG concept to temporal scales by capturing variations in hydraulic properties over different discharge frequencies.
(3) Sequence of consideration: MFDHG and AMHG differ in the order of considering cross sections and discharge frequencies. MFDHG first considers spatial properties (cross sections) and then incorporates temporal properties (discharge frequencies), while AMHG follows the opposite order.

Congruent hydraulics serve as a representation of shared characteristics within a group of studied cross sections, allowing for the characterization of the relationships between MFDHG and AMHG. The question at hand is whether the congruent hydraulics observed in AMHG and MFDHG represent two distinct approaches to capturing the same phenomenon. Taking the upper Yellow River and the Jinsha River as examples, we found that the fitted correlations between HG coefficients and exponents exhibited similar trends (Figure 8), e.g., fitted lines of depth-AMHG and depth-MFDHG of the Jinsha River showed nearly the same trend, although the MFDHG line occupied only a short portion of the AMHG line. Downstream adjustments of W, H, and V with Q were often smaller than those of individual cross sections. Specifically, the variations in Q, W, H, and V used to fit the DHG and MFDHG relations were 9.8–2830 $m^3\ s^{-1}$, 42.8–272 m, 0.51–7.21 m, and 0.45–3.45 $m\ s^{-1}$, respectively. The variations in Q, W, H, and V used to fit the AHG and AMHG relations were 5.2–3590 $m^3\ s^{-1}$, 11–346 m, 0.4–6.1 m, and 0.14–6.53 $m\ s^{-1}$, respectively. In addition, we calculated the congruent hydraulics based on AMHG and MFDHG. The results showed that the relative differences in congruent hydraulics estimated by AMHG and MFDHG showed small differences and were within the range of 1.5–48.0% (Table 4). River width-related congruent hydraulics, log (Q_{cW}) and log (W_c), showed the smallest relative difference. The similar AMHG and MFDHG trend lines and small relative difference of AMHG and MFDHG congruent hydraulics indicated that the spatiotemporal characteristics of HG can be interpreted by either the AMHG or the MFDHG. This might contribute to the establishment of an overall HG.

Table 4. Congruent hydraulics calculated by AMHG and MFDHG of the Yellow River and the Jinsha River.

River Systems	HG Types	log(Q_{cW})	log(Q_{cH})	log(Q_{cV})	log(W_c)	log(H_c)	log(V_c)
Yellow River	AMHG	6.8	7.8	8.3	5.3	1.4	1.3
	MFDHG	6.3	7.4	7.0	4.9	1.4	0.7
	Relative difference (%)	8.1	5.9	16.1	6.3	2.0	48.0
Jinsha River	AMHG	7.3	10.5	7.5	5.1	2.8	0.7
	MFDHG	8.2	8.4	8.2	5.1	2.2	1.0
	Relative difference (%)	10.9	20.2	8.3	1.5	22.6	36.0

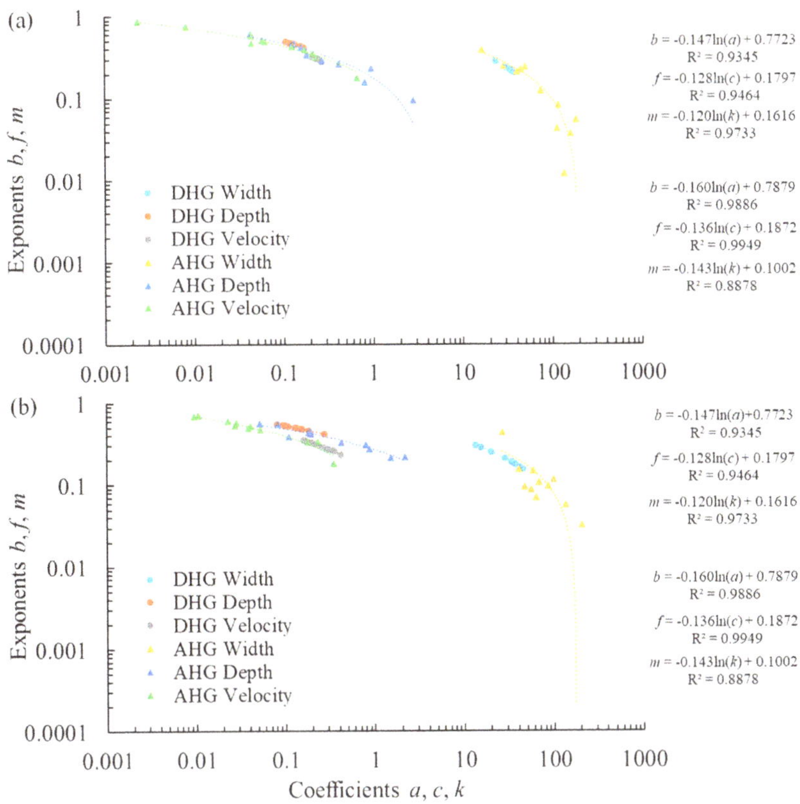

Figure 8. MFDHG and AMHG of the upper Yellow River (**a**) and the Jinsha River (**b**).

The overall HG relationship, although it has long been investigated by researchers [11,12,23,25], is expected to satisfactorily explain both AHG and DHG through AMHG and MFDHG. The ranges of applications of AHG and DHG are extended from the cross sectional scale and reach scale to the watershed scale by simultaneously considering multiple cross sections and discharge frequencies (Figure 7). Both AMHG and MFDHG are extensions of HG in the river network and reflect the applicability of hydraulic variable-discharge relations at the watershed scale. The concepts of AMHG and MFDHG provide methods and a theoretical basis for studying the spatial distributions of HG across river networks and reaching the goal of constructing an overall HG relationship.

6. Summary and Conclusions

Based on in situ measurements of six exoreic rivers and their tributaries originating from the QTP, this paper first verified the existence of DHGs in the study area, which is a prerequisite for studying MFDHGs, and then defined the MFDHGs according to mathematical expression and geomorphological significance. The MFDHG was finally verified both along a reach and for different reaches in the same river basin. The key findings and implications are as follows:

(1) DHG, as a frequency-specific expression of hydraulics and channel geometry, appeared widely in the six major rivers and their tributaries that originate from the QTP.
(2) The paired coefficients and exponents of DHG (a and b, c and f, and k and m) from multiple discharge frequencies along a given river reach or different reaches in the same

river basin are functionally related to one another, exhibiting a log-linear relationship. This is the mathematical expression of MFDHG.

(3) MFDHG, a mathematical construct arising from the use of power laws at a certain discharge frequency, is a novel geomorphic phenomenon after the discovery of AMHG. DHG rating curves reliably intersect at congruent Q_c, W_c, H_c, and V_c, indicating both geometric variability (to ensure rating curve intersection) and hydraulic self-similarity of river channels. This is the geomorphological expression of MFDHG.

(4) With increasing discharge frequency, the DHG coefficients and exponents showed opposite variations if both passed the M–K test at the 90% confidence level. Consistent variations in DHG coefficients and exponents with discharge frequency contributed to greater MFDHG strength.

(5) Although the empirical conclusion seemingly refutes previous decades of research defining DHGs as temporally independent, MFDHG relates individual DHGs under different discharge frequencies and contributes to the completeness of the HG system in terms of the spatiotemporal dimensions. This is a large step in refining the common features of DHG and will contribute to establishing an overall HG relationship considering multiple spatiotemporal dimensions across river reaches.

(6) Bed slope and bed material are important parameters that are adjusted based upon the incoming water and sediment. The future collection of additional data on bed slope, bed roughness, and materials in the study area using a combination of remote sensing, in situ measurements, and deep learning methods is proposed. These data can then be incorporated into the proposed MFDHG equations, further improving their accuracy and reliability.

Supplementary Materials: The following supporting information can be downloaded at https://www.mdpi.com/article/10.3390/w15112139/s1. Table S1: Cross sections studied in this research and their locations, slopes, upstream contributing areas, recurrence intervals of bankfull discharge, and bed/bank materials. Figure S1: Convergence of DHG curves of the main stream of the upper Yellow River (a–c), the main stream of the Huangshui River (d–f), the main stream of the upper Jinsha River (g–i), the tributaries of the Lantsang River (j–l), the main stream and the tributaries of the Nu River (m–o), and the main stream and the tributaries of the Yarlung Zangbo River (p–r) under multiple discharge frequencies. Red dots represent the congruent discharge, which indicates the convergence of individual DHG curves.

Author Contributions: Conceptualization, C.Q. and X.F.; data curation, Y.X. and G.W. (Ge Wang); funding acquisition, C.Q. and B.W.; methodology, C.Q.; software, Y.X. and G.W. (Ge Wang); validation, C.Q. and Y.X.; visualization, C.Q.; writing, review, and editing, C.Q., B.W., G.W. (Guangqian Wang) and X.F. All authors have read and agreed to the published version of the manuscript.

Funding: This research was funded by the Fund Program of the State Key Laboratory of Hydroscience and Engineering (2023-KY-02), the National Natural Science Foundation of China (Grant Nos. U2243218; U2243222; 52009061), the Postdoctoral Innovation Talents Support Program of China (Grant No. BX20190177), and the Shuimu Tsinghua Scholar Program (2020SM070).

Data Availability Statement: In situ-measured river width, average flow depth, flow velocity, and flow discharge data were provided by the Bureau of Hydrology at the Ministry of Water Resources of China, which is acknowledged here. These data can be sourced and made available from the Bureau of Hydrology at the Ministry of Water Resources of China (http://www.mwr.gov.cn/english/, accessed on 31 May 2023) upon reasonable request. Boundaries of the QTP are available at the National Tibetan Plateau Data Center (http://www.tpdc.ac.cn/en/data/61701a2b-31e5-41bf-b0a3-607c2a9bd3b3/, accessed on 31 May 2023). The SRTM 90 m DEM is available at http://www.tpdc.ac.cn/en/data/23e32e3e-8104-4798-b7a1-325df8fd1a95/, accessed on 31 May 2023.

Conflicts of Interest: The authors declare no conflict of interest.

References

1. Eaton, B. Hydraulic geometry: Empirical investigations and theoretical approaches. In *Treatise on Geomorphology*; Academic Press: San Diego, CA, USA, 2013; pp. 313–329.
2. Lindley, E.S. Regime channels. *Proc. Punjab Eng. Congr.* **1919**, *7*, 63–74.
3. Leopold, L.B.; Maddock, T. *The Hydraulic Geometry of Stream Channels and Some Physiographic Implications*; U.S. Geological Survey Professional Paper 252; U.S. Government Printing Office: Washington, DC, USA, 1953.
4. Park, C.C. Worldwide variations in hydraulic geometry exponents of stream channels: An analysis and some observations. *J. Hydrol.* **1977**, *33*, 133–146. [CrossRef]
5. Lee, J.; Julien, P.Y. Downstream hydraulic geometry of alluvial channels. *J. Hydraul. Eng.* **2006**, *132*, 1347–1352. [CrossRef]
6. Qin, C.; Wu, B.; Wang, Y.; Fu, X.; Xue, Y.; Li, D.; Li, M.; Zhang, Y. Dynamic variability of at-a-station hydraulic-geometry for mountain rivers in the southeast Qinghai-Tibet Plateau: The cases of Yalong River and upper Jinsha River. *Catena* **2020**, *194*, 104723. [CrossRef]
7. Qin, C.; Wu, B.S.; Wang, G.Q.; Wang, G. Spatial distributions of At-Many-Stations Hydraulic Geometry for mountain rivers originated from the Qinghai-Tibet Plateau. *Water Resour. Res.* **2021**, *57*, e2020WR029090. [CrossRef]
8. Stall, J.B.; Yang, C.T. *Hydraulic Geometry of 12 Selected Stream Systems of the United States*; Research Report No. 32; University of Illiniois Water Resources Research Center: Urbana, IL, USA, 1970.
9. Dingman, S.L.; Sharma, K.P. Statistical development and validation of discharge equations for natural channels. *J. Hydrol.* **1997**, *199*, 13–35. [CrossRef]
10. Parker, G.; Wilcock, P.R.; Paola, C.; Dietrich, W.E.; Pitlick, J. Physical basis for quasi-universal relations describing bankfull hydraulic geometry of single-thread gravel bed rivers. *J. Geophys. Res.* **2007**, *112*, F4005. [CrossRef]
11. Parker, G.; Wilkerson, G.V. Physical basis for quasi-universal relationships describing bankfull hydraulic geometry of sand-bed rivers. *J. Hydraul. Eng.* **2011**, *137*, 739–753.
12. Gleason, C.J.; Smith, L.C. Toward global mapping of river discharge using satellite images and at-many-stations hydraulic geometry-supporting data. *Proc. Natl. Acad. Sci. USA* **2014**, *111*, 4788–4791. [CrossRef]
13. Pelletier, J.D. Controls on the hydraulic geometry of alluvial channels: Bank stability to gravitational failure, the critical-flow hypothesis, and conservation of mass and energy. *Earth Surf. Dyn.* **2021**, *9*, 379–391. [CrossRef]
14. Song, X.; Zhong, D.; Wang, G.; Li, X. Stochastic evolution of hydraulic geometry relations in the lower Yellow River of China under environmental uncertainties. *Int. J. Sediment Res.* **2020**, *35*, 328–346. [CrossRef]
15. Andreadis, K.M.; Schumann, G.J.P.; Pavelsky, T. A simple global river bankfull width and depth database. *Water Resour. Res.* **2013**, *49*, 7164–7168. [CrossRef]
16. Bieger, K.; Rathjens, H.; Allen, P.M.; Arnold, J.G. Development and evaluation of bankfull hydraulic geometry relationships for the physiographic regions of the United States. *JAWRA J. Am. Water Resour. Assoc.* **2015**, *51*, 842–858. [CrossRef]
17. Rosgen, D.L. A classification of natural rivers. *Catena* **1994**, *22*, 169–199. [CrossRef]
18. Knighton, D. *Fluvial Forms and Processes*; Arnold: London, UK, 1998.
19. Wu, B.; Wang, G.; Xia, J.; Fu, X.; Zhang, Y. Response of bankfull discharge to discharge and sediment load in the Lower Yellow River. *Geomorphology* **2008**, *100*, 366–376. [CrossRef]
20. Hagemann, M.W.; Gleason, C.J.; Durand, M.T. BAM: Bayesian AMHG-Manning inference of discharge using remotely sensed stream width, slope, and height. *Water Resour. Res.* **2017**, *53*, 9692–9707. [CrossRef]
21. Raymond, P.A.; Zappa, C.J.; Butman, D.; Bott, T.L.; Potter, J.; Mulholland, P.; Laursen, A.E.; McDowell, W.H.; Newbold, D. Scaling the gas transfer velocity and hydraulic geometry in streams and small rivers. *Limnol. Oceanogr. Fluids Environ.* **2012**, *2*, 41–53. [CrossRef]
22. Raymond, P.A.; Hartmann, J.; Lauerwald, R.; Sobek, S.; McDonald, C.; Hoover, M.; Butman, D.; Striegl, R.; Mayorga, E.; Humborg, C.; et al. Global carbon dioxide emissions from inland waters. *Nature* **2013**, *503*, 355–359. [CrossRef]
23. Tabata, K.K.; Hickin, E.J. Interchannel hydraulic geometry and hydraulic efficiency of the anastomosing Columbia River, southeastern British Columbia, Canada. *Earth Surf. Process. Landf.* **2003**, *28*, 837–852. [CrossRef]
24. Rhodes, D.D. The b-f-m diagram: Graphical representation and interpretation of at-a-station hydraulic geometry. *Am. J. Sci.* **1977**, *277*, 73–96. [CrossRef]
25. Singh, K.P.; Broeren, S.M. Hydraulic geometry of streams and stream habitat assessment. *J. Water Resour. Plan. Manag.* **1989**, *115*, 583–597. [CrossRef]
26. Orlandini, S.; Rosso, R. Parameterization of stream channel geometry in the distributed modelling of catchment dynamics. *Water Resour. Res.* **1998**, *34*, 1971–1985. [CrossRef]
27. Gleason, C.J.; Wang, J. Theoretical basis for at-many-stations hydraulic geometry. *Geophys. Res. Lett.* **2015**, *42*, 7107–7114. [CrossRef]
28. Gleason, C.J. Hydraulic geometry of natural rivers: A review and future directions. *Prog. Phys. Geogr. Earth Environ.* **2015**, *39*, 337–360. [CrossRef]
29. Knighton, A.D. Variation in width-discharge relation and some implications for hydraulic geometry. *Geol. Soc. Am. Bull.* **1974**, *85*, 1069–1076. [CrossRef]
30. Chien, N.; Zhang, R.; Zhou, Z. *Fluvial Processes*; Science Press: Beijing, China, 1987.

31. Wolman, M.G. *The Natural Channel of Brandywine Creeck Pennsylvania*; U.S. Geological Survey Professional Paper; USGS: Reston, VA, USA, 1955; Volume 271, 56p.
32. Bray, D.I. Representative discharges for gravel-bed rivers in Alberta. *Canada. J. Hydrol.* **1975**, *27*, 143–153. [CrossRef]
33. Lewis, L.A. Some fluvial geomorphic characteristics of the Manati Basin, Puerto Rico. *Ann. Assoc. Am. Geogr.* **1969**, *59*, 280–293. [CrossRef]
34. Langbein, W.B. Geometry of river channels: Closure of discussion. *J. Hydraul. Div. Am. Soc. Civ. Eng.* **1965**, *91*, 297–313.
35. Parker, G. Hydraulic geometry of active gravel rivers. *J. Hydraul. Div. Am. Soc. Civ. Eng.* **1979**, *105*, 1185–1201. [CrossRef]
36. Chang, H.H. Geometry of gravel streams. *J. Hydraul. Div. Am. Soc. Civ. Eng.* **1980**, *106*, 1443–1456. [CrossRef]
37. Li, R.M.; Simons, D.B.; Stevens, M.A. Morphology of cobble streams in small watersheds. *J. Hydraul. Div. Am. Soc. Civ. Eng.* **1976**, *102*, 1101–1117. [CrossRef]
38. Williams, G.P. *Hydraulic Geometry of River Cross-Sections-Theory of Minimum Variance*; U.S. Geological Survey Professional Paper; U.S. Government Printing Office: Washington, DC, USA, 1978; Volume 1029.
39. Castro, J.M.; Jackson, P.L. Bankfull discharge recurrence intervals and regional hydraulic geometry relationships: Patterns in the Pacific Northwest, USA. *J. Am. Water Resour. Assoc.* **2001**, *37*, 1249–1262. [CrossRef]
40. Harvey, A.M. Channel Capacity and the Adjustment of Streams to Hydrologic Regime. *J. Hydrol.* **1969**, *8*, 82–98. [CrossRef]
41. Shen, C.; Wang, S.; Liu, X. Geomorphological significance of at-many-stations hydraulic geometry. *Geophys. Res. Lett.* **2016**, *43*, 3762–3770. [CrossRef]
42. Brinkerhoff, C.B.; Gleason, C.J.; Ostendorf, D.W. Reconciling at-a-station and at-many-stations hydraulic geometry through river-wide geomorphology. *Geophys. Res. Lett.* **2019**, *46*, 9637–9647. [CrossRef]
43. Wohl, E. *Mountain Rivers Revisited*; GEO Press: Washington, DC, USA, 2010.
44. Wohl, E.E.; Wilcox, A. Channel geometry of mountain streams in New Zealand. *J. Hydrol.* **2005**, *300*, 252–266. [CrossRef]
45. Wohl, E. Limits of downstream hydraulic geometry. *Geology* **2004**, *32*, 897–900. [CrossRef]
46. Eaton, B.C.; Church, M.; Millar, R.G. Rational regime model of alluvial channel morphology and response. *Earth Surf. Process. Landf.* **2004**, *29*, 511–529. [CrossRef]

hydrology

MDPI

Article

Application of the Iber Two-Dimensional Model to Recover the Water Quality in the Lurín River

Omayra Luzmila Mori-Sánchez [1], Lia Ramos-Fernández [2,*], Willy Eduardo Lluén-Chero [2], Edwin Pino-Vargas [3] and Lisveth Flores del Pino [4]

1 Experimental Irrigation Area, Universidad Nacional Agraria La Molina, Lima 15024, Peru
2 Department of Water Resources, Universidad Nacional Agraria La Molina, Lima 15024, Peru
3 Department of Civil Engineering, Universidad Nacional Jorge Basadre Grohmann, Tacna 23000, Peru
4 Center for Research in Chemistry, Toxicology and Environmental Biotechnology, Universidad Nacional Agraria La Molina, Lima 15024, Peru
* Correspondence: liarf@lamolina.edu.pe

Abstract: The Lurín River is one of the main sources of water for the city of Lima. However, the discharge of domestic wastewater, the presence of dumps, and long periods of drought cause the deterioration of the water resource. In this study, *DO*, BOD_5, *E. coli*, *T*, *EC*, *TSS*, *U*, and *h* were monitored at 13 monitoring points spread over 20 km of river influence. This information was used to calibrate the parameters of K_{dbo}, K_{aire}, K_{dos}, and K_{dec} in the Iber two-dimensional numerical model, obtaining values of 0.55 d^{-1}, [4.84 d^{-1}–80.65 d^{-1}], 10 g O_2 $m^{-2}d^{-1}$, and [1.49 d^{-1}–15.42 d^{-1}], respectively, with efficiencies ranging from "very good" to "satisfactory". In the hydraulic model, a discretization of the channel, banks, and plains of 3, 5, and 7 m, respectively, was considered, resulting in a computational calculation time of 4 days in each simulation. The greatest contamination occurs in July at km 5 + 400 up to the Pan-American bridge. Therefore, it is proposed to recover the river by optimizing the San Bartolo Wastewater Treatment Plant (WWTP) and a new WWTP in Pachacámac to avoid diffuse contamination, with discharge flows of 0.980 m^3s^{-1} and 0.373 m^3s^{-1}, respectively, and 4 mg L^{-1}, 15 mg L^{-1} and 1000 NMP/100 mL for *DO*, BOD_5, and *E. coli*, respectively.

Keywords: biochemical oxygen demand; *Escherichia coli*; Iber two-dimensional numerical model; dissolved oxygen

1. Introduction

The available water per capita in the world is reduced because of the increase in population, dumping of waste and chemicals from agricultural runoff, livestock, industrial and mining activities, and inadequate management of solid waste, which cause water shortages and deterioration of its quality. Contamination, without sufficient treatment, in rivers is a great concern, mainly because rivers are the source of water for domestic consumption; it is important to determine the quality of the water in rivers.

The Lurín river basin, which originates in the Andes Mountains, is located on the central coast of Peru, which has approximately two million inhabitants. Its climate is arid with scarce rainfall and inadequate management of water sources, leading to the inefficient development of agriculture and livestock, the main economic activities in the mentioned watershed [1]. Likewise, there is an accelerated process of unplanned urbanization, strong pressure for land and a risk of becoming a new industrial park. However, areas used for agriculture are still preserved, with tourism and archaeological potential resources, but with high poverty rates. In its lower section, the river is subject to important pressures derived from the extraction of water resources, the discharge of urban wastewater (treated and untreated), garbage disposal, livestock, and agricultural drainage. All this leads to an increasing scarcity of water and deterioration of its quality, negatively impacting aquatic ecosystems, and these factors are exacerbated during drought periods [2].

To solve this complicated problem, some tools are used to generate mathematical models of water quality to simulate working scenarios aimed at improving quality that can be used as a decision-making tool.

This study evaluated the current state of water quality using the two-dimensional Iber model, previously parameterized, to propose a recovery scenario for the Lurín River, to accomplish the environmental quality standard (ECAs) [3] and the Maximum Permissible Limit. (LMP) [4] of the Peruvian regulations.

2. Materials and Methods

2.1. Study Zone

The study was carried out in the lower part of the Lurín River (20 km), southeast of the city of Lima, located between 76°48′ and 76°54′ W longitude and 12°7′ and 12°16′ S latitude. This included the districts of Cieneguilla, Pachacamac, and Lurín from km 20 + 500 at the top of the Mototaxi bridge (L13) in the district of Cieneguilla, to km 0 + 578 at the top of the Panamericana South Bridge (L1) in the district of Lurín. Seven evaluations to obtain field information were performed between February and August 2019 (Figure 1).

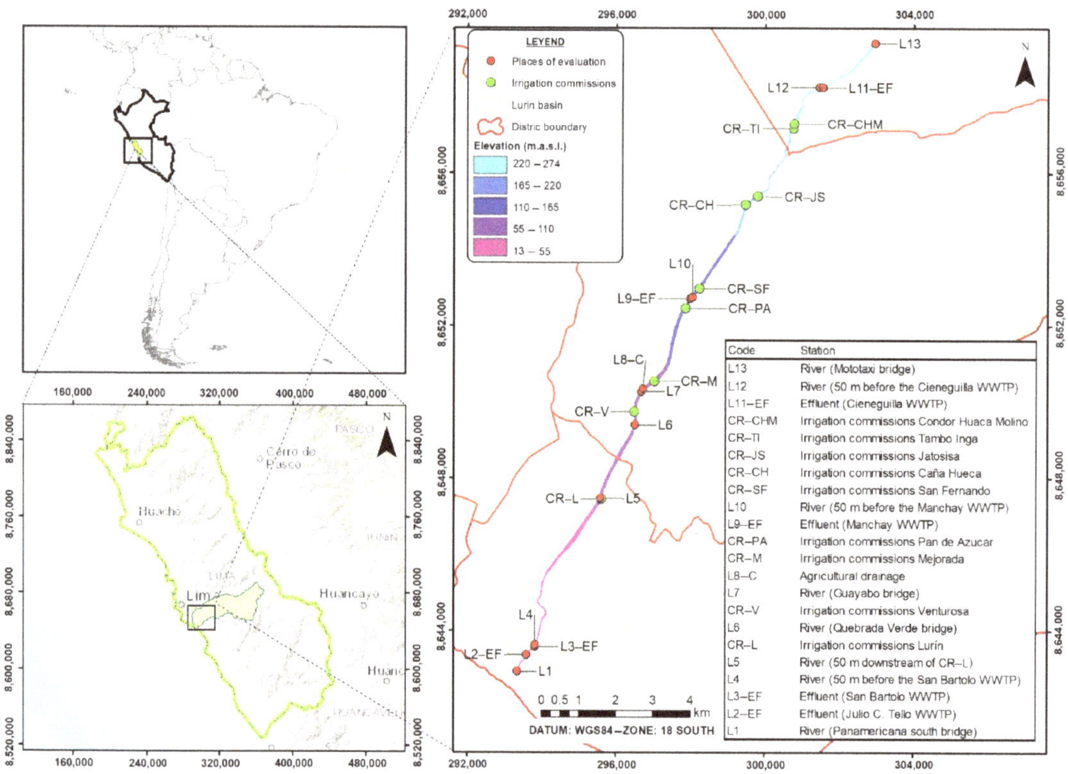

Figure 1. Geographical location of the evaluation stations and the intakes of the irrigation commissions in the lower part of the Lurín River.

2.2. Topographical Characterization

Images were captured with a DJI Zenmuse X4S camera attached to a Remotely Piloted Aircraft System (RPAS), Matrice 210 (DJI brand, Beijing, China), with an 80 m high flight plan, the lateral and frontal overlap of 75% and 70%, respectively, and a speed of 6 m s^{-1}. Further, the processing of images was done in the software photogrammetric Pix4Dmapper Pro (Pix4D SA, Prilly, Switzerland), and the topographical data was available for sections starting in km 9 + 300 up to km 15 + 100, sections that corresponded to the irrigation commissions of Mejorada and Jatosisa, respectively.

The Digital Elevation Model (DEM) was generated in Civil 3D, in tiff format, with a resolution of 1.0 m. Due to the heterogeneity of the topography, it was partitioned with an unstructured mesh, which got a better fit to the surface: in the riverbed with 3 m; the banks with 5 m; and the plains with 7 m, as shown in Figure 2. Additionally, it is worth mentioning that the mesh size influences the computational time of each simulation, and four days were used to simulate the 20 km.

Figure 2. (**a**) Mesh size, (**b**) Mesh with DEM interpolation in L3-EF and L4, (**c**) Mesh with DEM interpolation in L2-EF, (**d**) Mesh with DEM interpolation in L1, (**e**) n in the lower part of the study section from L4 to L1, (**f**) n in L3-EF and L1. n is the is Manning's roughness coefficient.

2.3. *Two-Dimensional Iber Model*

The Iber model of water quality includes a hydrodynamic model, with basic information on the depth of the water, the speed, and the turbulent viscosity. All are necessary to solve the convection-diffusion equation for each polluting substance [5]. Bladé et al. [6] indicated that the hydrodynamic behavior of the river, in a two-dimensional model, is solved with two-dimensional St. Venant Equation (1), where the effects of turbulence [7] and surface friction by wind are incorporated:

$$\begin{aligned}
&\frac{\partial}{\partial t}(hU_x) + \frac{\partial}{\partial y}(hU_xU_y) + \frac{\partial}{\partial x}\left(hU_x^2 + g\frac{h^2}{2}\right) = -gh\frac{\partial Z_b}{\partial x} + \frac{\tau_{s,x}}{\rho} - \frac{\tau_{b,x}}{\rho} + \frac{\partial}{\partial x}\left(\nu_t h\frac{\partial U_x}{\partial x}\right) + \frac{\partial}{\partial y}\left(\nu_t h\frac{\partial U_x}{\partial y}\right) \\
&\frac{\partial}{\partial t}(hU_y) + \frac{\partial}{\partial x}(hU_xU_y) + \frac{\partial}{\partial y}\left(hU_y^2 + g\frac{h^2}{2}\right) = -gh\frac{\partial Z_b}{\partial y} + \frac{\tau_{s,y}}{\rho} - \frac{\tau_{b,y}}{\rho} + \frac{\partial}{\partial x}\left(\nu_t h\frac{\partial U_y}{\partial x}\right) + \frac{\partial}{\partial y}\left(\nu_t h\frac{\partial U_y}{\partial y}\right)
\end{aligned} \quad (1)$$

where h is the water depth, U_x, U_y are horizontal velocities averaged in depth, g is the acceleration due to gravity, ρ is the water density, Z_b is the depth to the bottom, τ_s is the friction on the open surface due to friction produced by the wind, τ_b is the friction due to the friction at the bottom, and ν_t is the turbulence kinematic viscosity.

The τ_b is evaluated by Manning's Equation (2):

$$\tau_{b,x} = \rho g h \frac{n^2 U_x |U|^2}{h^{\frac{4}{3}}} \quad \tau_{b,y} = \rho g h \frac{n^2 U_y |U|^2}{h^{\frac{4}{3}}} \tag{2}$$

where n is Manning's roughness coefficient and $|U|$ is the resultant of the vectors U_x and U_y.

Cea et al. [5] mentioned the spatial and temporal evolution of species, and pollutant variables are modeled with a generic convection-diffusion equation [Equation (3)]:

$$\frac{\partial}{\partial t}(hC) + \frac{\partial}{\partial x}(hU_x C) + \frac{\partial}{\partial y}(hU_y C) = \frac{\partial}{\partial x_j}\left(h\left(r_i + \frac{\nu_t}{S_{c,t}}\right)\frac{\partial C}{\partial x_j}\right) + S_c \tag{3}$$

where C is the depth-averaged concentration of species, S_c is a generic source term which depends on the species considered, r_i is the coefficient of molecular diffusivity, ν_t is the turbulence kinematic viscosity, and $S_{c,t}$ (dimensionless) is the turbulence Schmidt number.

Cea et al. [5] indicated that the water quality module is completely paired with the hydrodynamic module. They share the same unstructured finite volume mesh adapted to topography, which allows defining parameters and visualization model outputs. Equation (3) is solved for each of the pollutants considered, where the reaction terms S_c for each species are as follows:

1. Temperature: four processes of heat transfer are considered, modeling the heat transfer between water and the atmosphere.

$$S_T = \frac{S_{Tem}}{C_\rho}(h) \tag{4}$$

where S_T is the source term of contributions or sinks of heat (W m^{-2}), S_{Tem} represents the heat transfer processes (1000 kg m^{-3}), and C_ρ is the specific heat of water (4180 J kg^{-1} °C^{-1}).

$$S_{Tem} = Q_{rad,in} + Q_{rad,out} + Q_{cond} + Q_{evap} \tag{5}$$

where $Q_{rad,in}$ is total net radiation absorbed by the water, $Q_{rad,out}$ is the long-wave radiation emitted by the water, Q_{cond} is heat transferred by conduction, and Q_{evap} is energy transferred by evaporation/condensation of the water.

1. Dissolved Oxygen *(DO)* and Biological Oxygen Demand (*BOD*): Cea et al. [8] indicated that one of the main uses of *DO* in a body of water is the degradation of organic matter (*BOD*); see Equation (6).

$$BOD_5 = BOD_u \cdot \left(1 - e^{-5 \cdot K}\right) \tag{6}$$

where BOD_5 is the *BOD* at 5 days, BOD_u is the ultimate *BOD*, which occurs at the maximum possible oxygen consumption when the substrate has been completely degraded, and K is the organic matter degradation constant at river temperature, expressed in d^{-1}.

The source term for *DO* and *BOD* are shown in Equations (7) and (8), respectively:

$$S_{DO} = K_{aeration} * \theta_1^{(T-20)} * (DO_{sat} - DO) - K_{bod} \cdot \theta_2^{(T-20)} \cdot F_{oxc} \cdot BOD_u - \frac{K_{sod}}{h} \tag{7}$$

$$S_{BOD} = -K_{bod} \cdot \theta_2^{(T-20)} \cdot F_{oxc} \cdot BOD_u - \frac{V_{sBOD}}{h} * BOD_u \tag{8}$$

where S_{DO} represents the reaeration source implemented in Iber, S_{BOD} is a term that represents the reaction due to degradation or reaction with other substances present in the water, $K_{aeration}$ is the aeration constant at 20 °C, T is the water temperature in °C, DO_{sat} is the *DO* saturation concentration, θ_1 is a correction coefficient for temperature (θ_1 = 1.024), K_{bod} is the carbonaceous organic matter degradation constant at 20 °C, θ_2 is the temperature correction coefficient (θ_2 = 1.047), F_{oxc} is a dimensionless attenuation factor due to low oxygen levels, K_{sod} is the rate of demand of oxygen by the sediment in kg $m^{-2}d^{-1}$, V_{sBOD} is the rate of sedimentation of organic matter in m, and h is the depth of water in m.

$$K_{aeration} = K_{airh} + \frac{K_{airw}}{h} \tag{9}$$

where K_{airh} is the reaeration constant at 20 °C based on the hydraulic characteristics of the river (h and U) without considering the V_{wind10}, and is calculated by the Covar method [8] (see Table 1). K_{airw} is the reaeration coefficient based on the V_{wind10}; it is calculated using Equation (10) proposed by Banks and Herrera [8]:

$$K_{airw} = 0.728V_{wind10}^{0,5} - 0.317V_{wind10} + 0.037V_{wind10}^{2} \tag{10}$$

where V_{wind10} is the wind speed measured at 10 m above the water level.

3. *E. coli*: a bacterium found in the gastrointestinal tract of homeothermic animals, such as humans, and therefore in urban wastewater.

$$S_d = -K_{dec} * C \tag{11}$$

where S_d is the term that represents the bacterial disappearance, K_{dec} is the constant of bacterial disappearance in $time^{-1}$, and C is the concentration of *Escherichia coli* (*E. coli*), expressed in the most probable number per 100 mL (NMP/100 mL).

Table 1. Equations of the Covar method to estimate the reaeration constant $K_{aeration}$.

Depth (h) in m Water Speed (U)) in ms^{-1}	Formula	Equation
If $h \leq 0.61$ m	Owens-Gibbs	$K_{airh} = 5.32\frac{U^{0.67}}{h^{1.85}}$
If $h > 0.61$ and $h > 3.45 * U^{2.5}$	O'Connor-Dobbins	$K_{airh} = 3.93\frac{U^{0.5}}{h^{1.5}}$
In other cases	Churchill	$K_{airh} = 5.026\frac{U}{h^{1.67}}$

SOURCE: Prepared with information from Cea et al. [8].

The degradation coefficient K_{dec} is estimated with the Mancini [9] empirical formula, based on temperature, salinity, and solar radiation [10].

$$K_{dec} = (0.8 + 0.2Sal)1.07^{(T-20)} + 0.086\frac{I_0}{K_e H_c}\left(1 - e^{(-K_e H_c)}\right) \tag{12}$$

where K_{dec} is the rate of disappearance averaged over a depth, $H_c(d^{-1})$, Sal is the salinity (g L^{-1}), T is the temperature of water (°C), I_0 is the incidence of solar radiation on the water surface (W m^{-2}), K_e is the coefficient of extinction of light in water (m^{-1}), and H_c is the depth of the vertical layer over which the *E. coli* spreads.

The rate of degradation of *E. coli* is largely dependent on the turbidity of the water, and is considered in the model through the light extinction coefficient K_e [11]:

$$K_e = 2.619 + 0.129 * NTU \tag{13}$$

where NTU is the turbidness of the water.

2.4. Hydrodynamic Characterization

Data on river flow were collected at eight points along the river and from four WWTP discharges (Cieneguilla, Manchay, San Bartolo, and Julio C. Tello); and in agricultural drainage. In addition, nine points were sampled in the streams of the irrigation commissions and the Tinajas dry ravine, which becomes active during periods of rain with flows of about 0.025 m^3s^{-1}. At hydraulic sections of the river, mean speed and the height of the water were measured.

Manning's roughness coefficient was delineated from the Red-Green-Blue (RGB) images collected by the RPAS in the visits to the fields, by observing changes in land use and, therefore, variation in roughness. The values were selected according to [12].

Information from stations L13 and L1 was assigned to the conditions of input and output at the boundary model requirements, respectively. In this case, the drains and sources of the river were considered as river basins of the irrigation commissions and Wastewater Treatment Plants (WWTPs) discharges, respectively.

2.5. Characterization of Water Quality

Data on water quality were collected in the field at 13 evaluation points: eight points in the river; four in the WWTP discharges (Cieneguilla, Manchay, San Bartolo, and Julio C. Tello); and in agricultural drainage. T, Electrical Conductivity (EC), DO, BOD_5, *E. coli*, and Total Solids in Suspension (TSS) were measured. The first three parameters were measured in situ and the last three were measured in the laboratory, and the salt concentration (Sal) was estimated from EC [13], see Equation (14):

$$Sal = -0.175 + 1.0053(EC) \tag{14}$$

Turbidness (NTU) was estimated from $TSSs$ [14], see Equation (15):

$$NTU = 1.0283 * TSS^{1.0282} \tag{15}$$

The information collected from these evaluations was compared to the Standards of Environmental Quality (ECA), for Category 3, to which the Lurín River belongs, where it is stated that concentrations for BOD_5 and *E. coli* in the river should not be over 15 mg L^{-1} and 1000 NMP/100 mL, respectively. In addition, DO concentrations should not be less than 4 mg L^{-1}, and the Maximum Permissible Limit (LMP), indicated for BOD_5, is 100 mg L^{-1}. Uncontrolled discharges were identified in Pachacamac, near km 5 + 400. Similarly, higher contamination values were found on the final stretch of the river, from km 5 + 500 to km 0 + 578 where stations L5 and L1 were located, respectively. These stations were assigned as input and output boundaries under the conditions required by the model, respectively. In addition, concentrations of discharges from the WWTPs were included.

2.6. Climate Characterization

Hourly data (0 to 24 h) of total net radiation (Ra_s), wind speed (V_{wind}), air temperature (T_{air}), and relative humidity (H_r) were collected from the Alexander von Humboldt weather station. The albedo (A_s), which depends on the type of surface where the radiation falls and the angle of incidence (α), was estimated with the expression $A_s = 1.18 \times \alpha^{-0.77}$, in which α was extracted from the webpage https://salidaypuestadelsol.com/sun/lima (accessed on 1 June 2021).

This information was used in Equation (5) to estimate the water temperature according to the dates of evaluation.

2.7. Model Calibration

Iber's two-dimensional model was used at a spatial scale of 1 m, the time in hours, and information for DO, BOD_u, *E. coli*, EC, Q, Ra_s, H_r, V_{wind}, and T_{air}.

The hydrodynamic model has been manually calibrated on 20 km of the river, with the Manning's roughness coefficients, after several trial and error simulations. Then, the simulated and observed U and h values were compared.

The water quality model was calibrated in the critical section from km 5 + 500 to km 0 + 578 (stations L5 to L1), with the parameters $K_{aeration}$, K_{bod}, K_{sod}, K_{dec}, and V_{sBOD}. It is important to remark that, from the information of h and U from the calibrated hydrodynamic model, $K_{aeration}$ was estimated using equations from Covar's method (Table 1). Values of K_{bod} were obtained in a range of 0.02 to 3.4 d^{-1}, as recommended by Brown and Barnwell [15]. After several simulations and comparing concentrations of DO and BOD_5, observed and simulated, K_{bod} was calibrated. For K_{sod}, a value of 10.0 g $O_2m^{-2}d^{-1}$ was inserted according to the type of sludge of urban origin [16]. K_{dec} was estimated according to Equation (12), and a zero value was given to V_{sBOD}, because turbulence and wind keep the particles in suspension.

The simulated concentrations were adjusted to those observed, and a trip back to the field with the data was organized to check for possible diffuse pollution, discharges, and/or undetected landfills. This is how a diffuse contamination was added at km 3 + 500 in Pachacamac.

After several simulations in March, May, and July, seven on each occasion, the model was calibrated, obtaining the Nash-Sutcliffe efficiency index (E), the ratio of the root of the mean quadratic error to the standard deviation of the variables of the observations (RSR), with limit values of E (−infinity to 1) and RSR (0 to +infinity), and rating the efficiency according to the scale proposed by Moriasi [17]. Furthermore, Pearson's correlation coefficient (R), which ranges from −1 to +1, was obtained with a *t*-Student test with a significance level (alpha) of 5%. The computational times for the hydrodynamic and water quality simulation were four days and one day, respectively. On the other hand, the use of the Iberplus plugin, which is an accelerated tool for flood modelling based on Iber, could be applied to reduce the computational time in the Iber water quality model [18,19].

2.8. Simulation of Scenarios for the Recovery of the River

Several simulations were performed for the critical month of July, including the optimization of the San Bartolo WWTP and the implementation of a WWTP in Pachacamac due to the discharge of uncontrolled domestic wastewater, with the result that the river complied with LMP [4] and ECA [3]. The procedure followed in the study is presented in a sequential diagram in Figure 3.

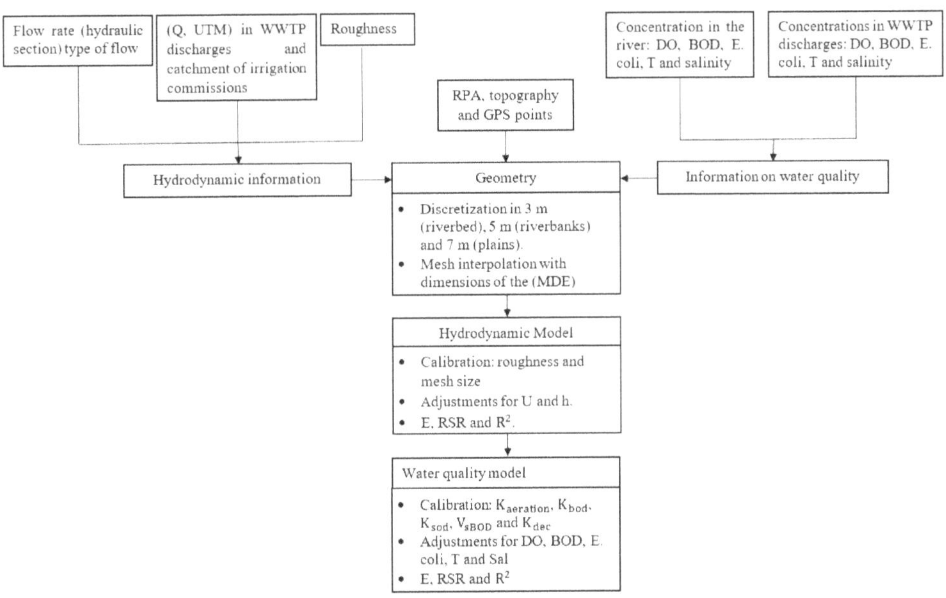

Figure 3. Flowchart of the methodology used in the study. WWTP is Waste Water Treatment Plant, RPAS is Remotely Piloted Aircraft System.

3. Results and Discussion

3.1. Hydraulic and Water Quality Characterization

The results of U, h, and Q, collected in the field at the evaluation points, indicated that U was less than 1.5 m s^{-1} and the values of h were less than 0.8 m, with a Froude number of less than 1, that is, the flow of the river shows to be at a subcritical stage. The lowest Q was recorded in L5 and L4, caused by the irrigation commission catchment upstream of these points, with mean $\underline{x}$ ± standard deviation S ($\underline{x}$ ± S) of 3.22 ± 4.92 m^3s^{-1} and 3.39 ± 4.99 m^3s^{-1}, respectively. In addition, the river was dry (Q = 0 m^3s^{-1}) in L5, on 19 June, 8 July, and 5 August 2019. There were high values of Q (4.35 ± 6.03 m^3s^{-1}) in L13 because of the limited catchment from irrigation commissions upstream from this point. In contrast, L1 had the highest value of Q (4.77 ± 5.50 m^3s^{-1}) caused by discharges from the San Bartolo WWTP (L3−EF) with values of 1.06 ± 0.57 m^3s^{-1}. The discharges to the river with the lowest Q were observed in L9-EF (0.01 ± 0.02 m^3s^{-1}) and L8−C (0.03 ± 0.048 m^3s^{-1}). The values ($\underline{x}$ ± S) of T, EC, TSS, DO, BOD_5, and *E. coli*, are shown in Table 2, where the following can be noted:

- The T and CE comply with current Peruvian regulations.
- The $TSSs$ have higher values (114.49 ± 159.31 mg L^{-1}) at station L1 due to $TSSs$ and higher Q value of the San Bartolo WWTP discharge. In contrast, station L13 shows the lowest values of TSS (29.24 ± 48.91 mg L^{-1}) caused by a lower flow.
- In the river, from the Mototaxi bridge up to 50 m after the collect point of the Lurín irrigation commission (L13, L12, L10, L7, L6, and L5), the DO complies with ECA even in the dry period (lower flow), probably due to the photosynthetic activity of the algae present; for example, in the Quebrada Verde Bridge, and the Guayabo bridge (L6 and L7). In sections of the river from L4 to L1, there are uncontrolled discharges (diffuse contamination from Pachacamac) at km 5 + 400. In addition, discharge from San Bartolo WWTP (L3−EF) that does not meet the LMP causes a decrease in DO. This is more evident in the dry period and fails to comply with the ECA.
- The discharge of agricultural drainage (L8−C) with a high content of organic matter shows values of BOD_5 (97.23 ± 139.75 mg L^{-1}), which does not comply with the LMP on some dates. It causes an increase in BOD_5 values (22.06 ± 22.52 mg L^{-1}) downstream (L7) that does not comply with ECA. Similarly, the discharge from San Bartolo WWTP (L3−EF) does not meet the LMPs, due to an irregular process and lack of quality control.
- The San Bartolo WWTP (L3−EF) does not comply with LMP for all parameters and dates of evaluation, while the Julio C. Tello, Manchay, and Cieneguilla WWTPs did not comply with the LMP for *E. coli* in April, May, and July.
- The discharge of agricultural drainage (L8−C) has *E. coli* values of (271, 142.86 ± 453, 937.74 NMP/100 mL), which does not comply with LMP. This causes an increase in *E. coli* downstream at (L7) and does not comply with ECA. Similarly, discharge from San Bartolo WWTP (L3−EF) that does not comply with LMP causes failure to comply with the ECA for *E. coli* downstream at (L1).

3.2. Model Calibration Results

From the calibration of the model, the parameter $K_{aeration}$ revealed values in the range of (4.84 d^{-1}–80.65 d^{-1}). For K_{bod}, good efficiency was accomplished in the model with a value equal to 0.55 d^{-1} at 20 °C. Figure 4a shows the values of E for DO and BOD_5 estimated with different values of K_{bod}, for seven simulations at each of the simulated dates (March, May, and July). The value of the parameter K_{dec} was in the range of (1.49 d^{-1}–15.42 d^{-1}).

Figure 4b–f show the observed and simulated concentrations for DO, BOD_5, *E. coli*, T, and EC, as well as the efficiency indices E, RSR, and Pearson's correlation (R).

The efficiency of the model was rated from "very good" to "satisfactory", according to the E values (0.75–1.0, 0.65–0.75, 0.50–0.65, <0.5) and the RSR (0–0.5, 0.5–0.69, 0.6–0.7, <0.7) with a rating of "very good", "good", "satisfactory" and "unsatisfactory", respectively, a scale proposed by Moriasi et al. [17] with the t- Student test, for a significance level (alpha) of 5%. Table 3 shows the E and RSR indices and Pearson's correlation (R) at the points of greatest river contamination.

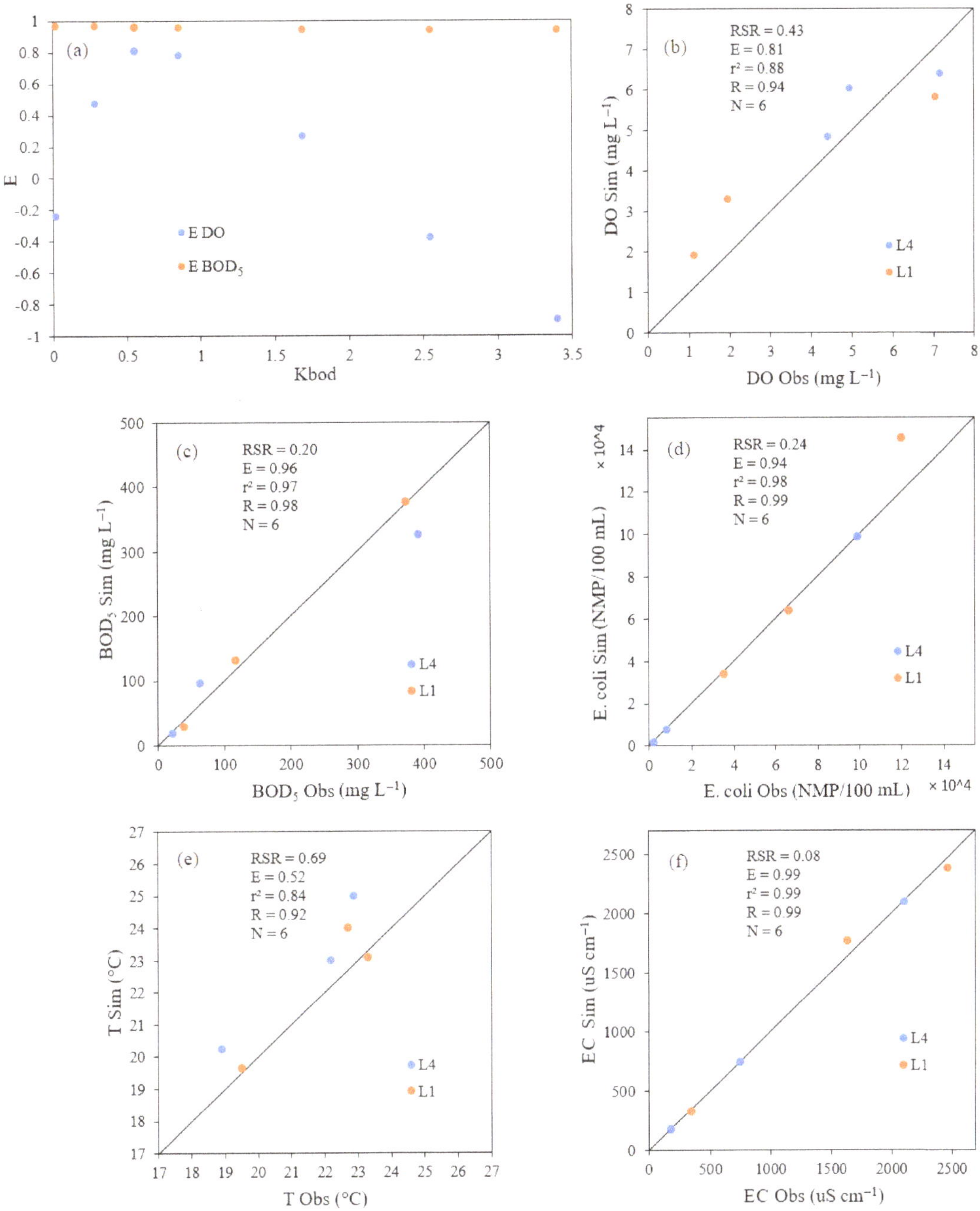

Figure 4. (**a**) Range of values of the E index due to changes in the parameter K_{bod} in March, May, and July. Comparison between observed and simulated values according to: (**b**) dissolved oxygen, (**c**) BOD_5, (**d**) *E. coli*, (**e**) temperature, and (**f**) electrical conductivity.

Table 2. Means and standard deviation ($\underline{x}$±S) for water quality from data obtained in evaluations completed between February and August 2019.

Stations	T (°C)	EC (uS cm^{-1})	TSS (mg L^{-1})	DO (mg L^{-1})	BOD_5 (mg L^{-1})	$E. Coli$ (NMP/100 mL)
L13	23.31 ± 2.58	363.80 ± 195.90	29.24 ± 48.91	7.23 ± 0.97	1.93 ± 0.48	143 ± 378
L12	24.37 ± 2.29	388.51 ± 222.50	27.9 ± 44.37	7.45 ± 1.61	4.22 ± 4.85	286 ± 488
L11−EF	25.00 ± 2.51	906.71 ± 156.05	3.70 ± 2.90	4.67 ± 0.35	2.83 ± 1.12	1714 ± 4536
L10	23.51 ± 3.92	430.84 ± 244.42	44.58 ± 74.48	6.41 ± 0.93	8.37 ± 9.25	143 ± 378
L9−EF	25.10 ± 12.31	1588.50 ± 775.11	20.82 ± 13.57	6.29 ± 3.07	45.83 ± 30.85	27,000 ± 20,410
L8−C	22.44 ± 3.76	1264.83 ± 908.29	56.93 ± 75.55	6.23 ± 1.69	97.23 ± 139.75	271,143 ± 453,938
L7	23.54 ± 3.24	757.49 ± 582.90	55.73 ± 84.51	5.75 ± 0.95	22.06 ± 22.52	60,714 ± 105,855
L6	23.06 ± 3.29	571.39 ± 373.53	56.51 ± 93.64	8.73 ± 2.87	21.19 ± 14.92	6143 ± 12,090
L5	24.78 ± 13.34	288.68 ± 192.83	163.92 ± 131.78	6.58 ± 3.54	7.97 ± 7.38	1500 ± 1069
L4	20.73 ± 2.44	1080.44 ± 892.35	107.74 ± 165.59	5.48 ± 1.51	142.12 ± 147.77	23,857 ± 36,108
L3−EF	21.80 ± 11.08	2440.00 ± 1195.64	28.76 ± 18.24	0.83 ± 0.61	307.31 ± 198.08	101,800 ± 61,576
L2−EF	23.59 ± 3.25	1843.00 ± 156.38	12.50 ± 6.22	5.57 ± 0.65	41.81 ± 18.93	5000 ± 7234
L1	21.27 ± 2.44	1549.43 ± 974.96	114.49 ± 159.31	3.77 ± 2.69	243.66 ± 225.86	72,428 ± 55,220

Table 3. Efficiency indices E and RSR, and Pearson's correlation (R) in the calibration of the Iber model for T, EC, DO, BOD_5, and *E. coli*.

Parameters	L4: 50 m before WWTP San Bartolo			L1: South Pan−American Bridge			L4 and L1		
	E	RSR	R	E	RSR	R	E	RSR	R
DO	0.546	0.674	0.805	0.806	0.440	0.974	0.813	0.433	0.940
BOD_5	0.932	0.260	0.989	0.995	0.070	0.998	0.959	0.202	0.983
E. coli	0.999	0.005	0.999	0.823	0.421	0.994	0.944	0.237	0.989
T	0.250	0.866	0.965	0.790	0.458	0.941	0.518	0.690	0.917
EC	0.999	0.003	0.999	0.988	0.108	0.994	0.994	0.076	0.997

3.3. Simulation in Present Conditions

Figure 5 shows the longitudinal profile from km 5 + 500 (L5) to km 0 + 578 (L1), indicating the distances at which each discharge is found (July). In L5, the river flow is zero, initiating diffuse contamination in Pachacamac, 2164 m downstream of L5, with concentrations of BOD_5 and *E. coli* of 341.68 mg L^{-1}, and 2.2 × 10^5 NMP/100 mL, respectively. At 2126 m downstream is the San Bartolo WWTP with discharges of DO, BOD_5, and *E. coli* of 0.65 mg L^{-1}, 469.95 mg L^{-1} and 1.4 × 10^5 NMP/100 mL, respectively, and at 300 m the Julio C. Tello WWTP is found with discharges of DO, BOD_5, and *E. coli* of 5.67 mg L^{-1}, 35.29 mg L^{-1} and 8 × 10^3 NMP/100 mL, respectively.

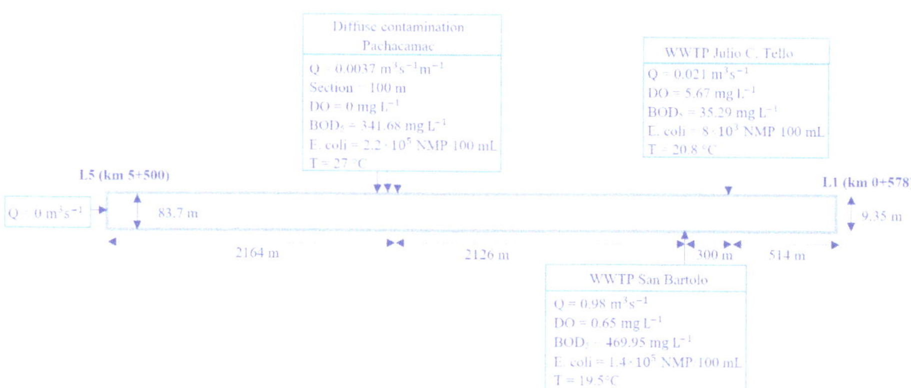

Figure 5. Schematic of the longitudinal profile from L5 to L1 with distances at which the discharges are found. Analysis date: July 2019.

Figure 6 shows the longitudinal profiles of the current simulation at the most critical condition time (July), where it is noted that the ECAs for BOD_5 and *E. coli* are not met in the section of the river from L5 to L1, because of the high concentrations of BOD_5 and *E. coli* from the diffuse contamination in Pachacamac. The *DO* does not comply with the ECAs between the San Bartolo WWTP and the Panamericana South Bridge because discharges of the San Bartolo WWTP with a water flow of 0.976m^3 s^{-1}, and concentrations of 0.65 mg L^{-1}, 469.95 mg L^{-1}, 1.4 × 10^5 NMP/100 mL of *DO*, BOD_5 and *E. coli*, respectively, do not comply with the LMP. In addition, the existence of solid waste landfills and other areas, such as informal latrines, were observed on the Panamericana south bridge.

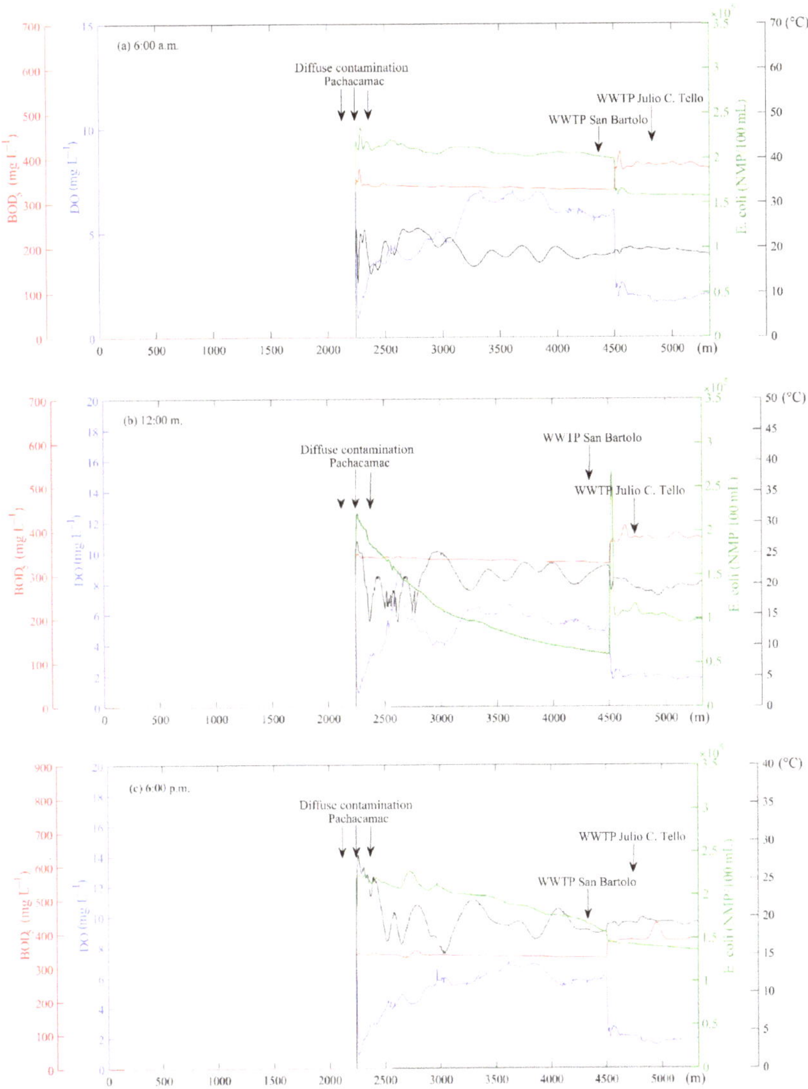

Figure 6. Longitudinal profile of the current situation of the river according to *DO*, BOD_5, *E. coli*, and *T* (from L5 to L1): 6 a.m. Table S1 (**a**), 12 m. Table S2 (**b**) and 6 p.m. Table S3 (**c**). Red line is BOD_5, blue line is *DO*, green line is *E. coli*, and black line is *T*.

3.4. Simulation of Recovery

The river recovery scenario was generated for July, with the improvement of the San Bartolo WWTP with a discharge flow of 0.98 m^3s^{-1}, and implementing the Pachacamac WWTP with a flow of 0.37 m^3s^{-1}, and both with concentrations of 4 mg L^{-1}, 15 mg L^{-1} and 1000 NMP/100 mL of *DO*, BOD_5, and *E. coli*, respectively, which allows them to follow the LMP and ECA that are required for the river.

From the calibration of the hydrodynamic model, the discharge flow for the new WWTP was estimated. After several trial and error simulations, it was identified that for the month of the lowest flow (critical), the estimated flow of wastewater dumped into the river in Pachacamac was 0.0037 $m^3s^{-1}m^{-1}$, in a section of 100 m, with a total flow of 0.37 m^3s^{-1}. The discharges must comply with LMP and ECA for a river of this category, and for that several simulations were performed. Values of 4 mg L^{-1}, 15 mg L^{-1} and 1000 NMP/100 mL were estimated for *DO*, BOD_5, and *E. coli*, respectively.

Figure 7 shows the longitudinal profiles of the polluting substances *DO*, BOD_5, *E. coli*, and *T*, for 6:00 a.m., 12:00 m., and 6:00 p.m. There, it is observed that after discharges take place in the new Pachacamac WWTP and San Bartolo WWTP, they continue for a segment of approximately 200 m in which ECAs are not met. However, this segment is considered a mixed one. It is also observed that as *T* increases, *DO* decreases, in an inverse relationship. Likewise, with increases in BOD_5, the *DO* decreases. Finally, *E. coli* at noon is reduced by 72%, because of higher solar radiation, with values up to 250 NMP/100 mL, as seen in Figure 7b, with higher values close to 1000 NMP/100 mL.

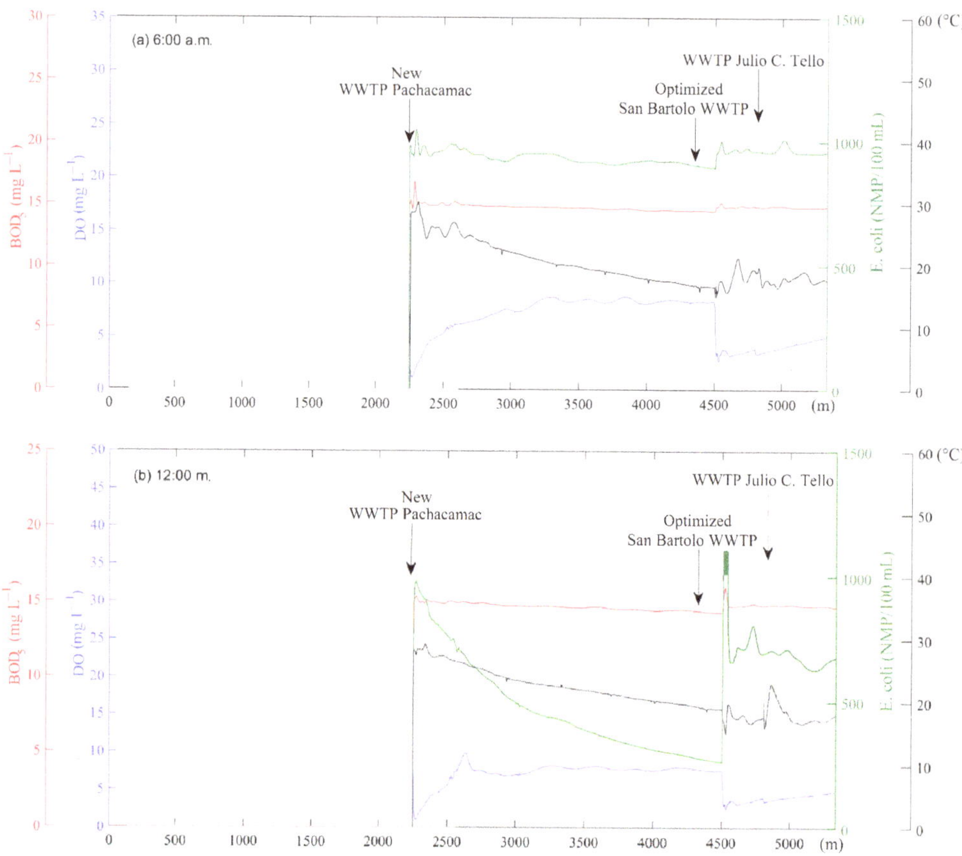

Figure 7. *Cont.*

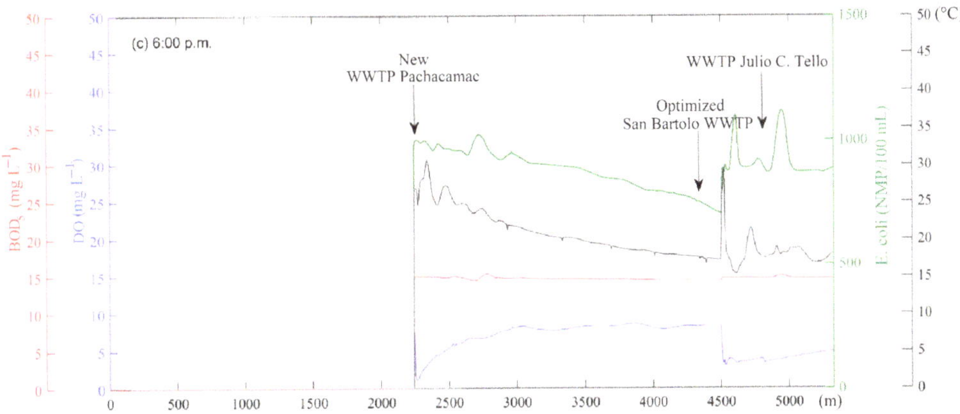

Figure 7. Longitudinal profile of river recovery according to *DO*, BOD_5, *E. coli*, and *T* (from L5 to L1): 6 a.m. (**a**), 12 m. (**b**) and 6 p.m. (**c**). Red line is BOD_5, blue line is *DO*, green line is *E. coli*, and black line is *T*.

To observe a two-dimensional behavior of the river, information was extracted from h, U, U_x, U_y, DO, BOD_u, *E. coli*, and T, in the cross section, 25 m after the San Bartolo WWTP (as shown in Figure 8), where it is noted that because of a greater influence of the discharges from San Bartolo WWTP, the difference in concentrations along the cross-sectional profile is high. It was also seen that on the right bank of the river, U reaches values of 0.27 ms $^{-1}$ and h to 0.09 m, which are lower with respect to the left bank, with values up to 0.49 ms^{-1} and 0.28 m of U and h, respectively, due to the contribution of discharges from San Bartolo WWTP. The parameter h influences the concentration of contaminants because, at a lower depth, it will be completely reaerated, generating a higher concentration of oxygen. For that reason, DO and h show an inverse relationship. At noon, T reaches values of 21.5 °C; on the other hand, at a higher concentration of BOD_u, there is a greater amount of organic matter to degrade and, therefore, the DO is lower.

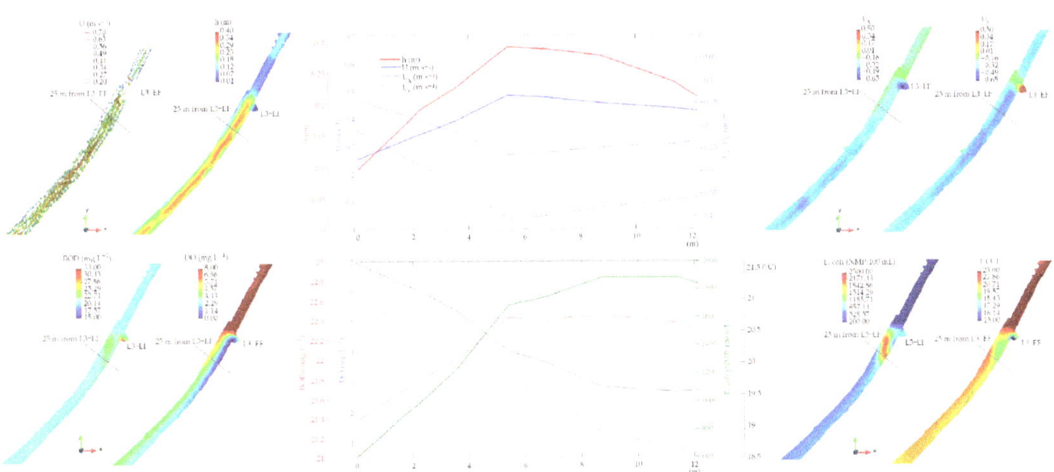

Figure 8. Cross-sectional profile of the recovery of the river according to DO, BOD_u, *E. coli*, T, h, U, U_x, and U_y, downstream San Bartolo WWTP, at 12:00 m. Top panel: red line is h, blue line is U, dotted blue line is U_x, dashed blue line is U_y. Bottom panel: red line is BOD_u, blue line is DO, green line is *E. coli*, black line is T.

In the cross sections, the highest concentration is found on the right side, due to the difference in speeds and heights, in addition to the great influence of the discharge from the San Bartolo WWTP, located at that end of the river.

4. Conclusions

The calibration of the Iber model revealed a performance ranging from "very good" to "satisfactory", with values of E, RSR, and R^2(0.813, 0.433, and 0.883) for *DO*, (0.959, 0.202, and 0.967) for BOD_5, (0.944, 0.237, and 0.979) for *E. coli*, and (0.518, 0.690, and 0.841) for *T*, with calibrated parameters of 0.55 d^{-1}, (4.84 d^{-1}–80.65 d^{-1}), 10 g O_2 $m^{-2}d^{-1}$, 0 m d^{-1}, and (1.49 d^{-1}–15.42 d^{-1}) for K_{bod}, $K_{aeration}$, K_{sod}, V_{sBOD}, and K_{dec}, respectively.

The most polluted area is located around the Panamericana south bridge, the critical month is July, with a flow of 1.2 m^{-3} s^{-1}, and values of 1.12 mg L^{-1}, 372.69 mg L^{-1}, and 1.2×10^5 NMP/100 mL for *DO*, BOD_5, and *E. Coli*, respectively. This does not comply with the ECA category 3 due to (i) diffuse contamination in the Pachacamac district, and (ii) inadequate operation of the San Bartolo WWTP, which fails to comply with LMP on all evaluated dates and for all substances, with values of 0.44 mg L^{-1}, 469.95 mg L^{-1}, and 1.4×10^5 NMP/100 mL for *DO*, BOD_5 and *E. Coli*, respectively. Diffuse contamination was estimated in Pachacamac with flows that go from 0.001 to 0.0037m^{-3} s^{-1} m^{-1}, and values of 0 mg L^{-1}, 888.37 mg L^{-1} and 2.2×10^5 NMP/100 mL, for *DO*, BOD_5 and *E. coli*, respectively.

It is proposed to recover the river by optimizing the San Bartolo WWTP and a new WWTP in Pachacámac to avoid diffuse contamination, with discharge flows for the critical month of July, of 0.980 m^3s^{-1} and 0.373 m^3s^{-1}, respectively, with concentrations that meet the ECA category 3: 4 mg L^{-1}, 15 mg L^{-1}, and 1000 NMP/100 mL for *DO*, BOD_5 and *E. coli*, respectively, with a flow of 0 m^3s^{-1} before diffuse pollution and a flow of 0.209 m^3s^{-1}, 50 m before the San Bartolo WWTP.

Supplementary Materials: The following are available online at https://www.mdpi.com/article/10.3390/hydrology10040084/s1.

Author Contributions: Conceptualization, E.P.-V. and L.R.-F.; methodology, E.P.-V., L.R.-F., O.L.M.-S. and W.E.L.-C.; validation, E.P.-V., L.R.-F., L.F.d.P. and O.L.M.-S.; investigation, E.P.-V., L.R.-F., O.L.M.-S. and L.F.d.P.; resources, L.R.-F., L.F.d.P. and W.E.L.-C.; data curation, E.P.-V., L.R.-F. and O.L.M.-S.; writing—original draft preparation, O.L.M.-S., L.R.-F. and W.E.L.-C.; writing—review and editing, L.R.-F., E.P.-V. and L.F.d.P.; supervision, E.P.-V., L.R.-F. and L.F.d.P. All authors have read and agreed to the published version of the manuscript.

Funding: This research was funded by The National Fund for Scientific and Technological Development (FONDECYT) of CONCYTEC-Peru, under the project "Surface and underground water resource management system for the safe use of water in climate change scenarios in the Lurín river basin" (Project No. 157-2017-FONDECYT-Peru).

Institutional Review Board Statement: Not applicable.

Informed Consent Statement: Not applicable.

Data Availability Statement: The data presented in this study are available on request from the corresponding author.

Acknowledgments: This paper was written as a result of collaboration that was initiated in the Project No. 157-2017-FONDECYT-Peru.

Conflicts of Interest: The authors declare no conflict of interest.

References

1. Osorio-Diaz, K.; Ramos-Fernández, L.; Velásquez-Bejarano, T. Projection of the impacts of climate change on the flow of the Lurín river basin-Peru, under CMIP5-RCP scenarios. *Idesia* **2022**, *40*, 101–114.
2. Momblanch, A.; Paredes-Arquiola, J.; Andreu, J.; Ramos-Fernández, L.; Baldeón, W.; Garcia Hernandez, J. Analysis of measures to improve water quality in the lower section of the Lurín River (Peru). In Proceedings of the IV Conference on Water Engineering, Córdoba, España, 21–22 October 2015; Volume B29. Available online: https://riunet.upv.es/handle/10251/138908 (accessed on 23 February 2020).

3. MINAM Peru. Environmental Quality Standard (ECA) for water. ECAs. D.S. 04-2017-MINAM. 2017. Available online: http://www.minam.gob.pe/wp-content/uploads/2017/06/DS-004-2017-MINAM.pdf (accessed on 6 August 2020).
4. MINAM Peru. Maximum Permissible Limits (LMP) for effluents. LMP. D. S. 03-2010-MINAM. 2010. Available online: https://sinia.minam.gob.pe/normas/limites-maximos-permisibles-lmp-efluentes-plantas-tratamiento-aguas) (accessed on 6 August 2020).
5. Cea, L.; Bermúdez, M.; Puertas, J.; Bladé, E.; Corestein, G.; Escolano, E.; Conde, A.; Bockelmann-Evans, B.; Ahmadian, R. IberWQ: New simulation tool for 2D water quality modelling in rivers and shallow estuaries. *J. Hydroinform.* **2016**, *18*, 816–830. [CrossRef]
6. Bladé, E.; Cea, L.; Corestein, G.; Escolano, E.; Puertas, J.; Vázquez-Cendón, E.; Dolz, J.; Coll, A. Iber: River modelling simulation tool. *Revista Internacional de Métodos Numéricos para Cálculo y Diseño en Ingeniería* **2012**, *30*, 1–10. [CrossRef]
7. Cea, L.; Puertas, J.; Vázquez-Cendón, M.-E. Depth Averaged Modelling of Turbulent Shallow Water Flow with Wet-Dry Fronts. *Arch. Comput. Methods Eng.* **2007**, *14*, 303–341. [CrossRef]
8. Cea, L.; Bermúdez, M.; Puertas, J. Two-dimensional water quality model for shallow flows in free sheet. Application to the mouth of the Ulla River. In Proceedings of the I Conference on Water Engineering, Madrid, Spain, 20–24 April 2009; pp. 27–28.
9. Mancini, J.L. Numerical Estimates of Coliform Mortality Rates under Various Conditions. *Water Pollut. Control. Fed.* **1978**, *50*, 2477–2484. Available online: https://www.jstor.org/stable/25040179 (accessed on 29 December 2020).
10. Cea, L.; Bermúdez, M.; Puertas, J. Uncertainty and sensitivity analysis of a depth-averaged water quality model for evaluation of Escherichia Coli concentration in shallow estuaries. *Environ. Model. Softw.* **2011**, *26*, 1526–1539. [CrossRef]
11. Parkhill, K.L.; Gulliver, J.S. Effect of inorganic sediment on whole-stream productivity. *Hydrobiologia* **2002**, *472*, 5–17. [CrossRef]
12. Chow, V.; Maidment, D.; Mays, L. *Open Channel Hydraulics*; McGraw-Hill Publisher: New York, NY, USA, 2004.
13. Bodelón, O.G.; Bernués, M.; Baltanás, A.; Montes, C. *Conductivity and Salinity in the Aquatic Ecosystems of the Doñana National Park (SO, Spain)*; Limnética Spanish Association of Limnology: Madrid, Spain, 1994; Volume 10, pp. 27–31. Available online: https://repositorio.uam.es (accessed on 14 October 2020).
14. Holliday, C.P.; Rasmussen, T.C.; Miller, W.P. Establishing the Relationship between Turbidity and Total Suspended Sediment Concentration. In Proceedings of the Georgia Water Resources Conference, The University of Georgia, Athens, Georgia, 23–24 April 2003; Available online: http://hdl.handle.net/1853/48455 (accessed on 6 December 2020).
15. Brown, L.C.; Barnwell, T.O. *The Enhanced Stream Water Quality Models QUAL2E and QUAL2E-UNCAS*; University of California: Oakland, CA, USA, 1987.
16. Thomman, R.V. *System Analysis and Water Quality Management*; McGrawm-Hill: New York, NY, USA, 1972; Available online: https://indico.ictp.it/event/a02286/contribution/64/material/0/0.pdf (accessed on 7 April 2021).
17. Moriasi, D.N.; Arnold, J.G.; Van Liew, M.W.; Bingner, R.; Harmel, R.; Veith, T. *Model Evaluation Guidelines for Systematic Quantification of Accuracy in Watershed Simulations*; American Society of Agricultural and Biological Engineers: St. Joseph, MI, USA, 2007; Volume 50, pp. 885–900. Available online: http://elibrary.asabe.org/abstract.asp??JID=3&AID=23153&CID=t2007&v=50&i=3&T=1 (accessed on 6 February 2021).
18. Garcia-Feal, O.; Cea, L.; Gonzalez-Cao, J.; Domínguez, J.M.; Gomez-Gesteira, M. IberWQ: A GPU accelerated tool for 2D water quality modeling in rivers and estuaries. *Water* **2020**, *12*, 413. [CrossRef]
19. García-Feal, O.; González-Cao, J.; Gómez-Gesteira, M.; Cea, L.; Domínguez, J.M.; Formella, A. An accelerated tool for flood modelling based on Iber. *Water* **2018**, *10*, 1459. [CrossRef]

water

MDPI

Article

Understanding the Planform Complexity and Morphodynamic Properties of Brahmaputra River in Bangladesh: Protection and Exploitation of Riparian Areas

Shiblu Sarker [1,*], Tanni Sarker [2], Olkeba Tolessa Leta [1], Sarder Udoy Raihan [3], Imran Khan [4] and Nur Ahmed [1]

1 Bureau of Watershed Management and Modeling, St. Johns River Water Management District, Palatka, FL 32177, USA
2 School of Planning, Design and Construction, Michigan State University, East Lansing, MI 48824, USA
3 Bangladesh Water Development Board, Dhaka 1215, Bangladesh
4 Institute of Water Modelling (IWM), Dhaka 1230, Bangladesh
* Correspondence: shiblu.buet@gmail.com

Abstract: The Brahmaputra River (BR) is a heavily braided river, due to various intricate paths, high discharge variability and bank erodibility, as well as multi-channel features, which, in turn, cause huge energy dissipation. The river also experiences anastomosing planform changes in response to seasonal water and sediment waves, resulting in a morphology with extreme complexity. The purpose of this study was to provide detailed and quantitative insights into the properties of planform complexity and dynamics of channel patterns that can complement previous studies. This was achieved by investigating the applicability of the anastomosing classification on the Brahmaputra river's planform, and computing disorder/unpredictability and complexity of fluctuations using the notion of entropy and uniformity of energy conversion rate by the channels, by means of a power spectral density approach. In addition, we also evaluated their correlation with discharge as a dynamic imprint of river systems on alluvial landscapes, in order to test the hypothesis that river flow may be responsible for the development of anastomosing planforms. The analysis suggests that higher discharge values could lead to less complex planform and less fluctuations on the alluvial landscape, as compared to lower discharge values. The proposed framework has significant potential to assist in understanding the response of complex alluvial planform under flow dynamics for the BR and other similar systems.

Keywords: anastomosing; Fourier transform; power spectral density; sample entropy; approximate entropy

1. Introduction

The Brahmaputra River (BR) is often characterized as a braided river due to its complex networking of channels, branches and bars, including high sediment loading, and significant variability in discharge and gradients [1–3]. The high variability in discharge and sediment loading from the Brahmaputra River Basin (BRB) are the main factors responsible for significant erosion–deposition processes [4–8], which also initiate the complex network along with bar dynamics. The morphological processes, such as soil erosion, deposition, channel movement and irregular planform structure, combined with significant stream power variability are critical for understanding such kinds of river systems [9–12]. As such processes frequently occur in braided river systems, they cannot be measured over a short period of time. Several previous studies have been conducted to better understand the planform process of braided rivers, that include the following: flume experiments [13–16], numerical modeling [17,18], satellite-based remote sensing products [19–24], and modern technology, such as digital photogrammetry and laser altimetry [25–28]. However, these previous studies have lacked detail analysis of morphodynamics and planform complexity of channels especially for braided river systems.

Since the BR is one of the world's most braided and sand-bedded river systems, and is ranked fifth in terms of its annual flow, it is an ideal river system for studying river planform complexity and randomness [29–33]. In addition, the BR system experiences a unique characteristic in terms of inter-seasonal variability of flow–sediment loading and morphological processes [32,34–36]. Furthermore, a series of large floods and major tectonic activity in the BR system have resulted in a complex morphodynamic environment [32,37,38]. The banks of the BR have been heavily eroded, including frequent changes in the channel courses. Such morphological dynamics and processes have tremendous impacts on the functioning of the riverine ecosystem and on the approximately 30 million people that live along the river banks [39,40]. Therefore, detailed analysis of the morphodynamics of the BR system, such as identifying the stable and unstable parts of the river's reaches, is of interest to enhance our understanding and so as to implement appropriate and sustainable mitigation measures for the system.

As previously discussed, the high variability of discharge and sediment loading from the BRB to the BR system are most likely responsible for significant erosion–deposition processes [4–7], which also initiate formation of complex network systems and dynamics along riverine and alluvial areas. Due to the high variability of discharge and sediment loading, and complex morphodynamics of the BR system, performing an analysis of the spatio-temporal variability of the planform and dynamic forcings are of interest for a better understanding of the BR system. In addition, it is also critical to investigate the effect of bar dynamics on the morphological reorganization combined with planform complexity. Our objective was to generate generic insights into the properties of planform complexity and dynamics of the channel patterns that complement previous studies. We address this by computing the planform complexity and randomness using complex network metrics and a theoretical approach that shed some light on understanding the dynamic behavior of the river and help to minimize the gap between previous studies on braided rivers.

2. Study Area

The Brahmaputra River Basin (BRB) outspreads in Bangladesh (5.47%), India (36.11%), Nepal (0.13%), Bhutan (7.20%), Tibet and China (51.08%), as shown in Figure 1. Geographically, the basin is situated between 82°1′ E and 97°46′ E, and 22°27′ N and 31°27′ N. Bounded by the Himalayas on the northwest, Tibetan plateau on the north, Ganges basin on the west, Meghna basin on the south, and Meghalaya subtropical forests on the south-east, the BRB drains an area of approximately 543,462 km^2 [41]. In India's territory, the basin stretches over the states of Arunachal, Sikkim, Assam, West Bengal, Nagaland, Meghalaya and Manipur, of which Arunachal represents the highest distribution of the drainage area (41.24%), followed by Assam (37.07%).

The BRB is characterized by an irregular shape with a maximum east–west length of 1540 km and a maximum north–south width of 835 km [41]. The BR originates from Kailash (the great glacier mass of Chema-Yung-Dung) of the Himalayas, at an elevation of about 5150 m, and flows for about 2900 km through Tibet, China, India and Bangladesh, eventually confluencing with the Ganges River. In China, the BR is known as the Yarlung Tsangpo, that flows east at an average elevation of 4000 m above MSL. Flowing eastward for a length of about 1130 km, the Yarlung Tsango River bends around Mt. Namcha Barwa, thereby forming the Yarlung Tsangpo Canyon, the deepest in the world. The river then turns towards the south, entering the Arunachal Pradesh State of India, wherein it flows for about 35 km and joins the Dibang and Lohit Rivers. From this confluence, the river becomes very wide and is referred as the Brahmaputra River, where the average slope of the river drastically reduces to 0.1 m/km compared to 2.82 m/km in Tibet, China [42] A few more tributaries join the main course of the river later, both on the left and right banks, before the river enters Bangladesh. At the border of Bangladesh, the river curves to the south and south–east directions. The branch curving towards the south continues as the

Jamuna and flows for a length of about 240 km to its confluence with the Lower Ganges at Aricha, locally called Padma. The Brahmaputra River has an average annual flood peak flow of 60,000 m^3/s at Bahadurabad of Bangladesh (around 150 km upstream of the confluence), with the monsoon flood typically occurring between July and August. When the bank of the BR is full, the discharge volume is reduced to approximately 44,000 m^3/s. The lowest discharge, approximately 5000 m^3/s, occurs in January and February. The discharge is quite steady during January to February but steadily increases from March to June and then gradually declines from September to December [7,43,44].

This study was accomplished by evaluating a huge section of the Brahmaputra River from geographical and temporal points of view. Therefore, we limited our focus to the Brahmaputra River's reach within the Bangladesh territory, as shown in Figure 2.

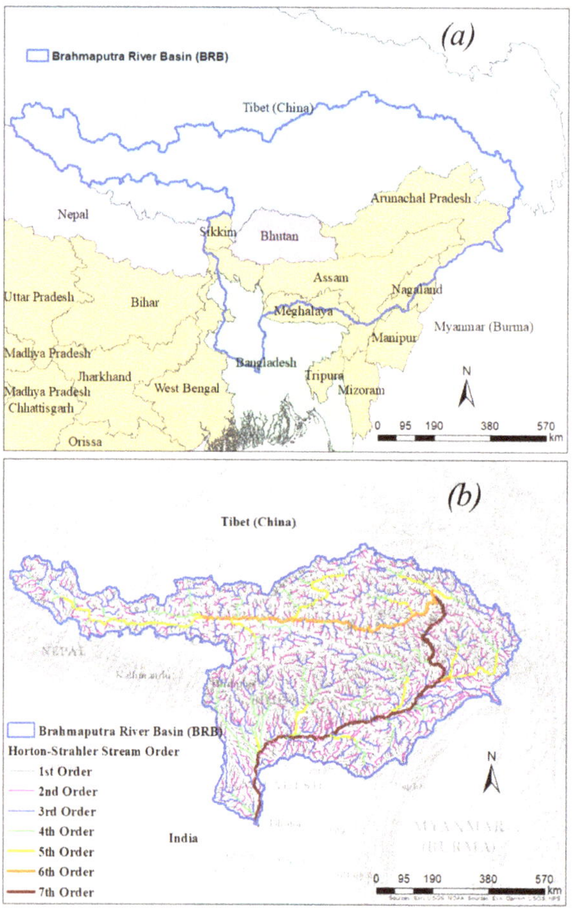

Figure 1. (a) The Brahmaputra River Basin (BRB) extent and (b) one-dimensional river network extracted from [4,6,8] for the BRB.

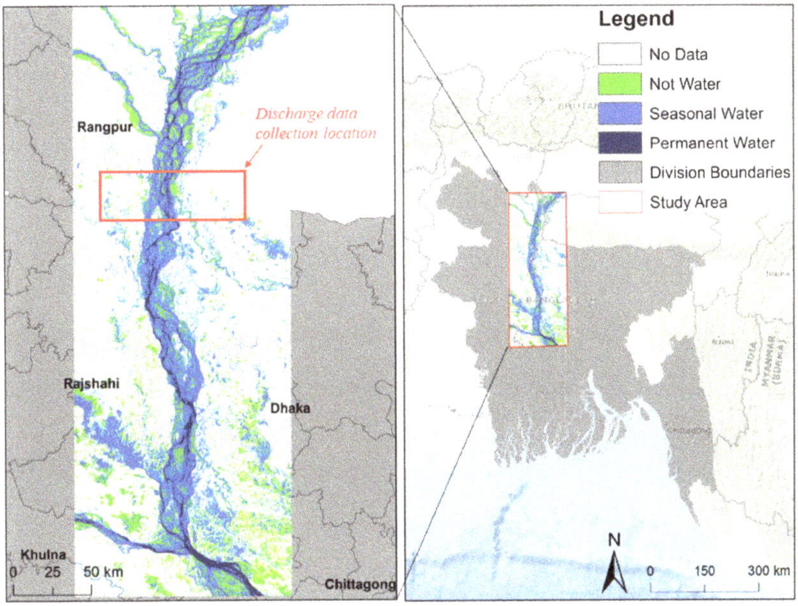

Figure 2. Brahmaputra River study region. The location of discharge data collection is depicted in red.

3. Methods

3.1. The Anastomosing River Principle

Anastomosing rivers usually occur in alluvial plains, where low energy and dense channel vegetation conditions persist, including floodplain geomorphology and structure. Based on the pattern of the channels, anastomosing rivers can be classed as having straight, meandering, and braided shapes. Avulsions or structures that divert flow and create a new channel pattern in the floodplain are commonly employed to build anastomosing rivers [3]. Such systems experience simultaneous multiple floodplain channels erosion, especially when bypasses are developed and older channel belt segments remain active for an extended period of time. The first type of anastomosis affects the entire floodplain, whereas the second type only affects a portion of it. Protracted anastomosis is generally caused by channel belt aggradation and/or channel capacity degradation as a result of in-channel deposition, both of which are facilitated by a low floodplain gradient [3]. There are numerous other reasons, including climate-related factors, such as catastrophic flood events, in-channel aeolian dunes or rapid base level rises. Based on the available information about the BR, the river can be considered an anastomosing river, which provides an ideal setting for hypothesizing its planform and characterizing its anastomosing nature [5]. Therefore, applying complex network theory is crucial to gain better understanding of the physical processes occurring in the alluvial landscapes of the river. In this study, the Brahmaputra River's high complexity permits us to hypothesize in regard to the meandering and braided categories, and, particularly, in regard to the anastomosing category of the river.

3.2. Channel Network Delineation

Several scholars have investigated the Brahmaputra River in detail over a long period of time. For this study, the data acquisition and the processing to delineate the channel network of the *BR* were performed using the online code editor of the Google Earth Engine (GEE). GEE provides free access to cloud-based large-scale geospatial data sets and analysis [45]. We used image collections or data stack, all available from the USGS Landsat surface reflectance products (TM, ETM+, OLI) and acquired during the dry season. We used yearly data for the period from 1987 to 2020. The dry periods were selected between 1 October and 30 December for each year. *GEE* was also used to intersect all the dry images within the study area and to combine them per year, creating a cloud-free yearly composite scene, as shown in Figure 3. We deployed ArcGIS to delineate the channel network of the BR, based on the annual seasonal and permanent water pathways.

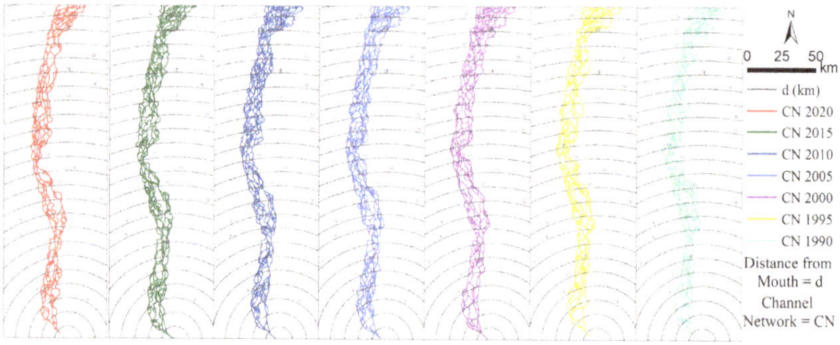

Figure 3. Delineation of the channel network for seven selected years from 1987 to 2020, based on the dry season of the Brahmaputra River.

3.3. Anastomosing Function

In this study, we proposed a series, called the Anastomosing Function (AF), to capture the one-dimensional spatial arrangement of the two-dimensional complex network planform of the Brahmaputra River. The River Network Width Function ($RNWF$) is commonly used by geomorphologists to characterize a river basin [4,46]. A $RNWF$ is a one-dimensional function that summarizes the river network's two-dimensional branching structure [47]. Moreover, it displays the distribution of travel distances within the network, as well as the probability distribution of travel duration under the assumption of constant flow velocity [47]. While $RNWF$ represents the number of channelized pixels or number of crossed channels that have the same distance from the basin outlet, the distances are measured along the flow path [4,46]. We developed the concept of AF, based on the similar notion of the $RNWF$ method previously used in many studies (see details in [4,46]). Similar to the $RNWF$ method, the AF computes the number of crossed channels as a function of the distance from the mouth of the Brahmaputra River, but not from the basin outlet (see details in [5]). Furthermore, since measuring distance along the flow path (longest channel) of braided rivers is very challenging, we, instead, proposed and adopted a radial distance method, with a constant interval from the mouth of the river, to generate AF. We applied this from the mouth of the Brahmaputra River where it meets the Ganges River. Mathematically, AF can be expressed as:

$$AF(d) = \#[Channelized\ I : d \leq R(I) \leq d + \delta d] \quad (1)$$

where $R(I)$ is the flow distance of channel intersection I from the mouth and δd is the scale of refinement. Usually, the distance d is normalized by R and $AF(d)$ is normalized by the total number of channel intersections rendered by their density. For a given anastomosing network topology, $AF(d)$ can be viewed as a stochastic process indexed by the distance d (similar to width function [47]). The concept of distance d was normalized by R for the seven selected years from 1990 to 2020, as depicted in Figure 3.

3.4. Discharge Data

We collected yearly maximum discharge data of the BR for the period from 1987 to 2020 at the Bahadurabad gauging station of Bangladesh, from the Bangladesh Water Development Board ($BWDB$), which serves as the national hydrological data provider of

Bangladesh (https://bwdb.portal.gov.bd/, accessed on 3 November 2021). The collected data is shown in Figure 4. The discharge data in this study was used as a dynamic imprint of the river to understand its relationship with planform complexity and fluctuation. Specifically, we tested our hypothesis using yearly maximum, mean wet, and mean dry discharge values as dynamic variables (see Figure 4).

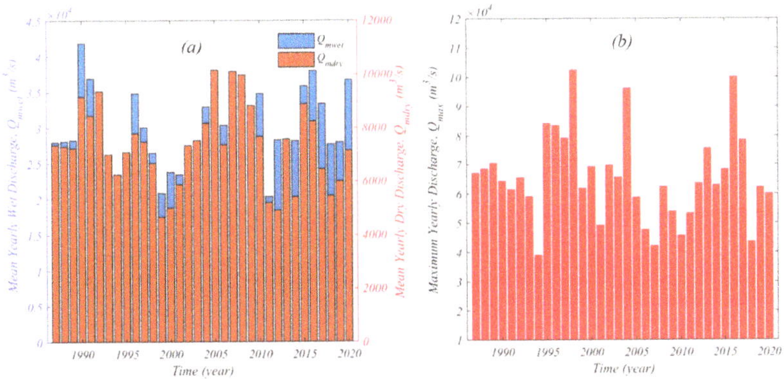

Figure 4. (**a**) Extracted mean yearly wet and dry discharge data and (**b**) Yearly maximum discharge data of the BR from 1987 to 2020.

3.5. Entropy

The entropy of a data series is a measure of its unpredictability. When moment statistics, such as mean and variance, are unable to distinguish between series, entropy is commonly utilized. Entropy measures the amount of information contained in a signal based on the probability of each signal value. In other words, entropy quantifies the degree of uncertainty associated with the occurrence of events across a space or time domain [48]. It can be expressed as:

$$En = -\sum_{i=1}^{N} p[af(i)] \log p[af(i)] \tag{2}$$

where $p[af(i)]$ is the probability of $af(i)$. The value $af(i)$ denotes each data value and N is the sample size of the corresponding signal represented by a vector $S = af(1), af(2), \ldots, af(N)$.

Approximate Entropy and Sample Entropy are two other algorithms commonly used to determine the regularity of data series, based on the appearance of patterns [49].

3.5.1. Approximate Entropy

Approximate Entropy ($ApEn$) is a form of Shannon Entropy and its calculation requires a large amount of data in a series. Steve M. Pincus created a statistical method to overcome the limitations of moment statistics by modifying an exact regularity statistic [50,51]. It was initially developed for the analysis of medical time series data, but its applications have since been expanded to other fields [4,50–52]. In this study, we were interested to compute $ApEn$ as a measure of disorder of a data series: for example, a data series S containing N data values, $S = af(1), af(2), \ldots af(N)$. From this data, a series of vectors can be constructed as:

$$AF(1) = af(1),\ af(2),\ \ldots\ af(m) \tag{3a}$$

$$AF(2) = af(2),\ af(3),\ \ldots\ af(m+1) \tag{3b}$$

$$\ldots\ldots\ldots\ldots\ldots\ldots\ldots\ldots \tag{3c}$$

$$AF(N-m+1) = af(N-m+1),\ af(N-m+2),\ \ldots\ af(N) \tag{3d}$$

The distance between two vectors $AF(i)$ and $AF(j)$ can be defined as the maximum difference in their respective corresponding elements.

$$\mathfrak{D}[AF(i), AF(j)] = \max_{k=1,2\ldots m}(|AF(i+k-1), AF(j+k-1)|) \tag{4}$$

where $i = 1,2,\ldots,N-m+1$ and $j = 1,2,\ldots,N-m+1$ and N are the numbers of data points in the series. For each vector $AF(i)$, a measure that describes the similarity between the vector $AF(i)$ and all other vectors $AF(j)$ $j = 1,2,\ldots,N-m+1$, $j \neq i$ can be constructed as:

$$C_i^m(r) = \frac{1}{(N-m+1)}\sum_{j\neq i}\theta(r - \mathfrak{D}[AF(i), AF(j)]) \tag{5}$$

where

$$\theta(af) = \begin{cases} 1, & af \geq 0 \\ 0, & af < 0 \end{cases} \tag{6}$$

In our context, the value of θ was 1. The symbol r specifies a filtering level and is related to the standard deviation of the series. Finally, $ApEn$ can be calculated by the following equation:

$$ApEn(m,r) = \varnothing^m(r) - \varnothing^{m+1}(r) \tag{7}$$

where

$$\varnothing^m(r) = \frac{1}{(N-m+1)}\sum_i In[C_i^m(r)] \tag{8}$$

The application of Approximate Entropy ($ApEn$) on the $AF(d)$ data is shown in the following flow chart (Figure 5).

3.5.2. Sample Entropy

Sample Entropy ($SampEn$) is a modified form of Shannon entropy that is used to evaluate the complexity of physical time series signals and physical states. While $SampEn$ is a measure of complexity, which is similar to approximate entropy, it excludes self-similar patterns [51,53].

Both $ApEn$ and $SampEn$ algorithms are based on the conditional probabilities (see details in [54]), and the first two steps (3) and (4) are similar to $ApEn$, as shown in Figure 5. After the second step, $SampEn$ is calculated for each template vector using Equation (9):

$$B_i^m(r) = \frac{1}{(N-m-1)}\sum_{j=1, j\neq i}^{N-m} \text{number of times that } \mathfrak{D}[AF(j) - AF(i)] < r \tag{9}$$

Then, summing all template vectors can be written as Equation (10):

$$B^m(r) = \frac{1}{(N-m)}\sum_{i=1}^{N-m} B_i^m(r) \tag{10}$$

Similarly, we can calculate each template vector using Equation (11):

$$A_i^m(r) = \frac{1}{(N-m-1)}\sum_{j=1, j\neq i}^{N-m} \text{number of times that } \mathfrak{D}[(AF+1)(j) - (AF+1)(i)] < r \tag{11}$$

and summing all template vectors can be calculated using Equation (12):

$$A^m(r) = \frac{1}{(N-m)}\sum_{i=1}^{N-m} A_i^m(r) \tag{12}$$

Finally, $SampEn$ can be calculated using Equation (13):

$$SampEn(m,r,N) = -log\left[\frac{A^m(r)}{B^m(r)}\right] \tag{13}$$

In this study, the values of m, r, etc. were determined based on the multi-scale phenomena of time series confirmed by previous research [4,55].

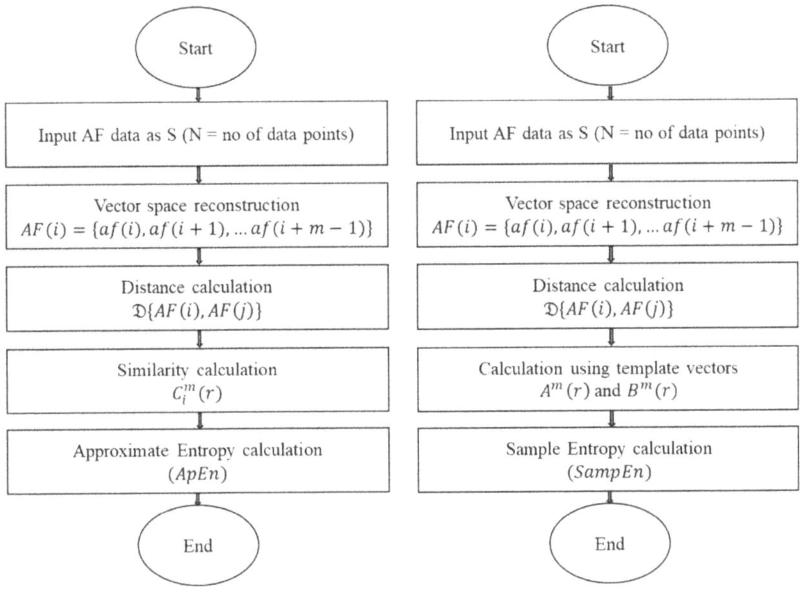

Figure 5. Details of the algorithms to compute Approximate Entropy ($ApEn$) and Sample Entropy ($SampEn$) on AF series.

3.6. Power Spectral Density

The Power Spectral Density (PSD) is a measurement of the signal's intensity or amplitude's frequency response. In general, it provides a standardized method for describing the distribution of energy in a signal across different frequencies. The PSD of $AF(k)$ as a discrete signal $AF(d)$ can be computed as the average magnitude of the Fourier transform squared [4,56], over a time interval and expressed as Equation (14):

$$PSD_{AF(k)} = \left|\frac{1}{2\pi}\sum_{d_1}^{d_2} AF(d)e^{-ikd}\right|^2 = \frac{\widehat{AF}(k)\widehat{AF}_*(k)}{2\pi} \tag{14}$$

where $\widehat{AF}(k)$ is the discrete Fourier transform of $g(d)$ and $\widehat{AF}_*(k)$ is its complex conjugate, and k is the wave number [4,57–59]. We analyzed this PSD in the power–law domain across the spatial frequency or wave number k as the Equation (15).

$$PSD_{AF(k)} \sim \frac{1}{k^\beta} \tag{15}$$

where β is the power–law exponent of the PSD. We referred to this β as the proxy of planform fluctuations of AF, which was computed using the slope of the linear regression fitted to the estimated PSD and plotted on log–log scales [4,60]. It is important to note that the entropy described in earlier sections was used to calculate complexity, while power spectral density was used to characterize planform fluctuations of the BR. In other words, the uniformity of discharge induced by the energy conversion rate of the planform can be measured by the β.

4. Results and Discussion

4.1. Extracted AF and Corresponding PSD of the Brahmaputra River's Planform

AF was generated using the method described in the method sections. We extracted AF, based on the dry season BR planform for each year between 1987 and 2020. Figure 6a illustrates the AF for the seven selected years between 1990 and 2020. The generated AF recreates the dynamic properties commonly observed in the BR's planform.

The PSD of AF was also generated using the procedure outlined in the method section. Figure 6b illustrates the PSD corresponding to the AF depicted in Figure 6a. The best-fitted slope was calculated and used as a measure of fluctuation to be analyzed in the subsequent section.

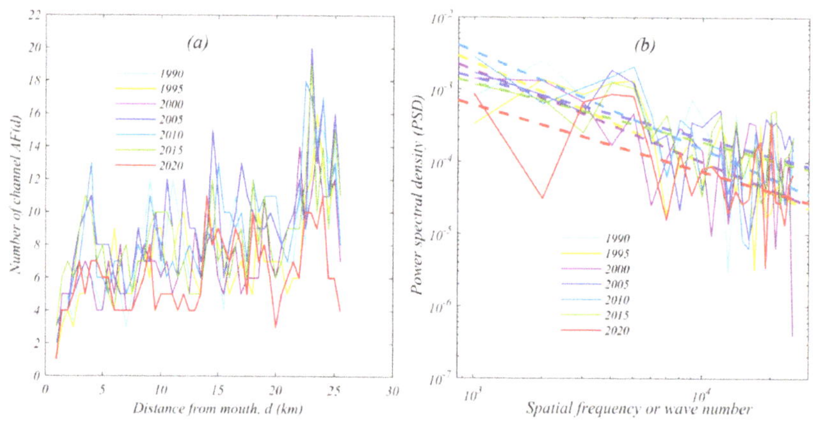

Figure 6. (**a**) Extracted AF for the BR for the seven selected years and (**b**) corresponding PSD of AF plotted on a log–log scale.

4.2. Disorder, Complexity and Fluctuation of the Brahmaputra River's Planform

As discussed in an earlier section, $ApEn$ and $SampEn$ represent the disorder and the complexity of the data series, respectively. Consequently, we used $ApEn$ and $SampEn$ to characterize the anastomosing river disorder and complexity nature of the BR. Figure 7a illustrates the BR's yearly value of $ApEn$ and $SampEn$ for the period 1987 to 2020. In addition to having a similar pattern, $ApEn$ and $SampEn$ were significantly linearly correlated with $R^2 \sim 0.17$, and p-value ≤ 0.05 at a 95% confidence interval. Consequently, we may conclude that the possibility of using AF to characterize the BR is enhanced as the complexity of channel patterns grows with disorder or unpredictability. Based on this argument, we expect that a correlation with river dynamic features may exist. To evaluate the dynamic imprint on river planform disorder and complexity, we further investigated the correlation between $ApEn$ and $SampEn$ to the discharge presented in the subsequent section.

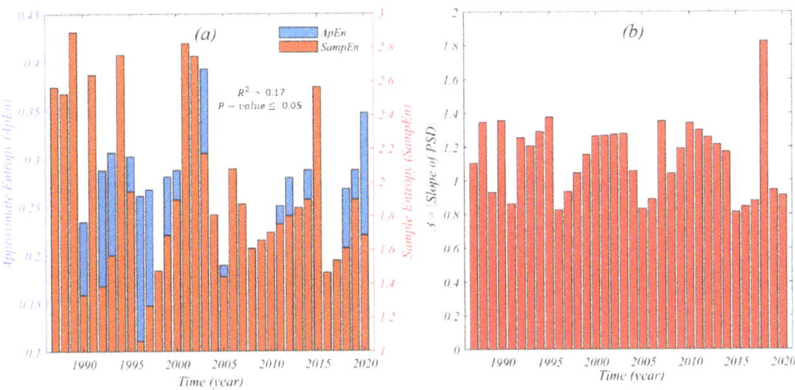

Figure 7. (**a**) Computed Approximate Entropy ($ApEn$) and Sample Entropy ($SampEn$) on AF series in a bar plot and (**b**) the corresponding β calculated by fitting the slope to the estimated PSD of AF series plotted on a log–log scale.

In addition to disorder and complexity, the uniformity of energy transfer rate was determined by calculating PSD of AF through $beta$. Figure 7b presents the yearly value of $beta$ for the period from 1987 to 2020. This can be used to describe the nature of the fluctuations of the BR's planform.

4.3. Association between River Discharge and Disorder, Complexity, and Fluctuation

Figure 8a–c exhibits the correlation between $ApEn$ with yearly maximum discharge (Q_{max}), mean yearly wet discharge (Q_{mwet}) and mean yearly dry discharge (Q_{mdry}). It was observed that the value of $ApEn$ increased as the Q_{max}, Q_{mwet} and Q_{mdry} decreased, as shown in Figure 8a–c. It was also observed that Q_{mwet} and Q_{mdry} contributed more to the formation of planform than Q_{max}. Hydraulically, higher discharge transports more sediment from the bed, and potentially widens the main channel [61], thereby reducing the properties of the anastomosing river planform and its complexity. On the other hand, reduced discharge facilitates more sedimentation in the river and formation of bars. The latter process eventually results in oblique flow phenomena that cause a complex network along the riverine landscape and increase complexity. Apart from physical intuition, the correlations between $ApEn$ and discharge values were found to be consistent with the value of $R^2 \sim 0.1$. Although the R^2 value was low, the t-test indicated the presence of significant correlation between $ApEn$ and discharge values within the 95% confidence interval (i.e., p-value $\leq\sim 0.05$).

Weak correlation was also noticed between observed discharge and $SampEn$ (see Figure 9a–c). However, the t-test indicated significant correlation existed within the 90% confidence interval (i.e., p-value $\leq\sim 0.1$) for $SampEn$ and the observed discharge. In addition, $SampEn$ was a fairly significantly complex metric in understanding the BR's planform. Therefore, we may conclude that $ApEn$ is a significant and consistent metric to measure the disorder properties of the BR's planform.

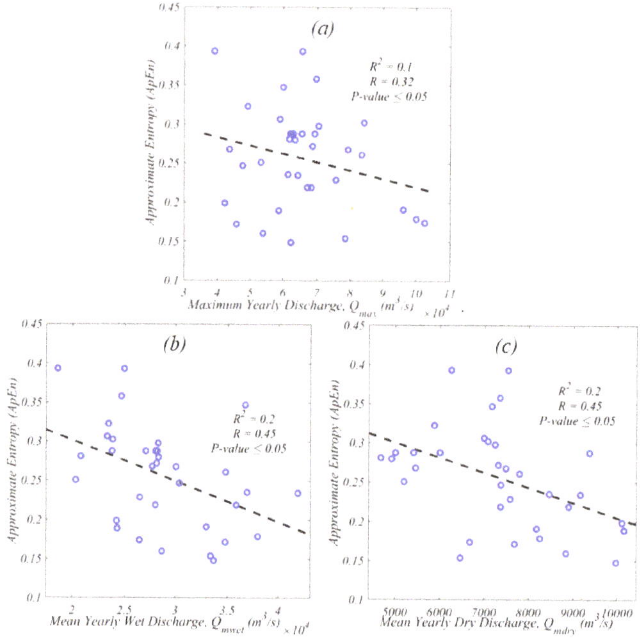

Figure 8. Correlation between Approximate entropy ($ApEn$) and (**a**) Yearly maximum discharge (Q_{max}), (**b**) Mean yearly wet discharge (Q_{mwet}) and (**c**) Mean yearly dry discharge (Q_{mdry}).

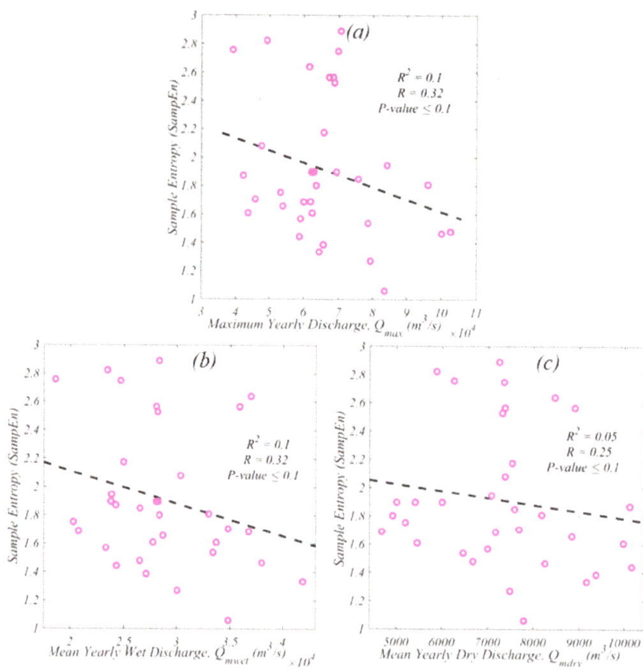

Figure 9. Correlation between Sample Entropy ($SampEn$) and (**a**) Yearly maximum discharge (Q_{max}), (**b**) Mean yearly wet discharge (Q_{mwet}) and (**c**) Mean yearly dry discharge (Q_{mdry}).

Figures 10a–c support our hypothesis that higher discharge results in less fluctuation on the BR's planform and vice versa. In addition, Q_{mwet} was more responsible for the fluctuation of the planform than Q_{mdry}. Hence, the BR's planform distributed energy at a non-uniform rate in the case of Q_{mwet}, and vice versa. In other words, the absolute value of the fitted slope of the *PSD* of *AF*, plotted on log–log scales, could also reflect fluctuation of the BR's planform, which was consistent with our disorder and complexity results.

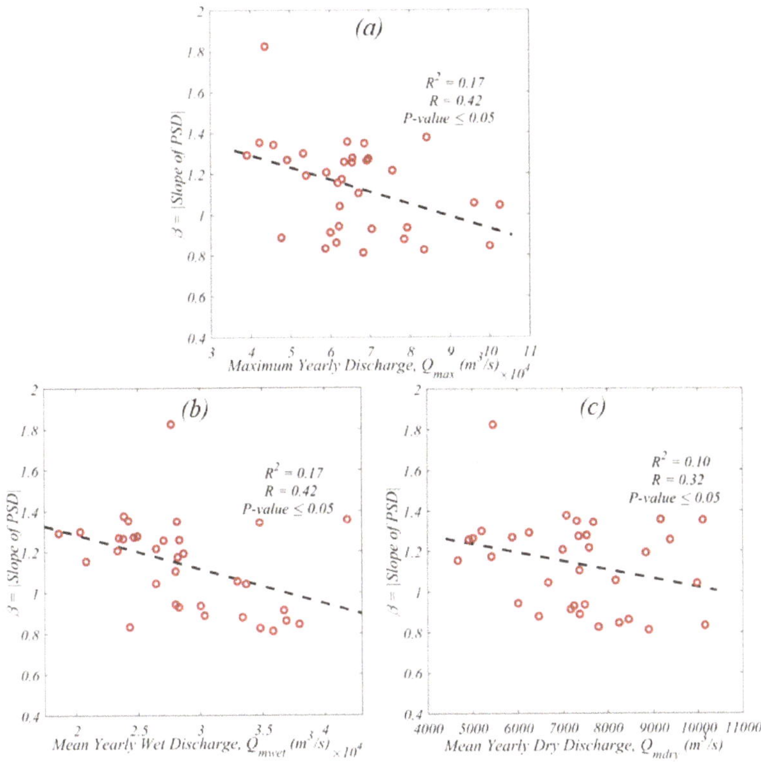

Figure 10. Correlation between β and (**a**) Yearly maximum discharge (Q_{max}), (**b**) Mean yearly wet discharge (Q_{mwet}) and (**c**) Mean yearly dry discharge (Q_{mdry}).

5. Potential Implications Towards Morphological Contexts

Although Leopold and Wolman (1957) proposed a three-part classification of channel planforms (straight, meandering, and braided) [1], there have been a great deal of physical processes responsible for channel changes and the classification of channel planforms. In addition, numerous channel patterns do not fit perfectly within these three categories of classification. In fact, there is higher complexity between meandering and braided categories, as well as within braided categories. As more research is conducted on river systems, new planform types, such as anastomosing, are being identified [62,63].

In this study, we investigated the channel network features of the BR. The *AF* was considered here as the one-dimensional signal of the two-dimensional channel pattern of the BR, meaning that every change in a channel had the potential to reshape *AF*. This planform network's dynamics was influenced, in part, by the evolution of individual bifurcations. We demonstrated the existence of at least one planform property with relevance to fluvial morphodynamics. This innovative characteristic could aid in describing flow curvature and bar dynamics. Hence, it is applicable to different planform networks in the geomorphology of rivers. Our findings indicate that defining a braided river as a network has the potential to enhance our knowledge of morphodynamics and results in a novel measure of complexity that recognizes the role of a channel as part of the entire planform network system.

6. Conclusions

In this study, a method of representing two-dimensional river planform as a one-dimensional signal is developed, which can be used as an objective metric to quantify planform disorder and complexity and to characterize and explore the anastomosing and planform nature of the Brahmaputra River. We achieved this by implementing a mathematical function, called the Anastomosing Function (AF). Additionally, we investigated the concept of entropy along with Power Spectral Density (PSD) to quantify the disorder, complexity, and fluctuation of the BR's planforms. The major findings can be summarized as follows:

- The generated and investigated AF is capable of accurately transforming a two-dimensional complex network into a one-dimensional spatial signal.
- The Approximate Entropy ($ApEn$) and Sample Entropy ($SampEn$) can be used to quantify the disorder and complexity of river's planforms, respectively, which confirms the reproducibility of the physical features of the river.
- Dynamic imprints, such as yearly maximum discharge (Q_{max}), have significant contributions to the river's planform complexity.
- Q_{max} also showed a significant and consistent contribution to the Brahmaputra River's planform fluctuation.

Overall, our findings reveal the potential use of AF, along with the concepts of entropy and PSD, to characterize a river under varying geomorphic and climatic conditions. The developed method could be used to quantify the climatic influence (i.e., change of discharge or discharge under extreme events [64]) on planform unpredictability and complexity of the BR. This could be an advantageous tool for engineers and urban planners in implementing sustainable urban development and management around the banks of braided and complex rives like the BR.

7. Limitations and Recommendations

This study aimed to understand the complexity of the Brahmaputra River (BR) system by utilizing and implementing information theory methods along with the available observed discharge data as one variable to explain the complexity and planform characteristics of the river. The study particularly used discharge data to understand if the response to discharge can be used as a proxy to extreme events on planform complexity. We found that discharge could signal the characteristics of the system and be used as a proxy in understanding planform complexity and geomorphic characteristics of the BR. The river's landscape, which has been experiencing intensive dynamics, the highest population densities and major economic interests, is indeed identified as being the most hydrographically complex and vulnerable area in the world.

Our study is very important to predict planform behavior for rivers that share similar characteristics with the BR. More importantly, our hypotheses were derived from satellite images and based on available discharge data (1987–2020) only. However, planform prediction is a complex phenomenon, requiring numerous hydrological variables and model-predicted outputs, especially to implement integrated water management strategies for braided rivers. While this study proposes a new framework for characterizing planform complexity and the use of observed/historical discharge in understanding the planform complexity of the BR, our method and findings should be used in conjunction with other field observations, hydrological variables, and numerical modeling for better water management and planning implementation. Additionally, although the correlation values between the methods and observed discharges were statistically significant, which led to the conclusion that discharge can be used as one proxy to describe the planform and complexity of the BR, the correlation values were low, which could be partially due to the use of short-term span satellite imagery and discharge data. Given the lack of additional very long-term data, including long-term observed discharge data, for the study area, we were not able to use additional data and introduce other indicators in characterizing the planform and complexity nature of the BR. Therefore, future work should focus on the use of additional indicators to provide further comprehensive characterization and understanding of the BR's planform and complexity.

Author Contributions: S.S. conducted the research, S.S. and T.S. wrote the initial draft, and analyzed the results. S.U.R. assisted in the data collection. O.T.L., N.A. and I.K. provided feedback on the interpretation of the results. Each author contributed to the manuscript's preparation. All authors have read and agreed to the published version of the manuscript.

Funding: This research received no external funding.

Conflicts of Interest: The authors declare no conflict of interest.

References

1. Leopold, L.B.; Wolman, M.G. *River Channel Patterns: Braided, Meandering, and Straight*; US Government Printing Office: Washington, DC, USA, 1957.
2. Charlton, R. *Fundamentals of Fluvial Geomorphology*; Routledge: London, UK; New York, NY, USA, 2007.
3. Makaske, B. Anastomosing rivers: A review of their classification, origin and sedimentary products. *Earth-Sci. Rev.* **2001**, *53*, 149–196. [CrossRef]
4. Sarker, S. *Investigating Topologic and Geometric Properties of Synthetic and Natural River Networks under Changing Climate*; University of Central Florida: Orlando, FL, USA, 2021.
5. Sarker, S. Understanding the Complexity and Dynamics of Anastomosing River Planform: A Case Study of Brahmaputra River in Bangladesh. *Earth Space Sci. Open Arch.* **2021**, 1. [CrossRef]
6. Sarker, S.; Veremyev, A.; Boginski, V.; Singh, A. Critical nodes in river networks. *Sci. Rep.* **2019**, *9*, 1–11. [CrossRef] [PubMed]
7. Khan, I.; Ahammad, M.; Sarker, S. A study on River Bank Erosion of Jamuna River using GIS and Remote Sensing Technology. *Int. J. Eng. Develop. Res.* **2014**, *2*, 3365–3371.
8. Gao, Y.; Sarker, S.; Sarker, T.; Leta, O.T. Analyzing the critical locations in response of constructed and planned dams on the Mekong River Basin for environmental integrity. *Environ. Res. Commun.* **2022**, *4*, 101001. [CrossRef]
9. Bridge, J.S. The interaction between channel geometry, water flow, sediment transport and deposition in braided rivers. *Geol. Soc.* **1993**, *75*, 13–71. [CrossRef]
10. Ferguson, R. Understanding braiding processes in gravel-bed rivers: Progress and unsolved problems. *Geol. Soc.* **1993**, *75*, 73–87. [CrossRef]
11. Klaassen, G.J.; Mosselman, E.; Bruehl, H. *On the Prediction of Planform Changes in Braided Sand-nd-Ed Rivers*; Delft Hydraulics: Delft, The Netherlands, 1993.
12. Pradhan, C.; Chembolu, V.; Bharti, R.; Dutta, S. Regulated rivers in India: Research progress and future directions. *ISH J. Hydraul. Eng.* **2021**, 1–13. [CrossRef]
13. Ashmore, P.E. Laboratory modelling of gravel braided stream morphology. *Earth Surf. Process. Land.* **1982**, *7*, 201–225. [CrossRef]
14. Ashmore, P.E. How do gravel-bed rivers braid? *Can. J. Earth Sci.* **1991**, *28*, 326–341. [CrossRef]
15. Ashmore, P. Anabranch confluence kinetics and sedimentation processes in gravel-braided streams. *Geol. Soc.* **1993**, *75*, 129–146. [CrossRef]
16. Young, W.; Davies, T. Bedload transport processes in a braided gravel-bed river model. *Earth Surf. Process. Land.* **1991**, *16*, 499–511. [CrossRef]
17. Murray, A.B.; Paola, C. A cellular model of braided rivers. *Nature* **1994**, *371*, 54–57. [CrossRef]
18. Murray, A.B.; Paola, C. Properties of a cellular braided-stream model. *Earth Surf. Process. Land.* **1997**, *22*, 1001–1025. [CrossRef]
19. Nykanen, D.K.; Foufoula-Georgiou, E.; Sapozhnikov, V.B. Study of spatial scaling in braided river patterns using synthetic aperture radar imagery. *Water Resour. Res.* **1998**, *34*, 1795–1807. [CrossRef]
20. Pradhan, C.; Bharti, R.; Dutta, S. Assessment of post-impoundment geomorphic variations along Brahmani River using remote sensing. In Proceedings of the 2017 IEEE International Geoscience and Remote Sensing Symposium (IGARSS), Fort Worth, TX, USA, 23–28 July 2017; pp. 5598–5601.
21. Pradhan, C.; Chembolu, V.; Dutta, S. Impact of river interventions on alluvial channel morphology. *ISH J. Hydraul. Eng.* **2019**, *25*, 87–93. [CrossRef]
22. Sapozhnikov, V.B.; Foufoula-Georgiou, E. Do the current landscape evolution models show self-organized criticality? *Water Resour. Res.* **1996**, *32*, 1109–1112. [CrossRef]
23. Sapozhnikov, V.B.; Foufoula-Georgiou, E. Experimental evidence of dynamic scaling and indications of self-organized criticality in braided rivers. *Water Resour. Res.* **1997**, *33*, 1983–1991. [CrossRef]
24. Walsh, J.; Hicks, D.M. Braided channels: Self-similar or self-affine? *Water Resour. Res.* **2002**, *38*, 1–6. [CrossRef]
25. Lane, S.N.; Westaway, R.M.; Murray Hicks, D. Estimation of erosion and deposition volumes in a large, gravel-bed, braided river using synoptic remote sensing. *Earth Surf. Process. Land.* **2003**, *28*, 249–271. [CrossRef]

26. Westaway, R.; Lane, S.; Hicks, D. The development of an automated correction procedure for digital photogrammetry for the study of wide, shallow, gravel-bed rivers. *Earth Surf. Process. Land.* **2000**, *25*, 209–226. [CrossRef]
27. Westaway, R.M.; Lane, S.N.; Hicks, D.M. Remote sensing of clear-water, shallow, gravel-bed rivers using digital photogrammetry. *Photogramm. Eng. Remote Sens.* **2001**, *67*, 1271–1282.
28. Westaway, R.M.; Lane, S.; Hicks, D. Remote survey of large-scale braided, gravel-bed rivers using digital photogrammetry and image analysis. *Int. J. Remote Sens.* **2003**, *24*, 795–815. [CrossRef]
29. Coleman, J.M. Brahmaputra River: channel processes and sedimentation. *Sediment. Geol.* **1969**, *3*, 129–239. [CrossRef]
30. Mosselman, E.; Huisink, M.; Koomen, E.; Seijmonsbergen, A. *Morphological Changes in a Large Braided Sand-Bed River*; John Wiley & Sons: Chichester, UK, 1995.
31. Thorne, C.R.; Russell, A.P.; Alam, M.K. Planform pattern and channel evolution of the Brahmaputra River, Bangladesh. *Geol. Soc.* **1993**, *75*, 257–276. [CrossRef]
32. Goswami, D.C. Brahmaputra River, Assam, India: Physiography, basin denudation, and channel aggradation. *Water Resour. Res.* **1985**, *21*, 959–978. [CrossRef]
33. Sarker, S. Essence of MIKE 21C (FDM Numerical Scheme): Application on the River Morphology of Bangladesh. *Open J. Modell. Simul.* **2022**, *10*, 88–117. [CrossRef]
34. Chembolu, V.; Dutta, S. An entropy based morphological variability assessment of a large braided river. *Earth Surf. Process. Land.* **2018**, *43*, 2889–2896. [CrossRef]
35. Dubey, A.K.; Chembolu, V.; Dutta, S. Utilization of satellite altimetry retrieved river roughness properties in hydraulic flow modelling of braided river system. *Int. J. River Basin Manag.* **2020**, *20, 1–14.* [CrossRef]
36. Karmaker, T.; Medhi, H.; Dutta, S. Study of channel instability in the braided Brahmaputra river using satellite imagery. *Curr. Sci.* **2017**, *112*, 1533–1543. [CrossRef]
37. Sarker, M.H.; Thorne, C.R.; Aktar, M.N.; Ferdous, M.R. Morpho-dynamics of the Brahmaputra–Jamuna river, Bangladesh. *Geomorphology* **2014**, *215*, 45–59. [CrossRef]
38. Valdiya, K. Why does river Brahmaputra remain untamed? *Curr. Sci.* **1999**, *76*, 1301–1305.
39. Dutta, S.; Medhi, H.; Karmaker, T.; Singh, Y.; Prabu, I.; Dutta, U. Probabilistic flood hazard mapping for embankment breaching. *ISH J. Hydraul. Eng.* **2010**, *16*, 15–25. [CrossRef]
40. Nayak, P.; Panda, B. Brahmaputra and the Socio-Economic Life of People of Assam. *The Mahabahu Brahmaputra*; Flood and River Management Agency of Assam: Guwahati, India, 2016; pp. 77–85.
41. Singh, V.; Sharma, N.; Ojha, C.S.P. *The Brahmaputra Basin Water Resources*; Springer Science & Business Media: Berlin/Heidelberg, Germany, 2004; Volume 47.
42. Government of Assam, Water Resources. Brahmaputra River System. Available online: https://waterresources.assam.gov.in/portlet-innerpage/brahmaputra-river-system (accessed on 1 October 2021).
43. Marra, W.A.; Kleinhans, M.G.; Addink, E.A. Network concepts to describe channel importance and change in multichannel systems: test results for the Jamuna River, Bangladesh. *Earth Surf. Process. Land.* **2014**, *39*, 766–778. [CrossRef]
44. Fischer, S.; Pietroń, J.; Bring, A.; Thorslund, J.; Jarsjö, J. Present to future sediment transport of the Brahmaputra River: reducing uncertainty in predictions and management. *Region. Env. Change* **2017**, *17*, 515–526. [CrossRef]
45. Sarker, T. Role of Climatic and Non-Climatic Factors on Land Use and Land Cover Change in the Arctic: A Comparative Analysis of Vorkuta and Salekhard. Ph.D. Thesis, The George Washington University, Washington, DC, USA, 2020.
46. Ranjbar, S.; Hooshyar, M.; Singh, A.; Wang, D. Quantifying climatic controls on river network branching structure across scales. *Water Resour. Res.* **2018**, *54*, 7347–7360. [CrossRef]
47. Lashermes, B.; Foufoula-Georgiou, E. Area and width functions of river networks: New results on multifractal properties. *Water Resour. Res.* **2007**, *43*. [CrossRef]
48. Shannon, C.E. A mathematical theory of communication. *Bell Syst. Tech. J.* **1948**, *27*, 379–423. [CrossRef]
49. Delgado-Bonal, A.; Marshak, A. Approximate entropy and sample entropy: A comprehensive tutorial. *Entropy* **2019**, *21*, 541. [CrossRef]
50. Pincus, S.M. Approximate entropy as a measure of system complexity. *Proc. Natl. Acad. Sci. USA* **1991**, *88*, 2297–2301. [CrossRef]
51. Sarker, S.; Sarker, T. Spectral Properties of Water Hammer Wave. *Appl. Mech.* **2022**, *3*, 799–814. [CrossRef]
52. Pincus, S.; Kalman, R.E. Irregularity, volatility, risk, and financial market time series. *Proc. Natl. Acad. Sci. USA* **2004**, *101*, 13709–13714. [CrossRef] [PubMed]
53. Richman, J.S.; Moorman, J.R. Physiological time-series analysis using approximate entropy and sample entropy. *Am. J. Physiol. Heart Circ. Physiol.* **2000**. [CrossRef] [PubMed]
54. Delgado-Bonal, A.; Marshak, A.; Yang, Y.; Holdaway, D. Analyzing changes in the complexity of climate in the last four decades using MERRA-2 radiation data. *Sci. Rep.* **2020**, *10*, 1–8. [CrossRef]
55. Ranjbar, S.; Singh, A. Entropy and intermittency of river bed elevation fluctuations. *J. Geophys. Res. Earth Surf.* **2020**, *125*, e2019JF005499. [CrossRef]
56. Sarker, S. A Story on the Wave Spectral Properties of Water Hammer. *engrXiv* **2021**. [CrossRef]
57. Stoica, P.; Moses, R.L. *Spectral Analysis of Signals*; Pearson and Prentice Hall: Upper Saddle River, NJ, USA, 2005.
58. Stull, R.B. *An Introduction to Boundary Layer Meteorology*; Springer Science & Business Media: Berlin/Heidelberg, Germany, 2012; Volume 13.

59. Gardner, W.A.; Robinson, E.A. *Statistical Spectral Analysis—A Nonprobabilistic Theory*; Prentice-Hall Inc.: Upper Saddle River, NJ, USA, 1989.
60. Pilgram, B.; Kaplan, D.T. A comparison of estimators for 1f noise. *Phys. D Nonlin. Phen.* **1998**, *114*, 108–122. [CrossRef]
61. Sarker, S. A Short Review on Computational Hydraulics in the context of Water Resources Engineering. *Open J. Modell. Simul.* **2022**, *10*, 1–31. [CrossRef]
62. Smith, D.G.; Smith, N.D. Sedimentation in anastomosed river systems; examples from alluvial valleys near Banff, Alberta. *J. Sediment. Res.* **1980**, *50*, 157–164. [CrossRef]
63. David Knighton, A.; Nanson, G.C. Anastomosis and the continuum of channel pattern. *Earth Surf. Process. Land.* **1993**, *18*, 613–625. [CrossRef]
64. Sarker, S. Fundamentals of Climatology for Engineers: Lecture Note. *Eng* **2022**, *3*, 573–595. [CrossRef]

MDPI AG
Grosspeteranlage 5
4052 Basel
Switzerland
Tel.: +41 61 683 77 34

MDPI Books Editorial Office
E-mail: books@mdpi.com
www.mdpi.com/books

www.ingramcontent.com/pod-product-compliance
Lightning Source LLC
Chambersburg PA
CBHW041017270926
51850CB00024B/971
* 9 7 8 3 7 2 5 8 4 7 5 5 6 *